AF540640

BIOLOGY OF AMPHIBIA

BIOLOGY OF AMPHIBIA

By

Dr. D.R. Khanna
Reader in Zoology
Gurukul Kangri University
Haridwar (Uttaranchal)
&
Dr. P.R. Yadav
Lecturer
Department of Zoology
D.A.V. College
Muzaffarnagar (U.P.)

DISCOVERY PUBLISHING HOUSE
NEW DELHI-110002

First Published-2005

ISBN 81-7141-932-1

Published by
DISCOVERY PUBLISHING HOUSE
4831/24, Ansari Road, Prahlad Street,
Darya Ganj, New Delhi-110002 (India)
Phone: 23279245 • Fax: 91-11-23253475
E-mail:dphtemp@indiatimes.com

Printed at
Arora Offset Press
Laxmi Nagar, Delhi–92

Preface

This is an introductory text for students of science, medicine, and allied researchers, and is designed to hold the reader's attention to stimulate his interest. The present title deals with the morphological, physiological and developmental aspects of Amphibians. Amphibians are the first land vertebrates evolved from fishes. They lead a dual life and carry on all physiological activities smoothly in water as well as on ground. The author has chosen those aspects of the subject which are essential for the student to understand, and arranged them in such a way that he can use them as pegs on which to hang more information. In order to achieve this, most of the instruction given in larger text books has been condensed and presented concisely in simple graphs and diagrams. The student can thus acquire an easily assimilated understanding of basic amphibian physiological activities which provides a firm foundation on which to build his further studies.

To make the work more comprehensive and informative, we have consulted many authoritative books, research journals, abstracts, monographs etc. We are grateful to al! those great scholars whose work are cited or substantially reproduced.

There can be no claim to originality except in the manner of treatment and much of the information has been obtained from the books and scientific journals available in the different libraries.

The authors express their thanks to their friends and colleagues whose continue inspirations have initiated them to bring out this book.

The authors are painfully aware of the shortcomings, errors and misprints that have crept in, and shall be greateful to receive suggestion for improvement of the next edition from all the readers.

The authors express their gratitute to Mr. Wasan and staff of M/s Discovery Publishing House for their whole hearted co-operation in the publication of this book.

Author

CONTENTS

1

INTRODUCTION

Scientists divide the animal kingdom into several major groups for classification purposes. By far the largest group is the invertebrates: it contains about 95 percent of the millions of known species of animals, including sponges, mollusks, crustaceans, and insects. Reptiles and amphibians are vertebrates—they belong to a group of animals characterized by having a bony backbone, a flexible but strong support column of articulated sections (vertebrae) to which the other body structures are attached. Vertebrates include not only amphibians (class Amphibia) and reptiles (class Reptilia), but also fishes (about three classes), birds (class Aves), and mammals (class Mammalia). The differences between reptiles and amphibians are generally more obvious than their similarities, although their joint study under the name "herpetology" (from the Greek *herpo*, to creep or crawl) is a scientific tradition dating back nearly two centuries.

Diverse Forms

The fossil record shows that both amphibians and reptiles were once much more diverse in size and structure than they are today (dinosaurs being the best known examples), and so there is a tendency to regard today's forms as a mere shadow of their former glory, having been displaced in dominance by the warm-blooded birds and mammals. However, the amphibians with 4,950 living species, together with 7,400 living reptiles, jointly outnumber the birds (9,000 species) and mammals (about 4,670 species). They are in no way more primitive than birds and mammals. Their behaviour and physiology are just as complex and equally well adapted to the astonishingly varied environments they inhabit.

Body Temperature and Metabolism

One significant factor in the distribution and behaviour of modern reptiles and amphibians is their inability to produce sufficient internal metabolic heat to maintain a constant body temperature, as do almost all birds and mammals. For this reason they are often termed "cold-blooded", or "ectotherms". Although some reptiles are able to generate enough metabolic heat to raise their temperature for a specific purpose for limited periods (for example, female pythons brooding their eggs), none can constantly maintain a temperature much higher than that of their surroundings without an external heat source. Their dependence on external sources of heat to reach and maintain an "active" body temperature greatly limits their ability to live and breed in colder climates, with the consequence that they are most abundant in tropical and warm temperate regions.

While being cold-blooded can have disadvantages, such as being more vulnerable to predators at low environmental temperatures, it also has a number of advantages. Amphibians and reptiles can simply "shut down" when conditions are unsuitable—for example, if it is too cold or food is scarce—and therefore do not have to use up large amounts of stored energy to keep themselves warm.

Skin and Scales

The skin of most amphibians is thin and highly glandular, and needs to be kept relatively moist to function effectively. It is also highly permeable, allowing water to be absorbed or lost, and is often

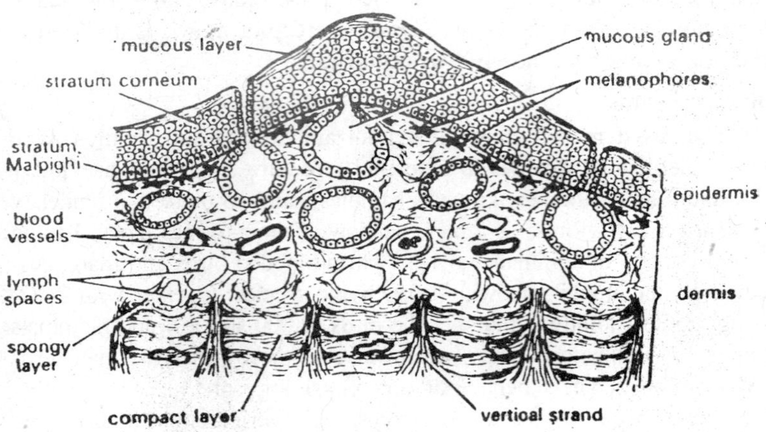

Fig. 1.1. V.S. of Skin.

involved in gas exchange as an adjunct to normal respiration through the lungs. This means that most amphibians are restricted to moist or humid environments.

The skin of reptiles differs from that of amphibians in that the outer layer is thickened to form scales, composed of keratin; some areas are further thickened to form tubercles and crests. In many reptiles (and a few amphibians) small bones develop in the skin just below the surface. These dermal bones, or osteoderms, add strength and protection to the skin and make it even more impervious to water loss. Skin colour and colour changes are largely determined by pigment cells just below the outer layers of skin.

Body Structure

From the muscular fins of their fish ancestors, the first amphibians evolved four limbs for moving about on land. Subsequent amphibians and reptiles share this basic tetrapod (four-limbed) vertebrate skeletal structure.

This basic skeletal structure has become highly modified in some amphibians and reptiles. An extreme adaptation is found in turtles, in which the greater part of the vertebral column, the ribs, and the limb girdles, have become fused with dermal bones to form a protective shell. In frogs, the development of an efficient jumping mechanism

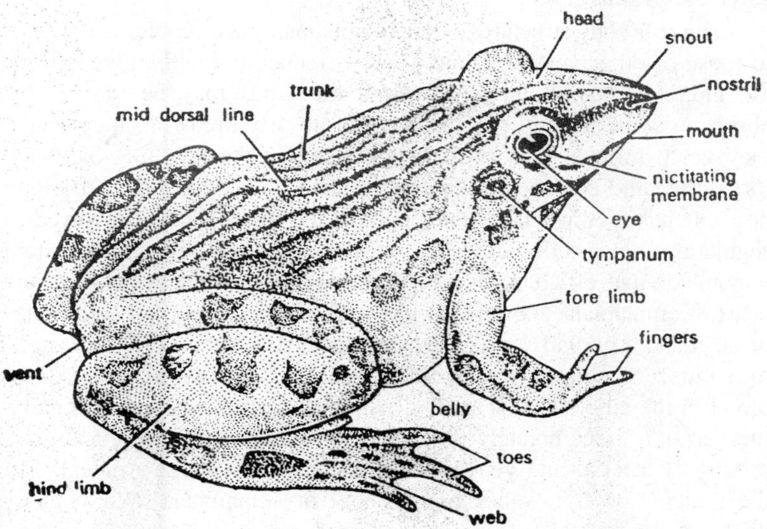

Fig. 1.2. An Indian frog, Rana tigrina.

has called for greater rigidity of the vertebral column and a solid structure to support the enormous mass of hind limb muscles. This has been achieved by a great reduction in the number of vertebrae and the incorporation of several vertebrae into a single long supporting bone called the *urostyle*.

Modern amphibians generally have simple pedicellate teeth, which are used to grasp the invertebrates on which they feed, but in many species the teeth are reduced or absent. Teeth in reptiles vary greatly in form and structure, from simple blunt teeth used to grasp and partially crush prey, to those with broad grinding surfaces or sharp shearing edges. Perhaps the most specialized are the fangs of snakes that have evolved grooves or hollows, making them efficient venom-delivering hypodermic needles.

In most amphibians and reptiles the tongue is an important adjunct to the teeth and jaws in catching prey. It is usually very muscular, flexible, and extrusible, and is often sticky with mucus to adhere to prey such as insects. It is also used to manipulate food within the mouth. In some groups of reptiles, such as monitor lizards and snakes, the tongue has become a specialized sense organ which senses and locates prey, but otherwise is not directly involved in the capturing of prey.

Internal Organs

The soft body structures of adult amphibians and reptiles are similar to those of other vertebrate animals. A trachea (windpipe) leads from the glottis to paired lungs, although the lungs may be reduced or absent in some aquatic species that can absorb almost all the oxygen they need through their skin. In most snakes the left lung has been lost during the course of evolution; and in many sea snakes the lung now extends forward along the length of the trachea and is involved in regulating the snake's buoyancy when diving. The circulatory system is complex and efficient, but with the exception of the crocodiles, the heart of amphibians and reptiles has only three chambers—two auricles or atria and a partially divided ventricle—although the flow is arranged to minimize mixing of oxygenated (arterial) and de-oxygenated (venous) blood in the single ventricle. In crocodiles, as in birds and mammals, the ventricle is completely divided by a septum, which prevents any mixing of arterial and venous blood and is much more efficient than the system found in other reptiles and in amphibians. While many different food specialists have developed among the amphibians and reptiles, once past the mouth the food is digested in a fairly standardized

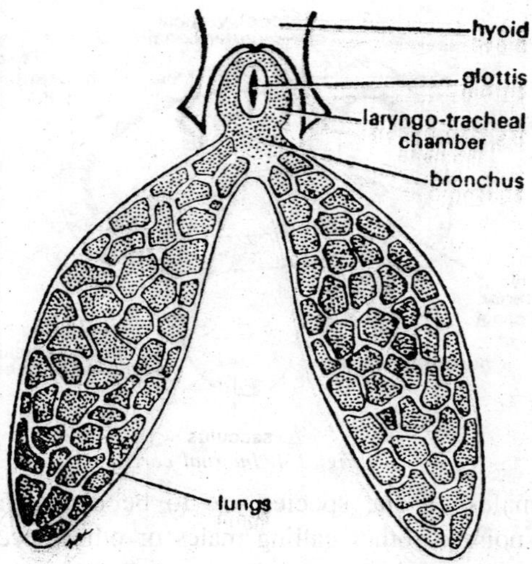

Fig. 1.3. Respiratory organs. (Dorsal view).

vertebrate digestive system. An important feature of both amphibians and reptiles—in which they differ from most mammals—is the cloaca, a large chamber just inside the vent, into which the gut and the ducts of urinary and sexual organs all enter.

For the majority of amphibians, living as they do in humid environments, water conservation is not a major problem. Consequently their nitrogenous waste products are converted into urea in the liver and then excreted by the kidneys and leave the body in a fluid urine. The same method is used by reptiles that live in moist or aquatic environments, but for reptiles in very dry environments such as deserts, where free water is scarce or rare, body water is too precious to be urine. Consequently in the majority of reptiles the liver converts nitrogenous waste into uric acid which, after excretion by the kidneys, passes to the cloaca where water is resorbed and solid uric acid crystals deposited; these leave the body as a moist paste, using only a fraction of the water that would be needed to excrete urea as urine.

Senses

For most amphibians and reptiles three senses are critiical to survival: sight, hearing, and olfaction (smell). The relative importance of each is strongly correlated with the lifestyles of any particular species.

In frogs in which the males attract females by calling to them, hearing clearly needs to be acute and discriminating if the calls of

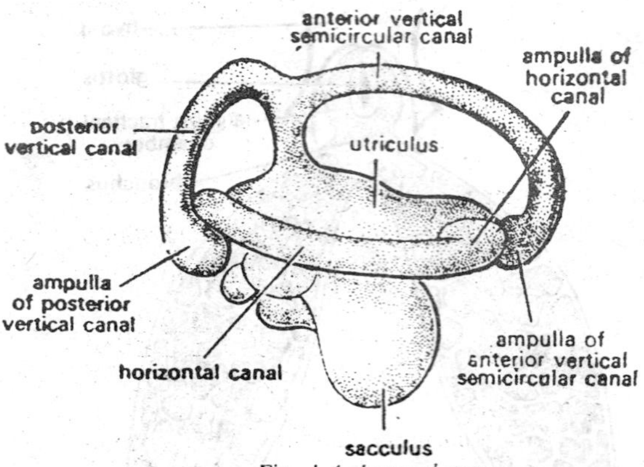

Fig. 1.4. Internal ear.

individual males of one species are to be distinguished from the background noise of other calling males or other species. The males of many lizards use brilliant colours to attract mates and to warn other males away from their territories. Clearly, acute vision which recognizes those colours is essential to such species.

The eyes of amphibians and reptiles are similar in general features to those of other land vertebrates, with characteristic cornea, iris, lens, and retina. In most amphibians there is an immovable upper eyelid and a movable lower eyelid, the upper part of which (the nictitating membrane) is usually transparent. In reptiles the lower eyelid is also usually movable and at least partly scaly, but in many lizards, and in all snakes, the lower and upper eyelid have become fused and the eye is covered by a large, fixed, transparent disk—the spectacle—which protects the eye from damage. The spectacle becomes scratched and dirty over time, and is shed with the skin at each sloughing cycle.

The ears are also typical of other land vertebrates. Sounds are received on a membrane on each side of the head—the tympanum—and transmitted by one or more fine stapedial bones to the inner ear. However, many amphibians and reptiles have limited reception of airborne sounds, as the tympanum is absent or covered by skin, and the inner bony structures are often modified to receive vibrations through other body tissues.

Olfactory senses (smell and taste) are also moderately to well-developed in most amphibians and reptiles. The most important olfactory organ in both amphibians and reptiles is a vomeronasal organ called Jacobson's organ, lying in the roof of the mouth. In amphibians and many reptiles it "tastes" the food in the mouth. In other reptiles, especially monitor lizards and snakes, the protruding tongue "tests" its environment by collecting minute particles which are passed for identification to Jacobson's organ by the tip(s) of the tongue. In this way such reptiles can follow the trail even of moving prey.

Reproduction and Life Cycles

The life cycle of an amphibian is fundamentally different from that of a reptile. Reptiles produce eggs within which the embryo develops and hatches (or is born, in the case of live-bearing species), generally as a miniature replica of the adult. Amphibians are characterized by a two-stage life cycle in which the eggs hatch into aquatic, gill-bearing larvae (for example, tadpoles) which eventually metamorphose into air-breathing, mostly terrestrial adults. Occasionally the entire tadpole stage takes place within the egg or within the body of the mother.

Almost all amphibians produce eggs, varying in number from a single egg to many thousands in a clutch, which are deposited in water or in humid sites on land. They are fertilized externally or internally. Most male amphibians lack any form of intromittant organ with which to introduce sperm inside the female, but some are able to protrude part of their cloaca into that of the female. In most frogs,

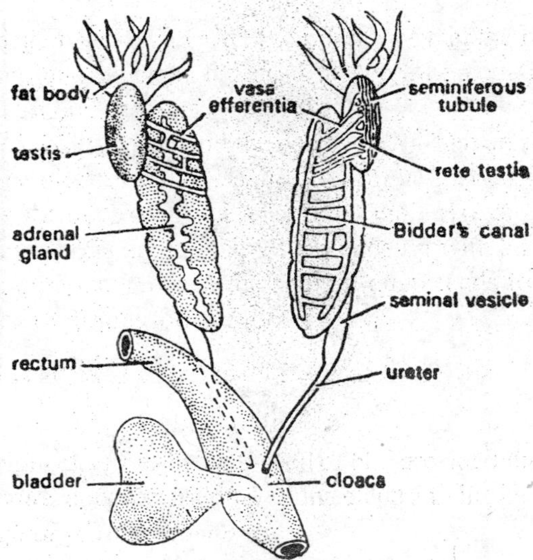

Fig. 1.5. Male urinogenital organs.

the male simply expels his seminal fluid over the eggs as they are extruded from the female's body. In salamanders, the male deposits a package of sperm (the spermatophore), which is picked up by the cloacal lips of the female and is held in her cloaca until needed for fertilization.

The egg of a reptile is fertilized internally. With the exception of the tuatara, in which sperm are exchanged through cloacal contact; reptiles have well-developed intromittant organs to deliver the seminal fliud. In lizards and snakes these organs are paired, so that each male has two functioning penises. Turtles and crocodilians have only a single penis.

The reptile egg is substantially more complex than that of an amphibian, and much better adapted to survival on land. In many reptiles, the egg does not develop a shell, and development occurs within the mother's body; the embryonic membranes are retained and in some cases form a primitive placental connection with the mother.

2

GEOLOGICAL TIME SCALE

To understand the complete evolutionary changes in the organisms (plants and animals, one has to study the organisms of the distant past which are now extinct. The extinct forms of life, which are also the connecting links between the most primitive forms and recent forms, are preserved in the rocks as *fossils*. The study of these fossils constitutes the subject-matter of *Palaentology*.

A number of estimates have been proposed from time to time about the age of earth. For examples, *Archbishop Ussher* of Ireland in 1645 declared that the creation of earth took place 26th October 4004 B.C., at 9 o'clock in the morning. This was the adopted version of the *Bible* for a long time. *James Hutton* believed that geological time was unimaginably long. *Lord Kelvin* (1897), from a study of the flow of heat from the interior to the surface of earth, estimated the age of earth from 20-40 million years.

The most accurate method of determining the age of a rock is by radioactive decay studies. Certain radioactive elements like uranium and thorium, found in certain rocks, decay into lead. The rate of this radioactive disintegration is constant and in unaffected by external physical conditions like temperature, pressure, etc. The formula may be written as $U^{238} = Pb^{206} + 8He^{4} +$ heat. It has been determined that on gram of uranium gives on decay 1/7,600,000 gram of lead in one year. The lead: uranium ratio in a rock would thus give the age of the rock. The maximum age of the earth has thus been estimated to be around 5500 m.y. (million years). Other methods of dating, *e.g.*, potassium—argon method, radiocarbon method, etc. are also known today.

By studying the fossils, present in the rocks, the geologists have divided the earth's past history into *Era* which are major divisions. The eras are divided into *periods*. The periods are subdivided into *Epoches*. These eras, periods and epoches are arranged on the time scale in an order of their age, and this arrangement is called, '*geological time scale*'. The first geological time scale was developed by *Giovanni Avduina*, an Italian Scientist, in 1760.

Eras

Based on the nature of the forms of life preserved as fossils in the rock, the earth's past history is divided into six eras.

1. Azoic
2. Archaeozoic or Eozoic (Primal life or dawn of life)
3. Proterozoic (Early life)
4. Palaeozoic (Ancient life)
5. Mesozoic (Mediaeval life)
6. Cenozoic (Recent life)

Periods

These are the fundamental units of the standard geologic time scale. There are 12 periods:

Periods of Palaeozoic era

There are 7 periods in Palaeozoic era:

1. Cambrian
2. Ordovician
3. Silurian
4. Devonian
5. Mississipian
6. Pennsylvanian
7. Permian.

Periods of Mesozoic era

There are three periods:

8. Triassic
9. Jurassic
10. Cretaceous.

Periods of Cenozoic era

There are two periods:

11. Tertiary
12. Quaternary.

Epoches

There are seven epoches. The epoches of cenozoic's period are:

Quaternary period

There are two epoches:

1. Recent
2. Pleistocene.

Tertiary period

It is subdivided into five epoches:

3. Pliocene
4. Miocene
5. Oligocene
6. Eocene
7. Palaeocene.

AZOIC ERA

This was the time, in which life was not present on the earth. During this time, earth was formed, cooled and underwent many changes. These changes made favourable conditions for the appearance of living organisms. The rock of this time are igneous type without any fossil.

ARCHAEOZOIC ERA

The manner in which life originated is not known. It is probable that sometime in early Archaeozoic Eras the earliest living matter that could reproduce itself got generated under very favourable conditions of temperature, humidity, light, etc., by the combination of certain chemicals. This first living matter may have been very similar to protoplasm. Filterable viruses which represent an intermediate stage between living and non-living entities, may have been the first to appear on the surface of the globe in the early Archaeozoic Period. Unicellular Protozoa and Protophyta may have been also present.

According to *Stirton* (1959), the oldest known fossils are blue-green algae, fungi, and probably a flagellate described from Ontario, Canada. The age of the rock in which these algae occur has been estimated to be about 2000 m.y. (million years) by radiometric methods.

PROTEROZOIC ERA

It may be called the '*Age of Primitive Invertebrates*'. Life was in the form of algae, precipitations, siliceous sponge spicules, remains of radiolarians, impressions of jelly fishes, a simple brachiopod (*Lingullela montana*) as well as deposits of graphite. These first organisms were quite possibly soft bodied. Reducing bacteria must also have been predominant during this era because the chief iron-ore deposits of the world are found in the Proterozoic era.

The fossil record from the rocks is much less probably due to absence of hard parts in the organisms to be preserved as fossils. The fossils which were formed must have been destroyed due to effect of pressure and high temperatures. This era is estimated to have lasted 3.5 billion years.

PALAEOZOIC ERA

This fossil record in this era are very extensive. Almost all the major invertebrate phyla are represented even in the early Palaeozoic era. The fossils of vertebrates also present late in the era because the early chordates were soft-bodied and could not be preserved as fossils. Palaeozoic era is divided into following periods:

Cambrian Period

The name Cambrian was proposed by *A. Sedgwick* in 1835. It is taken from *Cambria* the Latin name of City of Wales.

This period started by the melting of glaciers, slow rise of sea level and warm climate.

Almost all the main invertebrate phyla had existed during this period as also the main phyla of plants. Trilobite were the most dominant forms of life, the important genera being *Olenellus*, *Redlichia* in the early Cambrian times; *Paradoxides* in the middle Cambrian and *Crepicephalus* in late Cambrian. Small primitive brachiopods and gastropods with simple shells are found in early Cambrian rocks. Some foraminifers and radiolarians have also been found. A few graptolites are recorded in the Cambrian rocks. Echinoderms have been recorded from Cambrian rocks of Australia while some bryozoans from Russia and Canada. The first land plants also appear in this period.

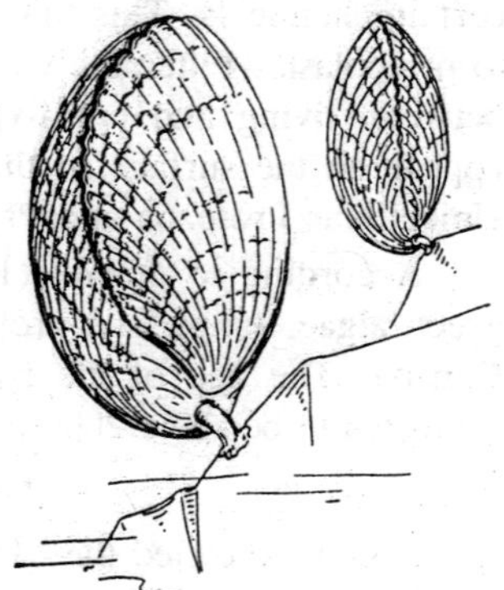

Fig. 2.1. Brachiopods, attached to rocks by their pedicles.

Ordovician Period

The name Ordovician was proposed by *Charles Lapworth* in 1879, on the basis of the ordovices, an ancient tribe, that formerly inhibited the City of Wales.

During this period, conditions became more favourable as the climate was uniformly warm with glacial activities in certain regions.

This was the "*Age of Giant Mollusca*". Graptolites, which started in late Cambrian, show their maximum development during Ordovician. They in fact serve as index fossils in Ordovician rocks. Trilobites were still abundant and reached the climax of their evolution. First ostracods appeared on the scene and brachiopods developed harder shells. Cephalopods, both straight-shelled and coiled type, were abundant. Echinoderms appeared in large numbers and show diversification, the starfish being very common.

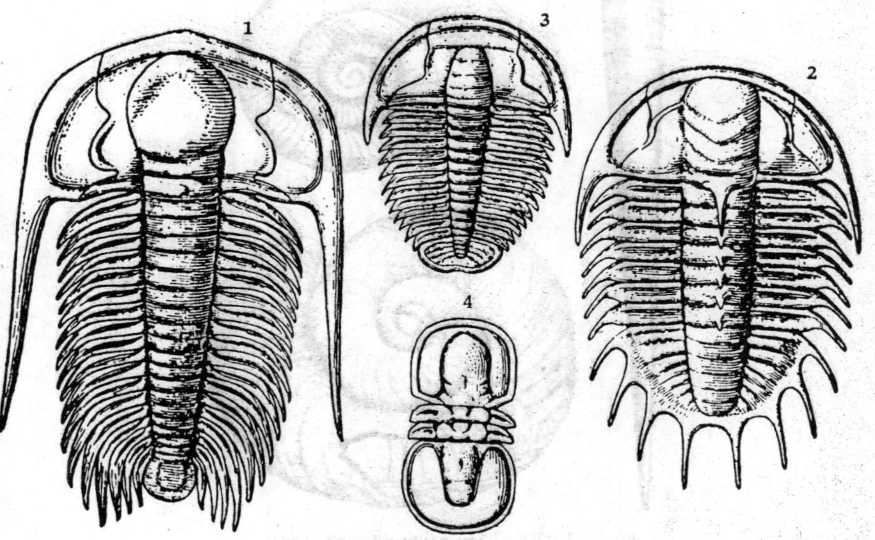

Fig. 2.2. Typical Cambrian trilobites.

Ordovician is also characterised by first appearance of conodonts which are very important marine environment indicators. Land plants are meagre. First appearance of fish-like animals Ostracoderms is also recorded in Ordovician rocks. This Period probably marks the first appearance of vertebrates.

Silurian Period

The term Silurian was proposed by *Murchison* in 1835 to designate the rocks present on the borders of Wales and England. This area was inhibited by Silures, an ancient tribe, inhibited Wales. The climatic conditions were mild and some areas were arid. Land elevation at places caused formation of deep continental seas. Rise of *Caledonian*

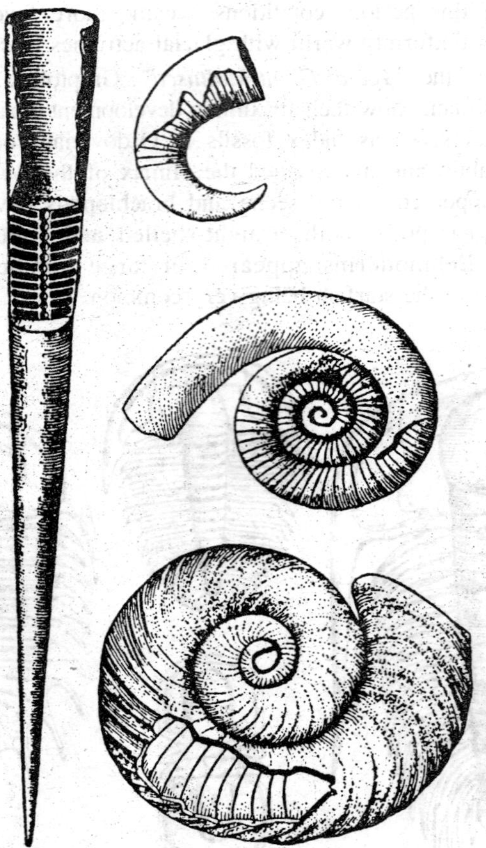

Fig. 2.3. Ordovician nautiloid cephalopods.

Mountains in British Isles, Scandinavia and N. Greenland and in N. Africa and East Central Asia.

The first air-breathing invertebrates on the earth were scorpions which appeared in the Silurian Period. Trilobites now started waning. Bryozoans were extremely abundant. Brachiopods continued their development and first *Spirifer* appeared in Silurian Period. Graptolites also started a decline at the end of Silurian Period.

The fish is found to have developed lower jaws in the Silurian Period. The organisation of the Silurian fishes is similar in many respects to that of the fishes of today. Although algae were still dominant but some terrestrial primitive ferns also evolved.

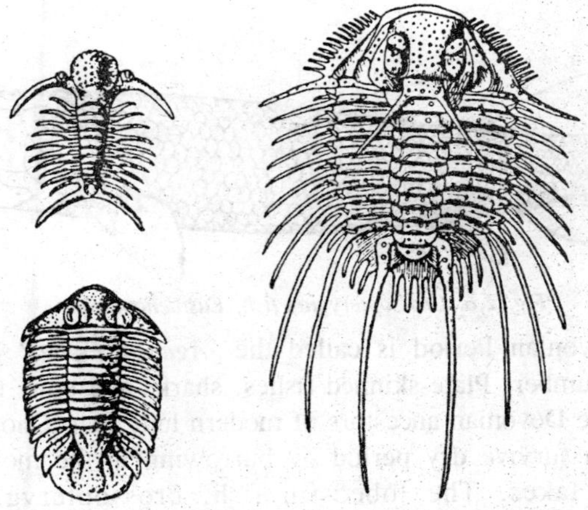

Fig. 2.4. Silurian trilobites.

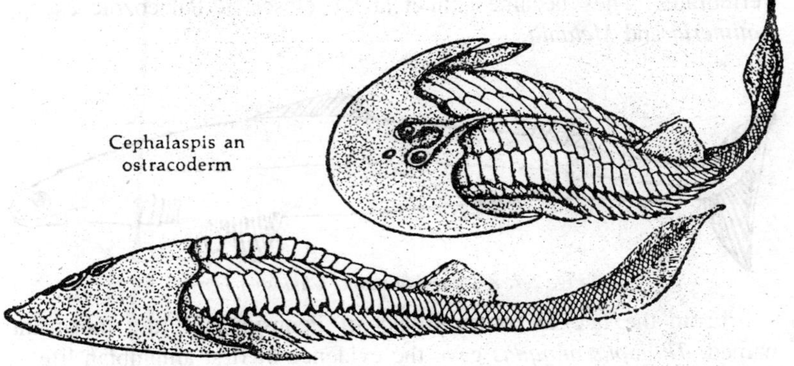

Fig. 2.5. An ostracoderm (armored fish), Cephalaspis; length one foot or less.

Devonian Period

The name Devonian was proposed by *Sedgwick* and *Murchison* in 1839 after Devonshire, a country of South-Western England, where rock formation of this period were first found.

The land elevation continued. Consequently larger seas were further divided into smaller seas. Due to fall of temperature, climate remained cold any dry. This period was marked by great volcanic activities and the formation of coal, oil and gas.

During the period, brachiopod developed wonderfully. Corals and bryozans were also abundant.

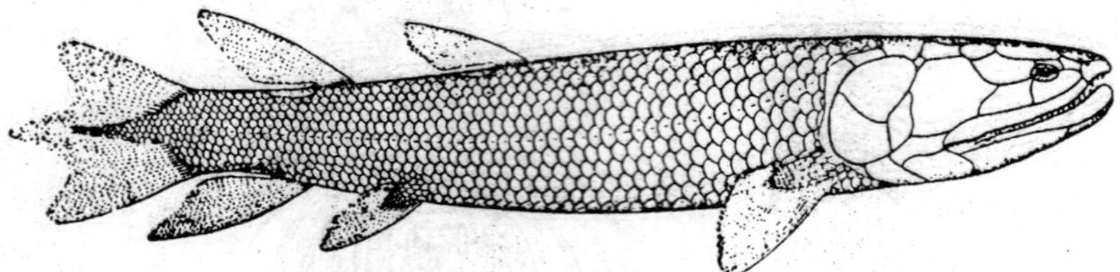

Fig. 2.6. Crossopterygian fish, Eusthenopteron.

The Devonian Period is called the '*Age of Fishes*' because of their great number. Plate-skinned fishes, sharks and bony fishes were common. The Devonian ancestors of modern lung fish, Choanichthyes, were able to survive dry period by burrowing in the moist beds of streams or lakes. The lobed-fin fish, crossopterygians were characterized by the presence of lobed rounded bases of the paired fins. These were on the direct line of evolution from fish to terrestrial vertebrates. They became extinct at the closer of Palaeozoic except *Latimeria* and *Melania*.

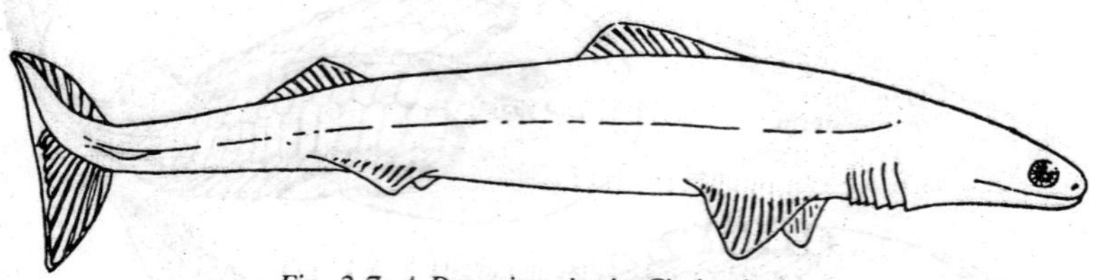

Fig. 2.7. A Devonian shark, Cladoselàche.

From the late Devonian Rocks of Pennsylvania, the foot-print named, *Thinopus antiquas* gave the evidence of first amphibian life.

Tall terrestrial Psilophytes, club-mosses and horse-tails occur in the rock of Scotland. First wingless insect also appeared in this period.

Mississippian Period

The name 'Mississippian' was given by *Alexander Winchell* in 1869 to designate the area in eastern Mississipi basin where rock formation of this period were found.

The climatic condition were hot and damp which caused formation of wide-spread swamps.

During this period, sharks were the dominant fishes in the seas and oceans. They were of many kinds but consisted principally of the more ancient shell-feeding types which were subsequently almost wholly blotted out.

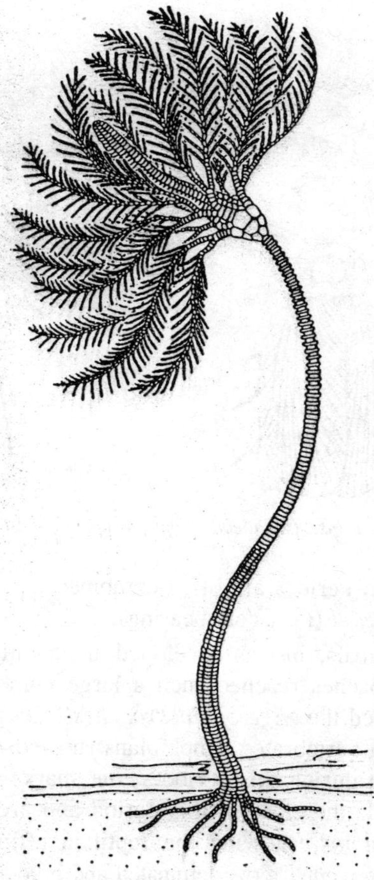

Fig. 2.8. A simple crinoid.

This period is called *Age of Crinoids* because they were most abundant and diversified. Amphibian spread and the insects evolved wings during Mississippian Period.

No evolutionary change in terrestrial plants in comparison to those of Devonian Period is observed in this period.

Pennsylvanian Period

Term 'Pennsylvanian' was proposed by *H.S. Williams* in 1819 which refers to the state Pennsylvania. The climatic conditions were disturbed so that the forests burried under the swamps and turned, in due course of time, into present-day coal mines. For this reason, Mississippian

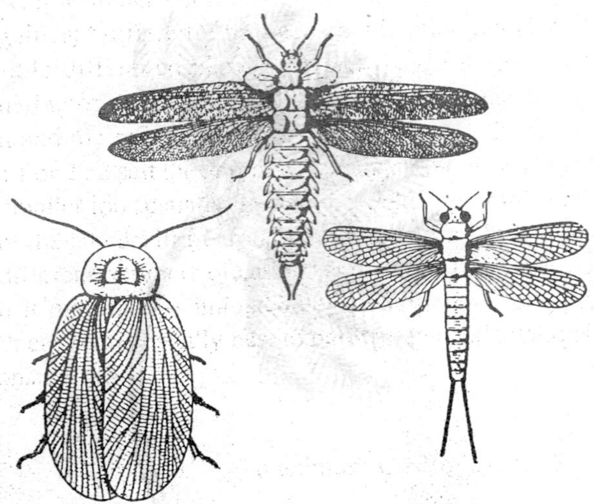

Fig. 2.9. Pennsylvanian insects. Paleodictyoptera (upper, and lower right). Cockroach (lower left).

and Pennsylvanian Periods are often combined together under a head, *Carboniferous Period* (i.e., coal bearing).

Amongst animals, insects developed abundently both in size and diversity. Cockroaches reached such a large number that sometimes this period is called the '*Age of Cockroaches*'. Dragon flies were also abundant. Among vertebrates, amphibians showed maximum diversity and specialization during this period. True sharks and sea-likes were common. Towards the close of this period appeared the first animal, *Tuditanus punctulatus*, showing the reptilian affinities, from Stegocephalians. Among plants, ferns dominated and bryophytes also appeared.

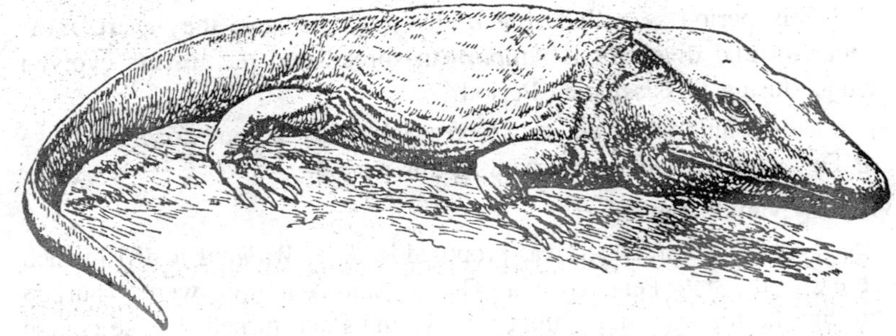

Fig. 2.10. A labyrinthodont amphibian, Diplovertebron.

Permian Period

The term 'Permian' was given by *Murchison* in 1841, because fossil record of the period were found in Russian province of Perm.

The cold and dry condition caused widespread glaciation. Extensive land elevation caused the formation of high mountains. Consequently, shallow seas were emptied and their bottom became exposed as salty deserts. Melting of glaciers gave rise to rivers upon the land.

Great diversification in reptiles took place and first mammal like reptiles appeared in the Permian times. Many important groups of marine invertebrates now dwindled. Permian marks the complete extinction of Trilobites which were very important and dominant during the early Palaeozoic. Due to the cold climates, the insect developed larval adaptations seen in the metamorphosis of present day insects.

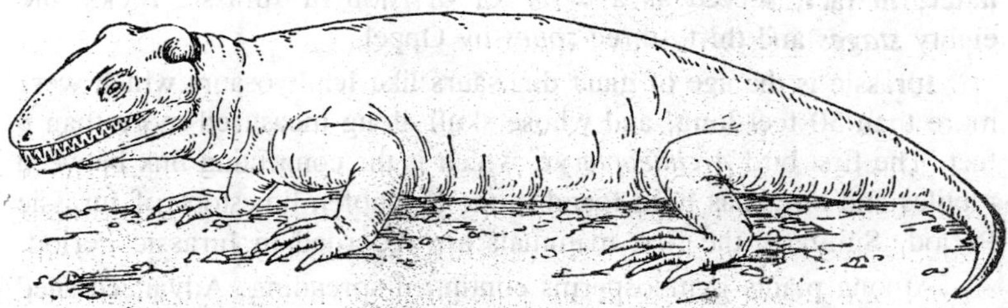

Fig. 2.11. A Permian vertebrate, Seymouria, combining characteristics of both amphibians and reptiles.

Among terrestrial animals, spiders, scorpions, centipedes, snails etc. were present.

Among plants swampy trees declined and were replaced by tall gymnosperms. The true conifers and cycads became most abundant and trees somewhat similar to date palms appeared.

MESOZOIC ERA

In contrast to the dominance of marine animals throughout Palaeozoic Era, the Mesozoic Era is characterized by the dominance of land animals, the most spectacular of which were giant dinosaurs which ruled the earth for thousands of years. That is why this era is known as '*Age of Reptiles*'. This Era is divided into three Periods.

Triassic Period

This name was give by *F. Von Alberti*.

This period was characterized by hot and dry climate and land elevation which helped in expression of deserts and continents.

Snails, bivalved molluscs, insects and sea-urchins diversified and spread. This period is characterized by the diversification of reptiles. Dinosaurs spread upon land; Ichthyosaurs (fish-like) in water; lizard-like Plesiosaurs in water; and bird like Pterosaus in air. The egg-laying mammals also made their appearance.

During this period horstails, clubmosses and seed fern declined, Gymnosperms progressed fast.

Jurassic Period

The name was given by *A. Von Humboldt* in 1799. This Period has been named after the Jura Mountains of Switzerland and France. The climates during this Period were mild throughout the world.

Consequently both invertebrate and vertebrate fauna thrived. Among the invertebrates, belemnites and ammonites were most abundant; the latter, in fact, served as a basis for division of Jurassic rocks into eighty *stages* and thirty-three *zones* by Oppel.

Jurassic is the age of giant dinosaurs like Ichthyosaurs which were more than 50 feet long, and whose skull along measured more than 8 feet. The first bird *Archaeopteryx*, which is the connecting link between reptiles and birds has been found in the Solnhofen limestone of Jurassic Period. Similarly the first mammals are recorded in Jurassic Period.

Among plants gymnosperms continued spreading. Advanced seed ferns gave rise to the first Angiosperms which were dicotyledonous.

Crataceous Period

The term was given by *J.J. d'omalus* d'Halloy in 1882 derived from Latin word *Creta*=chalk.

The climate was warm and somewhat uniform over the most earth surface, but became cooler later on. The seas spread widely over the continents in the earliest past of this period. The Larmide Orogeny changed the surface relief of the earth which affected the climates. The changed climate accelerated the evolution of mammals, birds, flowering plants etc. while caused complete extinction of certain groups of animals like the dinosaurs, ammonites and rudistids.

In the sea, sharks and bony fishes flourished. Snakes first appeared in this period. New varieties of birds appeared. Mammals were still small and less numerous. Marsupials diverged from the archaic forms. Placental mammals appeared.

During this period the dinosaurs became extinct because of their heavy body and small brain, lack of adaptability, coldbloodness, herbivorous food etc.

Among plants, the gymnosperms declined. Dicot angiosperms flourished and tall oak, maple etc. formed thick forests. The first monocots appeared.

CENOZOIC ERA

It is known as "*Age of mammals*". It was marked by great adaptive radiation in birds, insects and flowering plants. Wide variety of animals got food and protection as a result of development and spread of deciduous trees, herbs and grasses. The cenozoic era is divided into two periods, and each period into epochs.

Tertiary Period

This period includes five apoches.

Palaeocene epoch

In the beginning the climate was fairly warm but gradually it became colder.

After complete extinction of dinosaurs, nine new orders of terrestrial mammals evolved. Among these were the first primitive carnivores, lemurs, tarsiers etc. Modern groups of birds evolved. Plants also underwent a corresponding evolutionary change.

Eocene epoch

Climate remained less hot. These was a widespread erosion of mountains.

All the modern orders of mammals were present. The first horse *Eohippus* or Hyracotherium evolved also first camels which were originally of the size of a dog or even smaller evolved during this period. The horse had progressed to the *Epihippus* stage by this time. Pigs, rats, monkeys, whales and seals appeared.

Angiosperms spread and evolved, giving rise to several modern types. Heavy forest and limited grass land characterized in this period.

Oligocene epoch

Climate remained hot. The level of land decreased.

Most of the families of mammals were recognizable by this time. Larger brontotheres reached their climax in early Oligocene Period and suddenly died out before middle Oligocene-Period. The Oligocene camel *Poebrotherium* was of the size of a small sheep. The true flesh-eaters like dogs, cats, sabertoothed cats and other small animals, as well as rodents and insect-eating animals became most abundant. The horse had reached the *Miohippus* stage. It was of the size of a sheep with three toes in each foot. Tortoises of medium-size were abundant.

Hawks and vultures hovered in the sky and the 'American pigs' on the ground. Near the end of the Oligocene Period the first New World monkeys appeared.

Among plants tropical and subtropical forests, flowering plants and monocots became widespread.

Miocene epoch

In Miocene epoch, the Alps reached their hights and Himalayas were pushed up. So the climate remained somewhat cold. New mountain ranges such as Cascade, Sierra etc., formed upon America continent.

This Period saw the rise and rapid evolution of grazing mammals, long slender-limbed camels, *Oxydactylus*, bear dogs, small foxes, American pigs etc. were particularly abundant. The horse had reached *Parahippus* stage, a major step in evolution towards modern horse. A free interchange of old and new world terrestrial mammals took place among which cats, dogs, horses and mastodonts were especially common. In Late Miocene Period, the horse had reached *Merychippus* stage of evolution in North America. In Eurasia, the Miocene Epoch is marked by the appearance of hyenas, deer, giraffes, and true antelopes.

Terrestrial plants were more progressive.

Pliocene epoch

This period started with subsidence of land in many places and ended with the beginning of one or two Ice Ages. The climate became still cooler.

The mammals became more specialized and reached their peak of evolution. Camels, horses, rhinoceroses, etc. were abundant during this period. Camels had grown larger in size, and the horse had evolved into several genera. *Pliohippus*, which ultimately gave rise to modern horse, evolved during this period from its Miocene ancestors. *Hipparion* fauna was typical of Pliocene Period in Eurasia and has distributed from Western Europe and North Africa to South-western Asia, India, Burma and China. Characters found in the fossil jaws and teeth of *Dryopithecus*, found in Siwalik rocks of India, showed affinities with chimpanzee, gorilla and modern man. The lack of a land bridge between North and South America in early Pliocene Period inhibited the free migration and dispersal.

Thick forests declined due to cold. Tail and woody plants were replaced by soft monocot plants.

Quaternary Period

This period is divided into two epoches:

Pleistocene epoch

There are evidences in four *Ice Ages* alternating with warmer intervals. Widespread glaciation during this epoch lowered the sea level. Land connection between European continent and Britain, and Siberia and Alaska were established. Due to these land connections dispersal of horses, camels, deer, cats, saber, dogs, rodents etc., took place. The modern horse, *Equus* originated.

The first evidence of a primitive man *Homo heidelbergensis*, recognised by a massive mandible, was found in the middle Pleistocene rocks of Heidelberg, Germany. The Peking man, *Pithecanthropus pekinensis*, lived in the mildy cooler climates in parts of China and adjoining areas. The ape man of Java, *Pithecanthropus erectus*, found among Pleistocene fauna of Java, attests to the evidence of land bridge between Java an mainland of Asia. The entire Pleisotocene Period shows dominance of man.

Recent epoch

The climate has again become hot. Supremacy of man as the wisest and dominant species has been established. With increasing mental development, gradually he got control over water, air and land. He is trying to get control over plants other than earth. Thus we can say that this epoch is the '*Age of man*'.

AMPHIBIANS THROUGH THE AGES

The first backboned animals (vertebrates) evolved in the sea at the end of the Cambrian period about 500 million years ago. During the next 150 million years they continued to evolve in the world's oceans into amazingly diverse groups of fishes. But to colonize the land these aquatic vertebrates had to meet and overcome the tremendous challenges involved in exchanging an aquatic existence for a terrestrial one. These included breathing atmosphere rather than dissolved oxygen, abandoning the buoyancy of water, and modifying the body structure to meet the high gravitational forces encountered on land.

From Water to Land

The first vertebrates to make the transition from water to land were the amphibians. In the late Devonian period about 360 million years ago, they became the first tetrapods—that is, the first backboned animals with four articulated legs—from which all later vertebrates (reptiles, birds, and mammals) evolved. Indeed, once the amphibians had evolved the basic structures needed to colonize the land, the evolutionary opportunities, for vertebrates were immense. Only about

50 million years after the first amphibians appeared, a relatively short time in geological terms, the first reptiles evolved from an amphibian ancestor, and by the Triassic period about 240 million years ago they had become the dominant land-dwelling animals. This was the beginning of the era of the mighty dinosaurs.

From Fishes to Amphibians

The origin of early tetrapods is still a matter of scientific debate. Among the animals alive today, the closest relatives to the tetrapod ancestor are probably the lungfishes. They too have internal nostril openings (choanae), allowing air to be taken in while the mouth is closed, as well as lungs and a functionally divided heart.

Another living fish sharing similarities with tetrapods is the famous coelacanth, which was discovered in the seas near South Africa as recently as 1938, bearing testimony to the survival of a group thought to have died out 65 million years ago. The coelacanth's paired fins make the kind of movements that we assume tetrapod ancestors made, although its anatomical details are not closely related to those of tetrapods. Fossils of lobe-finned fishes of the Devonian period show closer similarities, both in skull structure and in the structure and in the structure of the pectoral fins.

Early Amphibians

The earliest known amphibians, *Ichthyostega* and *Acanthostega*, whose fossils were found in Greenland, lived about 360 million years ago. They belong to a group known as labyrinthodonts, named for the labyrinthine infoldings of the pulp cavity of their teeth. These early forms are interesting because they retain some fish-like characteristics, such as a fish-like tail and a rudiment of the bony gill cover in the cheek region of the skull. In fish the bone supporting the bony gill cover also supports the jaw articulation. Later, during the course of tetrapod evolution, this bone was eventually reduced to a slender rod (the stapes) capable of transmitting sound waves picked up by a tympanic membrane (the eardrum) to the pressure-sensitive inner ear.

In fish the pectoral or shoulder girdle forms the back end of the skull. In tetrapods such as amphibians and reptiles, this girdle is not attached to the skull but is separated from it by a neck which allows the skull to move more freely. The pectoral and pelvic girdles anchor and support the limbs, and the vertebral column becomes suspended between the pectoral and pelvic girdles like a bridge, requiring not only modification of soft tissues but also the evolution of special joints between the vertebrae to permit the backbone to bend sideways.

Simultaneous breathing and locomotion may not have been possible for the early amphibians if they ventured on land, because the rib movements needed to ventilate their lungs may haw conflicted with the muscle contraction required to undulate the vertebral column from side to side.

"Reptilomorph" Amphibians

One very primitive tetrapod from the early Carboniferous period about 340 million years ago was *Crassigyrinus*, whose fossils were found in Scotland. From its rather weak limbs, we assume it had aquatic habits. It too retained a rudiment of the original gill cover and had skull proportions similar to those of lobe-finned fishes. But the structure of the skull indicates that *Crassigyrinus* may be related to the Anthracosauria, one of the major divisions of labyrinthodont amphibians.

The anthracosaurs are now considered to belong to the "reptiliomorph" group, including the ancestor of reptiles, birds, and mammals—as opposed to the "batrachomorph" group, which included the ancestor of modern amphibians.

The reptiliomorphs were predominantly aquatic amphibians and retained traces of a lateral line canal system in the skull bones (used to detect changes in water pressure or water currents) and had an elongated body with short limbs adapted for eel-like locomotion. A few had well-developed limbs, such as *Proterogyrinus*, a predominantly terrestrial animal up to 60 centimeters (24 inches) long.

During the Permian period 285 to 245 million years ago, the evolutionary trend in some amphibians seems to have been toward adaptations for terrestrial life. Fossils have been found in "red beds" (red sandstone) in locations that in the Permian were dry coastal plains. *Seymouria*, from the Permian red beds of Texas, was about 60 centimeters (24 inches) long, rather massively built, and would have lived mainly on land; it was a reptiliomorph in a number of features, most notably the skull and the vertebral column. *Diadectes*, the earliest tetrapod that can be positively identified as a plant-eater, was up to 3 meters (10 feet) long. The fact that this reptiliomorph amphibian was for a while classified as a reptile highlights the difficulty scientists have in separating reptiles from amphibians on their skeletons alone.

"Batrachomorph" Amphibians

Among the earliest labyrinthodonts were two representatives of the "batrachomorph" amphibians (Greek *batrakhos* means "frog"), which lived during the early Carboniferous period about 360 million years ago. They are classified in the order Temnospondyli. *Caerorhachis* was predominantly terrestrial, and unlike the reptiliomorphs (and all modern reptiles and amphibians) it did not have the "otic notch", a concavity in the skull above the cheek region which accommodates the tympanic membrane. *Greererpeton*, an aquatic animal, also had no sign of this otic notch; in its skull the stapes was a massive element (similar in that in fossils of finned fishes) supporting the braincase. The development of a middle- ear structure occurred only later in batrachomorph evolution, among the Temnospondyli, apparently independent from the reptiliomorph group. This suggests that the middle ear of modern amphibians has quite a different origin from the middle ear of reptiles and mammals.

Dendrerpeton from the late Carboniferous period of North America, about 290 million years ago, was one of the earliest representatives of the Temnospondyli with an otic notch and predominantly terrestrial habits. In later representatives of this group, such as the larger, massively built *Eryops* and the heavily armored *Cacops*, both found in North America, the skull was still relatively high and narrow, with the eyes on the sides but quite high; the vertebrae and the ribcage were well developed; and strong girdles and limbs supported the animals on their predatory excursions. During the Triassic period when seas inundated the northern continents, many amphibians grew to astonishing size, the body was elongated, and the limbs reduced; in *Metoposaurus* and *Cyclotosaurus* the skull was large (up to 1 meter, or $3^1/_4$ feet long) and flat, with the eyes on the top. Among other representatives of the Temnospondyli were the Branchiosauridae, a diverse family whose fossils are abundant in Europe, including not only the aquatic larvae of predominantly terrestrial adults, but also creatures that reached sexual maturity as an aquatic larva, like the axolotl today (order Caudata).

It is generally accepted, on the basis of a number of characteristics (most importantly the structure of the inner ear and the pedicellate teeth), that all modern amphibians have a common ancestral origin within Temnospondyli.

Modern Amphibians

The class Amphibia, subclass Lissamphibia, is represented by the three living orders of amphibians: the Anura (frogs and toads), the Caudata (salamanders and newts), and the gymnophiona (caecilians).

Perhaps the ancestors of today's amphibians belonged to one particular branch of the Temnopondyli. *Doleserpeton* was a small terrestrial amphibian of the early Permian period about 285 million years ago. Fossil bones found in Oklahoma, United States, share with modern amphibians many common features in vertebral structure. *Doleserpeton* also had teeth of a type known as pedicellate, which occur in today's amphibians, in which the tooth is divided horizontally by unmineralized tissue—quite unlike the labyrinthodont infolding described earlier. Most fossil skulls of *Doleserpeton* are smaller than 12 millimeters ($^1/_2$ inch) in length, so it is possible that they are the skulls of immature animals, and the peculiar tooth structure may be a juvenile feature.

The fossil record of lissamphibians is rather poor, frogs being he notable exception. *Triadobatrachus* from the early Triassic of Madagascar, 245 million years ago, resembled modern frogs in a number of features such as the skull and the elongated hind limbs.

The earliest "true" salamander fossil was found in Russia and lived during the late Jurassic, about 150 million years ago. The earliest caecilian fossil is from the late Cretaceous, more than 65 million years ago, but only a vertebra was found. However, evidence suggests that the caecilian group is as old as the bread-up of the Gondwana landmass into the southern continents, which occurred some 200 million years ago.

3

ORIGIN OF AMPHIBIA

The origin of the amphibians was a momentous event, since attainment of the land opened vast new territories to the vertebrate animals and permitted them to evolve their most advanced forms. Despite the importance that terrestrial vertebrates were to have, however, their initial evolution was not in any way unusual or spectacular. The amphibians were not the last survivors of a lesser class but one of a number of new forms produced as the early bony fishes diversified rapidly in the Devonian period. At their first appearance, they gave the impression less of a revolutionary new group than of fishes peculiarly adapted for special habits of life. Outwardly, except for their legs, they resembled the rhipidistian fishes from which they sprang. Very likely, they continued to swim in the shallows, as their sharp-toothed forebears had, preying upon the abundant placoderms and early paleoniscoids to be found there. Paleontologists are quite certain of the relationship between the rhipidistians and the amphibians even though they have not discovered the animals intermediate between the finned and limbed forms. The remains of the oldest tetrapods in their collections leave no doubt about the derivation of the axial skeleton from fishes of the rhipidistian group. Since the fossil material provides no evidence of other aspects of the transformation from fish to tetrapod, paleontologists have had to speculate how legs and aerial breathing evolved and why a group of fishes produced forms that habituated themselves little by little to life on land.

The Devonian Environment

Since the evolution of any group of living things is a progressive response to the demands of the environment, investigators have discussed at length the conditions that prevailed at the time of the rhipidistian-amphibian transition. By pooling interpretations based upon geological, botanical, and zoological data, they have been able to visualize the earth as it was then and to guess what might have driven the incipient amphibians on their new course. In the Devonian period, they reason, primitive green plants were advancing tentatively beyond the dampest ground at the edges of freshwater lakes and streams. Harbingers of the great tree-fern and horsetail forests of Carboniferous time, these early land plants sheltered the many-legged arthropods which were also acclimating themselves to life in dry air. Beyond the water's edge, the land stretched away unprotected by vegetation. Rain fell and ran off in torrents, crumbling igneous and metamorphic rock into pebbles, sand, and clay. In some areas, the wind picked up the finer particles and blew them into dunes, while in other places loads of sediment were washed into the water and layered over the bottom. The red colour of much of this transported material has provided scientists with an important clue to the climatic conditions that existed in the Devonian years. Geologists know that the ruddiness of soils is due to hematite, a compound containing iron in ferric, or highly oxidized, form. Iron in this state is produced when sediments containing ferrous silicates are laid down in warm and moist areas where the surface is subject to drying from time to time. Ferric iron is stable once it appears, and deposits containing it are compressed into layers of rock called red-beds.

Behaviour of the Earliest Amphibians

As long ago as 1916, Barrel pointed out that the existence of extensive Devonian red-beds signified widespread semiarid conditions during the period when the amphibians arose. He believed that lungs evolved as an adaptation to the droughts that intervened between weeks or months of steamy rainfall. Lull concurred and emphasized the role of natural selection in preserving a population of vertebrates that was developing the ability to sustain itself out of water. Romer described graphically how the earliest tetrapods might have saved themselves from death by leaving a drying pond and hitching themselves overland to a deeper, better-filled pool. Critics of Romer's theory challenged it on several bases. Inger questioned the idea that the Devonian climate was surely semiarid, stating that red soils are being formed today in parts of the world where tropical rains are uninterrupted. Even if dry seasons did exist, he maintained, it was hardly likely that the first amphibians behaved as Romer thought they did. Since they were dependent upon small fish for food, they would have lived in permanent waters rather than in ponds which became periodically uninhabitable. A drought severe enough to destroy the pond or lake in which they had established themselves would certainly evaporate similar bodies of water within a radius of many miles. An attempt to migrate any distance under such conditions would result in the animals' dying of desiccation on the bare ground. Inger imagined that lungs were selected for, not because their possessors ventured onto dry terrain, but because these organs enabled fishes in warm, oxygen-poor water to survive by breathing air at the surface. He thought that the animals' first excursions beyond the water's edge were prompted by increasing population pressures in old communities; during humid nights, individuals driven by lack of adequate food or space might have squirmed over the wet earth in search of less crowded ponds much as the fighting fish *Betta* and the catfish *Clarias* do today in tropical Borneo.

Goin and Goin seconded Inger's suggestion, reporting that this kind of behaviour can also be observed in certain modern frogs. These amphibians do not attempt to migrate in the dry season. Sharp reduction in the water level of a pond, in fact, causes the inhabitants to gather in the center, where the remaining water is deepest. Landward movement occurs when rain wets the ground but has not yet restored the pond to its former size. Young frogs leave, then, and colonize less densely populated pools nearby. The Goins agree further with Inger that protoamphibians, impelled to quit their home waters under similar conditions, probably began to eat the newly terrestrial arthropods as a supplement to their regular diet of fish. Once accustomed to a terrestrial food supply, the air-breathing former fishes were launched on their land career.

Romer insists that amphibians remained aquatic animals much longer than Inger and the Goins suppose. Although he holds to his idea that the earliest amphibians traveled over land to escape their drying pools, he believes that well into the Carboniferous period their normal life was carried on in the water. Not until that time, when insects and land plants radiated with amazing rapidity, would there have been sufficient food material on land to sustain a population of vertebrates. The climate in Devonian years was a far harsher one, he still thinks, than Inger envisions. That seasonal droughts were the rule rather than uninterrupted humidity seems proved by the mud cracks and evaporites that appear in the rocks of the age. Transition to terrestrial feeding and a more or less terrestrial habitat would have been unlikely during a period when desert dryness occurred in alternation with earth-soaking rains. Though the first tetrapods were able, in his opinion, to survive an occasional forced march to new waters, the rigorous conditions which necessitated their migration would have precluded their dallying to catch any scorpions or other arthropods that might have crossed their path. During their early history the amphibians probably stayed in the water, protected from the extremes of climatic change, and developed their taste for invertebrates by eating the aquatic insect larvae which hung upside down from the surface of the pond. Romer adduces more direct evidence for his contention that amphibians emerged from the water relatively late by turning to the fossil record; the oldest members of the class retained the sinuous body of the swimmer and had, with few exceptions, small legs that would hardly have allowed easy locomotion on land. Not until the end of the Carboniferous period did a number of amphibians appear whose strong limbs and stout body were adapted for terrestrial. rather than aquatic life.

Structural Modifications Necessary for Living on Land

Why the amphibians left the water and when constitute only a part of the puzzle to be solved by those who are studying the origin of the first tetrapods. Paleontologists tracing amphibian history have also to analyze the sary to transform a fish into an animal viable on land before they can explain the steps by which rhipidistian forms evolved into the earliest terrestrial species. It is possible to draw such guide lines, because observation of living animals makes patent the different demands on the vertebrate body of watery and aerial environments. Underwater, animals are supported by the medium, kept moist, and supplied with dissolved oxygen. Once they come out into the dry air, they are faced with the necessity of holding themselves up and moving in a much less buoyant substance, of extracting oxygen from a gaseous mixture, and of conserving the water that forms the basis of their protoplasm. The physical forces operative in the terrestrial environment make certain mechanical arrangements a necessity and others an impossibility if the animal is to succeed in these tasks. Determining the time and the order in which the required structural changes took place is a more speculative matter. Although many of the modifications were correlated and so must have occurred simultaneously, paleontologists are not certain of how much alteration took place as protoamphibians were acclimating themselves to land life and how much occurred preadaptively at an earlier time when the animals were still wholly aquatic.

Breathing Mechanisms in Early Amphibians

Although several Devonian fishes possessed lungs, it is not at all clear that elaboration of these structures was an early event in amphibian evolution. Ever since its initial appearance as an outgrowth from the pharyngeal region of the gut, the lung seems to have functioned as a respiratory membrane. Its usefulness to aquatic forms confined in waters of low oxygen content surely explains the perpetuation of the organ in several lines of bony fishes. Although much later in vertebrate evolution the lung itself would increase in complexity as advances in the design of internal organs made possible a higher rate of metabolism, a simple internal respiratory sac not too different from that of the fishes would very likely have sufficed for the cold-blooded protoamphibians. A land vertebrate could not survive, however, without some mechanism for assuring the passage of air from mouth to lungs. In fishes this transport may be a passive process: a fish can gulp air and then plunge head downward, causing the bubbles to rise through the pharynx and so pass backward into the lung. A land animal which remains horizontal must have some way of forcing or drawing air through its respiratory tract. Certain structural innovations which appear in early tetrapods could have arisen in conjunction with the requirement for a respiratory current. The increased length and stoutness of the ribs characteristic of the first known amphibians afforded added surface for the attachment of muscles which could have produced a rise and fall in the body wall, rhythmically changing the pressure around the lungs. Air, sucked into the respiratory sacs when their internal pressure fell below that of the outside, would have been expressed when the pressure on the lungs rose. This kind of respiratory mechanism, which survived in higher vertebrates, was surely not the only one which evolved in the ancient amphibians. The gradual broadening of the head in many forms can be interpreted as evidence for the existence in Paleozoic times of the force-pump apparatus that remains in the ribless amphibians of today. In these animals, the floor of the mouth is lowered with the nostrils open to draw air into the mouth and then raised with the nostrils closed to press the air backward into the lungs. The wider the floor of the mouth, the larger its underlying muscle sheets and the greater the efficiency of the mechanism that depends upon their action. The tendency toward flattening and widening of the head evident in several lines of Paleozoic amphibians could be explained by postulating use of the force-pump process in these animals and the continuing selection of types whose skull structure made possible its more effective operation.

In the fossil record there is evidence to suggest that reliance on either one of these respiratory mechanisms developed slowly, both phylogenetically and in the life of the individual. The young of the early amphibians, like the young of modern forms, grew gills first and then developed functional lungs as they matured. During their metamorphosis, the animals almost certainly depended to some extent upon the thin, moist skin (if they were not heavily armored) and the mouth lining as auxiliary respiratory surfaces, as existing amphibians do. Whether the older amphibians were able to continue to respire through the skin as adults is not clear. Although fossil evidence is scanty, it seems that many, if not all, of them retained a covering of bony scales that would have limited gaseous exchange at the surface of the body. The scales over the back and flanks were thinner than those of the ancestral rhipidistians, however, and possibly not present everywhere, so that some cutaneous respiration might have persisted throughout life in these animals.

One paleontologist believes that the appearance of legs facilitated respiration as well as locomotion in terrestrial vertebrates. I. I. Schmalhausen points out that a fish lying on its side on the ground bears the weight of the body wall upon its viscera despite the presence of ventral ribs. When the internal organs are compressed in this way, the fish can force air through the pharynx only with great difficulty. For air to pass easily into the lungs, the trunk has to be propped up so that the respiratory organs hang suspended in the body cavity. A rhipidistian fish could have used its muscular fins to lift its body from the surface temporarily, but in the absence of legs the body would have dragged or flopped when the animal moved. Although there might have been a rhipidistian that could expend the large amounts of energy necessary to make the lungs work under these conditions, a form with a fish-like trunk would have been better adapted for living on land if the body was kept clear of the ground.

Changes in the Skeleton Associated with Terrestrial Life

Elevating the body was also desirable as preparation for easy locomotion over dry ground. A fish moves against the water which buoys it up, generating a minimum of friction, but a tetrapod in contact with the substratum would be considerably hindered by the scraping of its body along the uneven surface unless, like the snakes, it was specially modified for that kind of locomotion. Raising the trunk and tail eliminates at once the necessity of slithering along every rise and fall in the terrain and the danger of scraping the epidermis to shreds in doing so. The transformation of the fins from steering devices to piers for the suspension of the body involved changes in every part of the appendicular skeleton. Besides reorienting at least a portion of the limb in a vertical direction and exchanging for the fin rays feet that could be planted flat on the ground, the relationships of the girdles to the axial skeleton had to be substantially modified. In fishes, the pelvic girdle always consisted of a pair of bony plates embedded in the ventral body wall. A tetrapod limb, pressing upward against such a structure, would cause it to sink inward and compress the soft organs in the posterior part of the body cavity. If the hind legs were to hold up the animal and push it forward, there had to emerge a rigid connection between the pelvic girdle and the vertebral column. With such an arrangement, the thrusting force of the foot against the ground could be transmitted with little loss to the main axis of the body.

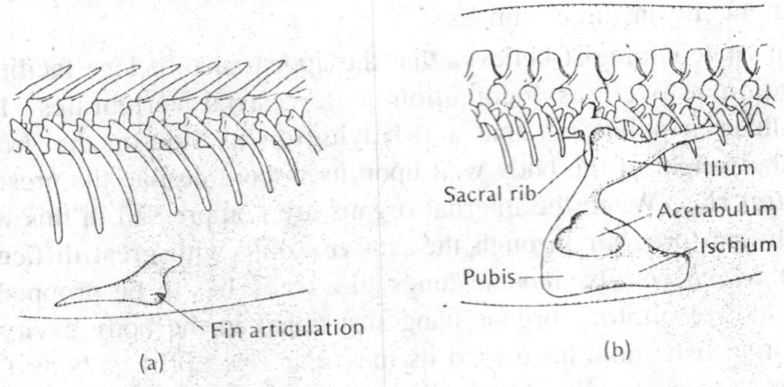

Fig. 3.1. The relationship of the pelvic girdle to the vertebral column in fish and early tetrapods. (a) Left lateral view of the column and pelvic girdle of a fish, showing the girdle embeded in the musculature of the ventral body wall. (b) The same view of the column and girdle in a primitive amphibian.

A union of the appendicular and axial parts of the skeleton in the pectoral region already existed in fishes: from earliest times, the dermal shoulder shield had articulated firmly with the back of the skull. This association, while it was undoubtedly advantageous for fishes, was unsuitable for a tetrapod. The forelegs are a sort of landing gear in a terrestrial animal: they receive the weight of the body as it is propelled forward by the hind limbs. To absorb the shock generated by the descent of the trunk, the pectoral girdle has to be disconnected from the head. Even if an amphibian had been able to withstand the tension on the skull that would have been created at every step so long as the articulation between girdle and skull remained, another difficulty would

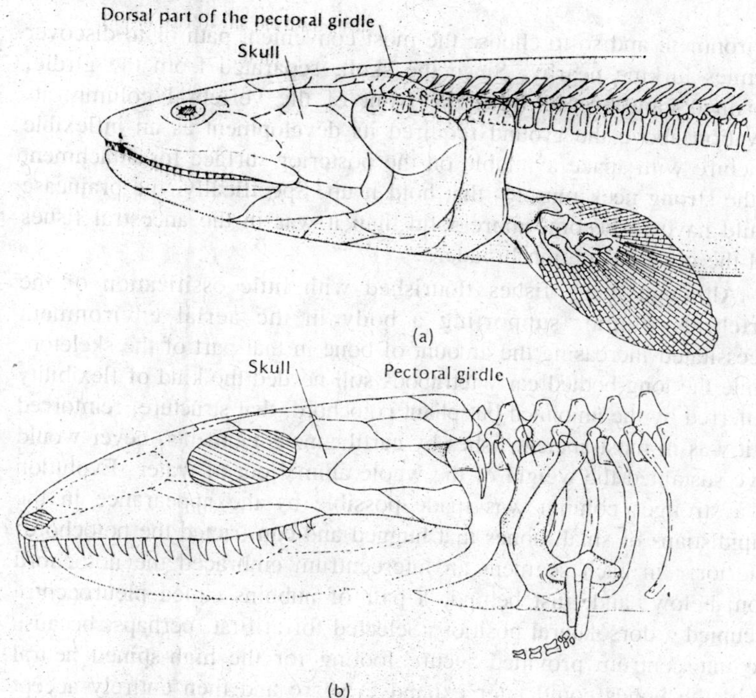

Fig. 3.2. The relationship of the pectoral girdle to the skull in fish and early tetrapods. (a) Skull and pectoral skeleton of a rhipidistian fish. (b) Skull and pectoral skeleton of an early amphibian.

have ensued. Because the scapulocoracoid element which receives the limb bone abuts the cleithrum of the shoulder shield along its vertically oriented medial surface, pressure from the leg would set up a shearing stress between the endochondral part of the girdle and the rigidly held dermal portion. Since the skeleton is most vulnerable to shearing forces, the resulting weakness of the suspensory apparatus would limit the weight, and thus the size, of terrestrial vertebrates. Transformation of the girdle into one strong enough to support a walking animal of any mass demanded a loss of the connection of the dermal bones with the skull and a concomitant expansion of the endochondral part to which the legs are attached.

Freeing the skull from the pectoral girdle surely brought with it other advantages for an animal making its way on land. For the first time, a vertebrate would be able to turn its head and to lift it above the level of the body without tilting its tail downward. This ability made it possible to see beyond small obstacles in the immediate environment and so to choose the most convenient path or to discover enemies lurking nearby. Since the skull, separated from the girdle, remained cantilevered from the front of the vertebral column, its elevation above the ground required its development as an inflexible structure with space available on the posterior surface for attachment of the strong neck muscles that hold it up. Specifically, the braincase would have to become more solid than it was in the ancestral fishes and the occipital region broader.

Although many fishes flourished with little ossification of the vertebral column, supporting a body in the aerial environment necessitated increasing the amount of bone in that part of the skeleton. While the long-bodied early tetrapods still needed the kind of flexibility conferred by the unjointed but pliant notochord, that structure, reinforced as it was in most ancient fishes by cartilaginous elements, never would have sustained the weight of the whole animal out of water. Evolution of a stronger column was made possible by the appearance in the rhipidistians of small bones that hugged and constricted the notochord: anteriorly in each segment an intercentrum embraced the notochord from below, and, just behind, a pair of nubbins called pleurocentra occupied a dorsolateral position. Selected for at first, perhaps, because the intercentrum provided secure footing for the high-spined neural arch, the bones could later expand to share and then entirely accept the stresses projected through the vertebral column of the land animal. Contact between these central elements was apparently insufficient to produce stability, however, for no tetrapod evolved without an accessory articulation more dorsally between the neural arches. The outgrowth fore and aft of paired zygapophyseal, or yoking, processes from the roof of each arch resulted in the development of gliding joints which added strength to the column but interfered little with its flexibility.

Protecting the Body from Desiccation

The structural modifications which enable an animal to support itself and to respire in an airy medium would be of no benefit unless the animal could protect its body from desiccation. Since vertebrates, like all living things, are composed of watery protoplasm, this is a formidable task. Eventually, in the tetrapod line, it was to be accomplished in a way similar in principle to that utilized by all organisms adapted for terrestrial life: the body became enclosed in a nonliving coat impervious to water, which exposed moist tissues in as few places as possible. Whereas other forms of life manufactured cork and cuticle, shell, slime, and exoskeleton to retard the loss of water, tetrapods evolved a layer of dead, keratin-filled cells at the surface of the epidermis which confined internal fluid reasonably well. Gills, which have to remain moist and exposed if they are to function, had to be abandoned, and respiratory membranes developed in more protected parts of the body. The formation of internal nares was preadaptive for animals that had to restrict contact between the dry air of the environment and their respiratory organs. With a pair of narrow passages that led from the exterior to the front of the oral cavity, protoamphibians could admit in two fine streams the air necessary for respiration rather than exposing the moist interior of the mouth broadly by parting the jaws. In higher tetrapods, protection of the internal tissues would be increased by further isolating the respiratory stream within the oral cavity and lengthening the distance through which air had to travel before reaching the lungs.

If the condition of modern amphibians is representative of the level to which the first tetrapods had brought their waterconserving abilities, it is obvious that the bare beginnings of protection against desiccation were sufficient to allow vertebrates to gain the land. In living amphibians, the skin exhibits a keratinized layer but one thin enough in most species to be permeable to water under certain circumstances. The delicacy of the protective layer enables it to serve a double function: covered with mucus, it slows the loss of water when the animal is exposed to the air but permits fluid to enter when the animal submerges itself after a terrestrial excursion. The increase in cutaneous permeability shown on the latter occasion is under hormonal control. Although it is not known whether endocrine activity of this sort is a specialization in modern amphibians or a legacy from ancient forms, it is safe to suppose that physiological mechanisms must have appeared early to supplement the structural alterations which adapted vertebrates for land life. If hormonal regulation of cutaneous permeability did exist in the first amphibians, it would have been effective combined (as it is in extant forms) with endocrine control of the kidney. If the emergent tetrapods secreted hormones that acted to conserve water by inhibiting filtration of fluid from the bloodstream into the urinary tubules or by encouraging the recovery of water by resorption through the tubule wall, the ability of these animals to survive in the atmosphere would have been enhanced.

Scales disappeared, and other structural defenses against drying out remained limited among the amphibians. These animals assured their viability by behavioural adaptations rather than by paralleling the higher tetrapods in the development of increased physical resistance to desiccation. They frequented moist areas adjacent to streams or lived in marshes, where the heat of the sun made the air steamy rather than dry. When drought was inescapable, they burrowed or lay dormant underground. Undoubtedly, the earliest amphibians fertilized their eggs externally, as many living forms do, laying them in the water, where they were safe from drying if not from the depredations of hungry fishes. Surviving embryos, unprotected by extraembryonic membranes, became gilled larvae that passed through an obligatory aquatic stage before they could step out on land. Those later amphibians which became more terrestrial evolved a variety of special reproductive habits designed to accommodate or abbreviate the larval requirement for water. Their eggs are laid in damp moss, gelatinous froth, or temporary puddles or fertilized and held for a time within the body; the embryos may mature with unusual speed, acquiring adult characteristics before they leave the egg. None of these peculiar inventions permitted the degree of independence from fresh water enjoyed by higher vertebrates, which enclose their embryos in a fluid-filled sac and protect them further with a shell or by housing them internally until they can survive in dry air.

Modifications of the Sensory Organs in Protoamphibians

Although tetrapods were able to spend time on land before they evolved a truly impervious skin and a way of reproducing in the absence of water, they could hardly have managed if their sensory organs had not soon become adapted to functioning in terrestrial surroundings. The eyes of bony fishes, their chemoreceptors, and their lateral-line organs for sensing vibrations in the environment were designed to work in a watery medium. In air, the surface of the lidless eyes would dry in a short time, as would the soft tissue of the lateral-line canals. The olfactory epithelium, usually folded loosely in fishes, would quickly lose its moisture and shrivel. In addition to glands that secreted mucus over the skin, the first vertebrates which were even partially terrestrial required others that continuously bathed the exposed parts of the sense receptors. The nasal lining, recessed within the respiratory passage, was protected sufficiently by this method alone. The eye acquired a lid that could be passed over the surface periodically for instant rewetting of the cornea. With these minor alterations and certain others in the shape of the cornea and lens, the olfactory epithelium and the eye became immensely valuable for the new vertebrates. Since light travels and substances diffuse more rapidly through air than through water, tetrapods could sense changes in their environment at a much greater distance than fishes ever could. That they relied increasingly upon these sensory organs can be inferred from the gradual enlargement of the cranial spaces for the olfactory and optic lobes of the brain, to which impulses from these receptors pass.

No structural modification, however, could turn the lateralline system to advantage in a land animal. Its action depended upon the passage through superficial canals of a current of water to deform a gelatinous coating around the cilia projecting from its sensory cells. Even if the gummy coating could have been kept wet, it is unlikely that pressure waves in air would have been forceful enough to disturb it. Because sensitivity to aerial vibrations allows an animal to detect the presence of enemies it cannot see or smell, it was to be expected that a substitute for the lateral-line system, if one appeared, would be selected for. Actually, no new organ arose in the protoamphibians, but by a series of changes in old structures, an arrangement was achieved whereby cells in the inner ear closely related to the sensory cells of the lateral-line system were made responsive to external vibrations. Of chief importance in this renovation was the change that took place in the hyomandibular bone. In its position behind the spiracular gill slit in ancient fishes, this element of the branchial skeleton had served as a prop for the jaws, bracing the quadrate bone against the otic region of the braincase. When, in the evolution of the protoamphibians, the upper jaw became attached firmly to the skull, the hyomandibula relinquished its articulation with the quadrate and expanded in a lateral direction. A process that already extended outward to the operculum in rhipidistians broadened and, after the disappearance of the gill cover in the earliest tetrapods, came to rest against a membrane that spanned the space left between the skull roof and cheek. At its medial end, the hyomandibula remained pressed against the cranial bones that enclosed the inner ear. When eventually the cranial wall became membranous instead of bony where the hyomandibula touched it, the old suspensory element was transformed into a functional stapes or columella. Enveloped by an extension of the spiracular pouch, now the cavity of the middle ear, the bone was free to vibrate when the external membrane, or tympanum, against which it rested, was disturbed by vibrations in the air. The new stapes, as it moved, transmitted pressure waves through the membrane at its medial end to the fluid around the inner ear. There, the sensory cells were stimulated, as they always had been, by the agitation of the liquid in which their gelatin-covered cilia were bathed.

Oldest known Amphibians: the Ichthyostegids

Although it is not possible to tell from the remains of the first known amphibians how far evolution of the ear had progressed, it is clear that by late Devonian times a number of other structural adaptations for land life had been realized. Paleontologists have found in what is now eastern Greenland fossils of vertebrates that were even then robust tetrapods. These animals were not adapted to withstand the arctic conditions that currently prevail in that part of the world. The climate at the end of the Devonian period was far milder. For most if not all of the year, rain water that fell in the highlands ran freely down to the sea, filling the land with streams, shallow backwaters, and rivers that housed a rich variety of lobe-finned fishes. It was an environment conducive to fossilization: not uncommonly, dead fish were carried along with the current until they were buried in sediment at the bottom, often in estuarine areas, where their bones rested side by side with brackish-water placoderms, acanthodians, and sharks. The amphibians that appeared in eastern Greenland were transported downstream in the same way. Alive, they shared the freshwater habitat of their rhipidistian cousins, leaving it from time to time to make forays out onto solid ground.

The ichthyostegids, as paleontologists named these early tetrapods, already had well-developed legs. The limb skeleton exhibited, fully defined, the pattern that was to remain characteristic of land vertebrates. A single, large bone extended through the upper part of the leg to meet, at knee or elbow, two elements which supported the lower limb. These bones articulated distally with a number of small ones that gave flexibility to the wrist or ankle region. The five-toed foot stretched out flat, its metapodial bones bracing those of the digits. Although the proximal bone extended laterally from the side of the body like the fin of a fish, the lower part of the leg was turned in a vertical direction in such a way that it could lift the body from the ground.

The girdles of the ichthyostegid amphibians were as completely adapted for bearing weight as the limbs. The dermal part of the pectoral girdle had lost its connection with the back of the skull: the bones through which the attachment was made in ancestral fishes, the posttemporal and supracleithrum, had disappeared entirely. Instead of being braced against the skull, the remaining dermal bones on each side abutted a new bone formed in the ventral midline, the interclavicle. The dominant element in each half of the pectoral girdle was now the scapulocoracoid, with which the upper limb bone articulated. It rose dorsally to form a strong blade for the attachment of muscles which bound the girdle against the trunk. The pelvic girdle had evolved the rigidity and the contact with the axial skeleton that allowed it to serve as a firm anchor for the hind legs. The two loosely associated ventral plates of fishes had given way to a structure built of left and right halves almost immovably joined to each other in the ventral midline. In its basic design, this girdle was like that of later tetrapods. The three bones of each side met laterally in the acetabulum, the depression into which the head of the proximal limb bone fits. Two of them, the pubis and the ischium, spread from this point ventromedially to meet their counterparts in the midline, affording a surface for the origin of muscles that moved the leg downward and back and forth. The third bone, the ilium, extended upward to reach the sacral rib and vertebra, to which it was securely tied. Besides bracing the girdle against the vertebral column, this bone provided anchorage for the muscles which raised the limb. A posteriorly projecting process of the ilium in ichthyostegids apparently gave attachment to muscles less important in terrestrial.locomotion than the levators of the leg, for it appeared in only one later group of amphibians before vanishing forever.

Although fossil evidence for terrestrial adaptation of soft organs is not direct, as it is for the changes in the appendicular skeleton, it is possible to make some inferences about the status of the respiratory organs in the ichthyostegids. The structural requirements for pulmonary breathing were present: internal nares existed at the edge of the palate, and broad ribs framed a capacious chest cavity, in which the lungs could have expanded freely. That the animals may have relied to a great extent upon their lungs is suggested by the limited space which remained for the gills. The branchial chamber, squeezed behind the long jaws, no longer extended posteriorly under a large opercular cover. Its opening apparently fitted under the vestigial subopercular bone, the only part left in ichthyostegids of the movable gill shield of fishes. It seems certain that at this stage in the evolution of amphibians the skin was important as a respiratory organ. The presence of lateral-line canals attests to its having been kept wet, and the reduced scale cover implies exposure of a vascular surface through which the exchange of oxygen and carbon dioxide could have taken place.

Behaviour of the Ichthyostegids

The persistence of the lateral-line system in ichthyostegids signifies that these animals still lived primarily in the water. They retained a strong resemblance to fishes and continued to swim and to feed in the same manner as their ancestors. Though heavy-bodied, they were elongated and streamlined for movement below the surface. From the remains of the vertebral column, paleontologists can tell that it was supple enough to undulate as the axial muscles pulled upon it. The large notochord, reinforced but apparently not restricted by the adjacent pleurocentra and intercentra, could still be flexed laterally despite the development of neural arches which articulated with each other in tetrapod fashion. If any doubt existed about the ability of the ichthyostegids to swim, it would have been dispelled by a glance at the form of the tail. It was flattened from side to side like that of any fish and supported by radial elements that were braced against the neural and hemal arches. Such a structure acts only to propel an animal through the water and does not appear in vertebrates which rely primarily upon their legs for locomotion. The ichthyostegids glided about in the shallows as easily as their finned relatives, no doubt, and like them lived on small fishes which they snapped up and impaled upon their conical teeth.

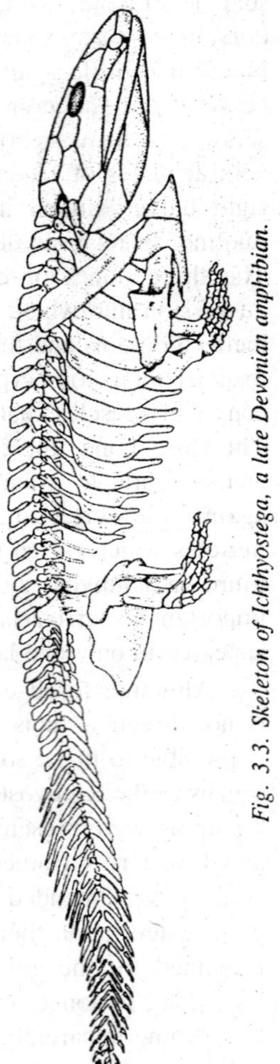

Fig. 3.3. Skeleton of Ichthyostega, a late Devonian amphibian.

Structure of the Skull in Ichthyostegids

Outwardly, the head of these early tetrapods looked not very different from that of the contemporary lobe-finned fishes. Under the skin, however, the skull had assumed the form which was to remain basic for later terrestrial vertebrates. The front and rear parts of the braincase joined each other immovably behind the recess for the pituitary gland, and the ensheathing dermal bones of roof and palate exhibited an arrangement so similar to that of other Paleozoic tetrapods that the homology of the corresponding elements is unquestionable. The head was protected by a cover of plate-like bones, which joined each other tightly at sutures. As in later tetrapods, the dorsal region was shielded by a series of paired elements which met at the midline, equivalents, most paleontologists agree, of the nasal, frontal, parietal, and postparietal bones. Beside them were the orbits, surrounded by the prefrontal, postfrontal, postorbital, jugal, and nearby lacrimal plates, five ossifications which persisted in land vertebrates of many lines. Anterior to the lacrimal, a small bone, the septomaxilla, bordered the nasal aperture dorsally, and the premaxilla edged the front of the upper jaw. The latter element bore marginal teeth, as did the maxilla which extended from the nostril back to the quadratojugal bone over the corner of the mouth. The side of the head behind the jugal and postorbital bones was sheathed by squamosal, tabular, and supratemporal elements, the first two of which, with a small preopercular plate, rimmed the otic notch, the cleft in the posterior margin of the skull that was closed in the living animal by the tympanic membrane. The undersurface of the skull was as heavily armored as the roof and cheek. The dermal palate that lay beneath the braincase within the arch of the upper jaw consisted of a lateral series of paired bones, the vomers, palatines, and ectopterygoids, all of which bore teeth, and a more medial pair of large pterygoid elements that fitted closely against a central parasphenoid bone. Each pterygoid was prolonged posteriorly into a curved flange that formed the inner side of an opening through which muscles passed to insert on the lower jaw. The bones of the lower jaw-a narrow element bearing teeth and several that supported it from below-were also easily comparable to those of later tetrapods.

Skull of Ichthyostegids and Rhipidistians Compared

The similarity of the arrangement of the dermal bones in the skulls of ichthyostegids and rhipidistians is one of the strongest pieces of evidence linking the amphibians to that particular group of fishes. No other Devonian vertebrates exhibited a pattern from which that of the tetrapods could have been derived. The elements in the skull of the rhipidistians were not identical to those of the ichthyostegids in every way, but the differences do not seem great enough to preclude a phylogenetic connection between the two kinds of animals. The snout region of the rhipidistian, much shorter than that of the ichthyostegid, was covered by a number of small bones rather than a single pair of broad nasals. The two large bones of the tetrapod could have resulted from fusion of the little plates present in the fish, however, since the tendency to combine neighboring dermal bones in the skull roof is common among vertebrates. The ichthyostegids had left in front of the paired nasal bones an internasal nubbin, which could have been a last surviving remnant of the mosaic of small plates that once existed in that area. Loss of bones is another well-documented kind of evolutionary change which could account for discrepancies between the skulls of rhipidistians and ichthyostegids. The tetrapods lost the large opercular of the fishes as well as a row of extrascapular elements at the rear of the skull and the small intertemporal bone which in rhipidistians lay anterior to the supratemporal.

The apparent difference in the location of the opening for the median eye proved the most formidable obstacle for those who would relate the ichthyostegids to the rhipidistians. In amphibians and other tetrapods that retained the rudimentary eye in the midline, light reached it through a foramen between the parietal bones. In rhipidistians the foramen existed, but it appeared at first to lie farther forward, between the frontal elements. Paleontologists of Stensio's school, convinced as

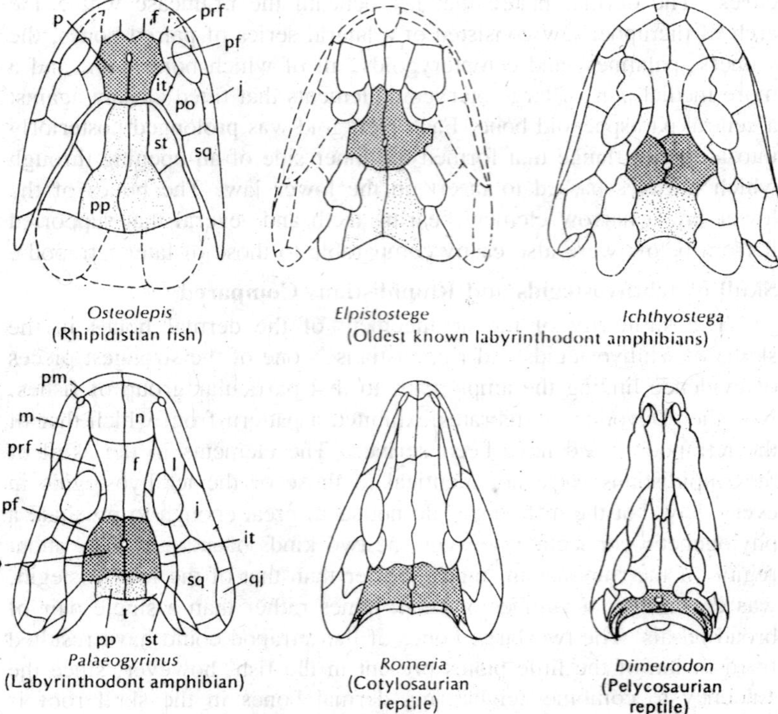

Fig. 3.4. Skull of a rhipidistian fish and several Paleozoic tetrapods in dorsal view, showing posterior migration of the dermal bones of the roof and the progressive shortening of the postparietal elements.

they all were that the rhipidistians were ancestral to the tetrapods, suggested that the foramen had changed its position relative to the bones of the skull roof. Unlike fusion or loss of bones, however, this sort of change would have been an extraordinary one. Despite Jarvik's contention that movement of the parietal foramen had occurred in certain instances, other workers remained skeptical. Finally, Westoll proposed what he considered a more probable explanation: the medial eye grew upward from the brain beneath the same pair of bones in the rhipidistians and their tetrapod descendants, but the bones themselves seemed differently placed because of a change in the proportions of the skull. In the short-snouted rhipidistians, the bones between the orbits that flanked the median eye were really the parietals, not the frontals as Jarvik thought. As the tetrapods evolved, the bones in the front of the skull fused and elongated. The true frontals extended posteriorly between the orbits, and the parietals, with the foramen between them, retreated over the skull roof. In supportt of his hypothesis, Westoll pointed out that in *Elpistostege*, a Devonian animal that he regarded as transitional between the rhipidistians and ichthyostegids, the size and position of the frontal and parietal bones were intermediate. Whether Westoll is correct in using the arrangement of the bones in *Elpistostege* as evidence for his theory is uncertain, since only a partial skull of this form is known and its relationship to the ichthyostegids is not surely established. Support for Westoll's idea has come from other research: Romer has substantiated the possibility of a gradual lengthening of the skull anterior to the orbits and a shortening of the region behind by demonstrating that this trend continued beyond the amphibian level through the first reptiles and into the mammal-like reptile line.

The advantage of the posterior extension of the parietal bones that accompanied this change in early tetrapods may have been that the skull was strengthened thereby. In rhipidistians, the boundary between the parietal and postparietal bones constituted a division of the dermal roof that coincided with the intracranial articulation of the braincase below. The fishes were able, therefore, to entire anterior part of the skull as the mouth was opened. Whatever advantage there was in an enlarged gape was apparently overshadowed by the necessity for rigid construction of the head in an animal that supported itself out of water. In ichthyostegids, the intracranial joint was immobilized, and the overlying dermal bones interlocked in such a way that the roof was no longer effectively transected. Continued posterior displacement of the parietal bones brought them to a position directly above the suture between the two halves of the braincase. The solid dermal cover across the old intracranial division, coupled with the expansion of the parasphenoid bone beneath it, reinforced the interior part of the skull significantly.

Evolution of the Internal Nares

If solidification of the skull occurred before tetrapod status was attained, it must have evolved only in the rhipidistians that were directly antecedent to the ichthyostegids. By contrast, internal pares, structures which also seem to be adaptations for land living, characterized rhipidistians generally and constitute another basis for recognizing these fishes as ancestors of the first amphibians. Schmalhausen attributes the presence of openings from the nasal passages into the front of the mouth, not to incipient terrestrial habits on the part of all rhipidistians, but to the advantage these openings conferred upon the rhipidistians underwater. In developing his thesis, Schmalhausen points out that these fishes swam and hunted among the weeds, where their sense of smell was especially important in detecting their prey. He supposes that at an early time in their evolution rhipidistians relied upon currents, as other fish did and still do, to carry water in and out of blind-ended nasal sacs. Like other bony fishes, they must have had at least two external nostrils on each side of the snout: one for entrance and the other for exit of the water. Their olfactory acuity depended upon the rapidity with which the environmental water could be made to flow over the sensitive nasal epithelium. Since the rhipidistians were not built for speedy swimming which would force a steady stream of water through the nostrils, they depended upon their respiratory current to change the water in the nasal cavities. According to Schmalhausen, water traveled more quickly through the nasal sac and out at the back as the posterior nostril became displaced toward the edge of the upper jaw, where suction created by water entering the mouth was strongest. The spread through the population of the genetic factor responsible for the new location of the posterior nostril was encouraged by another advantage besides increased powers of olfaction. Once the nostril had crossed the maxilla to open into the mouth through the palate (and that it did so Schmalhausen believes is indicated by forms like *Panderichthys*, in which the nostril seems to straddle the bone), the fish could take water into the oral cavity without moving the jaws. When a rhipidistian waiting in the weeds to ambush its prey was able to dispense with the rhythmic opening and closing of its mouth, it no longer produced the turbulence that must have betrayed the presence of its ancestors to many of their intended victims. Although Schmalhausen's speculations concerning the hunting habits of the rhipidistians can never be confirmed, his supposition that the internal nares, or choanae, served first for the passage of water and only later for the intake of air is a reasonable one. Internal nares, like many other vertebrate structures, may have been selected for because they were advantageous in one way and then retained because they became useful in another.

Structure of the Teeth in Ichthyostegids

In addition to the internal pares, the teeth of the ichthyo stegids were a legacy from the rhipidistians. The teeth which edged the jaws and studded the palate in both forms were conical and sharp, like those of most fish-eating vertebrates, but they bore distinctive striations from the base to the tip of the crown. When a tooth is cut in cross section, each striation can be seen to mark an infolding of the tooth wall. In the interior of the tooth, the enamel is further convoluted to

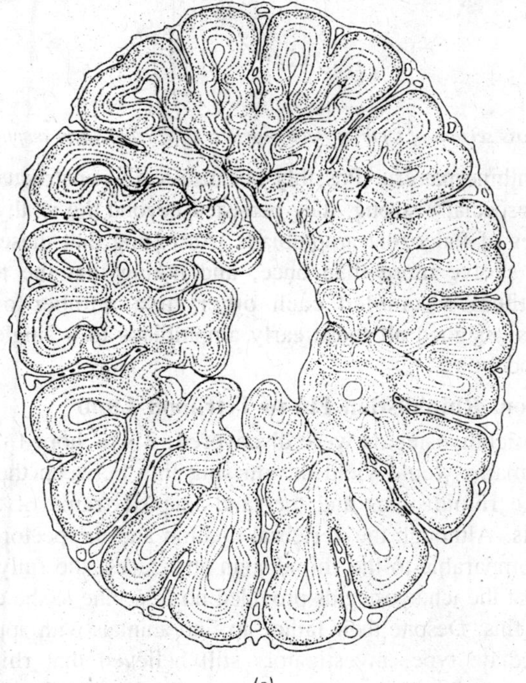

(a)

Fig. 3.5. Cross section of teeth of Polyplocodus, a rhipidistian.

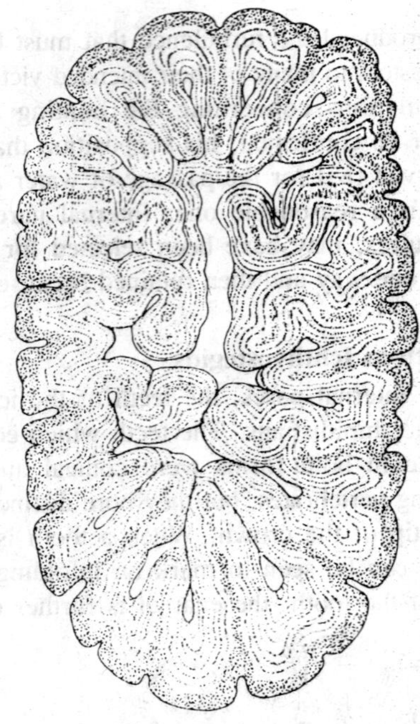

(b)

Fig. 3.6. Cross section of teeth of Benthosuchus, a labyrinthodont amphibian.

form a labyrinthine pattern. Possibly because teeth constructed in this manner were especially strong, they persisted beyond the ichthyostegids in several lines of Paleozoic amphibians, Assuming that these curious teeth did not evolve more than once, and noting that the tetrapods which bore them resembled each other in a number of ways, paleontologists assigned all these early amphibians to a single group, the Labyrinthodontia.

Transition from Rhipidistian Fin to Tetrapod Limb

When the ichthyostegids were discovered, it was hoped that their postcranial remains would yield the missing link between the piscine skeleton of the rhipidistians and the land-adapted frame of the later labyrinthodonts. Although the vertebral column and the pectoral girdle were easily comparable to the rhipidistian structures, the fully formed five-toed legs of the ichthyostegids provided no new clue to the evolution of limbs from fins. Despite their failure to find animals with appendages of an intermediate type, investigators still believed that rhipidistian fins were the most likely precursors for the legs of land animals. In contrast to the fins of the actinopterygian fishes, in which several short basal elements articulate with the girdle, rhipidistian fins attached to the girdle through a single bone, as legs do. This most proximal bone was followed distally by a series of smaller ones, making a supportive axis that projected some distance beyond the trunk into the fin. Each axial elementt had braced against its anterior side a radial bone that broadened the surface available for the insertion of appendicular muscles. These elements, interposed between the girdle and the fin rays, seemed a better source for the bones of the tetrapod limb than the transverse rows of basal rods present in the ray-finned fishes. Taking into account anatomical and embryological evidence, Westoll was able to correlate each of the rhipidistian ossifications with a counterpart in the pentadactyl appendage. The proximal element, which he called the first mesomere, he considered homologous to the humerus in the pectoral limb. It projected directly outward from the body in the early limb, just as it did in the ancestral fin. The second mesomere, or axial element, became the ulna and twisted forward to lie beside the first radial, now the radius. The change in the orientation of the second mesomere produced the sharp bend at the future elbow joint and also crowded the second radial element, which subsequently disappeared. The third mesomere with its radial formed the most proximal of the carpal bones, and the remaining mesomeres and radials became more distal carpal elements bracing digits I and II. The lateral carpal elements, Westoll thought, were derived from postaxial processes of the central mesomeres which eventually became independent bones. The femur, fibula, tibia, and tarsals in the pelvic appendage arose through the same kind of transformation. According to Westoll's theory, the bones of the digits, their supporting metacarpals (or metatarsals), and the most distal row of carpals (or tarsals) were not derived from any elements identifiable in the rhipidistian fin but evolved as wholly new structures as the fin rays were lost. Westoll defended this concept by pointing out that in the embryonic development of the limb in many terrestrial vertebrates the bones of the hand and foot originate separately from the more proximal elements of the appendage.

Arguments against Diphyletic Origin of Lepospondyls and Labyrinthodonts

If sufficient evidence is ever accumulated to show that the various lepospondyls descended separately from lobe-finned fishes or, as has been argued more strongly, that the lepo spondyls as a group are derived from a different piscine stock than the labyrinthodonts, the proponents of a diphyletic origin for the amphibians will be proved correct. The possibility that the Carboniferous tetrapods arose from two sources has been Judged unlikely, however, by several workers. H. Szarski maintained that two lines of animals could not have established themselves simultaneously in the same ecological niche. Furthermore, he refused to believe that labyrinthodonts and lepospondyls gained their virtually identical appendicular skeletons by convergent evolution. Adaptation for a similar function might have resulted in bent limbs and feet with spread-out toes, but the evolution in both lines of a pelvic girdle with three paired bones rather than two or four he regarded as an improbable coincidence. E. E. Williams interpreted the similar evolutionary tendencies shown by lepospondyls, labyrinthodonts, and modern amphibians as evidence that all of them emanated from a common ancestor. In undergoing loss of limbs, reduction of bone, and solidification of vertebrae before the middle of the Carboniferous period, lepospondyls exhibited precociously the kinds of change that took place later in the other lines. In Williams' opinion, vertebral structure was dissimilar in lepospondyls and labyrinthodonts, not because the two groups were derived from different sorts of fishes, but because the lepospondyls had given up the two-part centrum of the rhipidistians completely while the early labyrinthodonts had not. The cylindrical centrum of lepospondyls was evolved, Williams believed (like the similarly shaped centrum of salamanders long afterward), by expansion of the pleurocentrum at the expense of the intercentrum. Williams based his inference on a review of vertebral development in amphibian embryos: he found that in the living forms centra arise in a location equivalent to that of the pleurocentrum in the early tetrapods. The close correspondence between the appearance of the spool-shaped centra in lepospondyls and urodeles and the underlying homology between the structures of the latter and those of other amphibians increased the probability of a single stem for all the amphibians.

Jarvik's Support of Amphibian Diphyly

Jarvik regards the suppositions of Szarski and Williams as misleading. From his own research on fossil fishes and modern amphibians, Jarvik has concluded that there is significant anatomical evidence in favour of the diphyletic or even polyphyletic origin of tetrapods. He agrees with Holmgren that the urodeles came from a different line than other tetrapods, but he traces them to a separate branch of the rhipidistians rather than to the Devonian lungfishes. Jarvik recognizes a cleavage of the early rhipidistians into two orders, Porolepiformes and Osteolepiformes, and believes that each group had given rise independently to terrestrial forms. His studies show that modern urodeles still retain structural patterns inherited from the porolepiform fishes, and living anurans (the frogs and toads), comparable ones derived from the osteolepiforms. In examining the cavities and foramina in the cranium of the rhipidistian *Porolepis*, he found that the area for the anterior end of the brain, the passageway for the olfactory nerves, and the form of the nasal cavities were fundamentally like the corresponding parts of the urodelan head. His reconstruction of the cranial nerves and blood vessels furthered the resemblance of *Porolepis* to salamanders. In the porolepiform palate, Jarvik found a pair of deep recesses between the premaxillary bones and the vomerine elements which he believed held intermaxillary glands like those of modern urodeles. By contrast, *Eusthenopteron*, an osteolepiform rhipidistian, showed none of the peculiar specializations of *Porolepis* and the salamanders. Instead, the spaces for the brain, nerves, and vessels conformed to those existing in anurans and the tetrapods of more advanced classes.

Additional evidence for Jarvik's hypothesis came from the anatomy of the branchial skeleton. From the ventral parts of it preserved in *Eusthenopteron* and in the porolepiforms *Glyptolepis* and *Holoptychus*, Jarvik distinguished, again, two different structural designs. In the porolepiform fishes, a single median element, the basibranchial, had braced upon it all but the most posterior branchial arches as well as a long rod-like bone, the urohyal, which had a bifurcated end. In *Eusthenopteron* the basibranchial was divided into two segments, which together braced all the ventral arch elements. The urohyal lacked the bifurcated end and had in front of it an elongated sublingual rod, which did not appear in the porolepiforms. Jarvik pointed out the similarity of the branchial skeleton of salamanders to that of the porolepiforms and suggested that inheritance of the porolepiform design determined the peculiar mode of tongue development observable in modern urodeles. In these animals the tongue arises anterior to the branchial skeleton from a crescent-shaped glandular field present in the larva. According to Jarvik, such a glandular area existed in the porolepiform rhipidistians, flanked by denticles, of which vestiges appear in certain young salamanders. The region from which the tongue develops in anurans and higher tetrapods is that which was supported by the sublingual rod in the osteolepiforms. A homologue of that median element is present in the modern animals in the form of an anteriorly projecting skeletal process which underlies the tuberculum impar, the earliest part of the tongue to appear in the embryo. The embryonic derivation of the tongue is regarded by Jarvik as important because he believes that the developmental process does reflect accurately the separate history of the urodeles and the anurans. For this reason, he also included in his argument for the diphyletic origin of the amphibians the resemblance of the developing anuran limb to the osteolepiform appendage. He pointed out that the embryonic skeleton of the urodelan appendage exhibits a pattern that is different from the osteolepiform-anuran one. Since the fins of the porolepiforms are known only in outline, he had to rest his case without demonstrating that the endoskeleton of the urodele limb was in any way like that of those fishes.

Thomson's Rebuttal of Jarvik's Hypothesis

Few paleontologists take issue with Jarvik's meticulous anatomical descriptions, but many argue that his interpretation of the facts is open to question. Thomson does not accept Jarvik's conclusions and has attempted to show that none of his objections to monophyly is valid. If the development of the urodelan limb is thoroughly reviewed, Thomson maintains, it will be seen that the skeletal elements of the foot do not arise differently. They form and shift their position during growth, much as they do in other tetrapods. Certain rearrangements of the bones occur earlier in urodeles, but the mature skeleton of the foot resembles that of any unspecialized amphibian or reptile. Similarly, Thomson thinks that in the development of the tongue in urodeles, all trace of a sublingual element may be missing, not because the urodeles arose from rhipidistians which lacked it, but because the sublingual was reduced earlier in the urodelan line than in other amphibian stocks. Since anurans and higher tetrapods have only a rudimentary element under the tuberculum impar, Thomson supposed that the tendency toward reduction of the large osteolepiform sublingual rod was inherited in common by all the tetrapods. Rather than signifying the diphyletic origin of urodeles and anurans, the absence of the sublingual element in one group and its presence in the other may indicate simply that the rate at which the bone diminished differed in the immediate ancestors of the modern animals.

Thomson thinks that Jarvik has based his argument for diphyly on insufficient evidence and an exaggerated view of the dichotomy between the two groups of rhipidistian fishes. The porolepiforms especially are poorly known: it cannot be ascertained whether the structures Jarvik has described for *Porolepis* and *Glyptolepis* are distinguishing characteristics of all the porolepiforms or peculiarities of these forms. In Thomson's opinion, Jarvik has made the former assumption too readily and has overestimated the divergence these traits imply. Thomson insists that the differences now recognized between the porolepiforms and osteolepiforms are of minor significance. He minimizes the importance of the variations in the course of nerves and vessels through the snout and suggests that the urodele-like bulge in the anterior end of the cranial cavity of *Porolepis* may be an artifact. The anterior palatal recesses described by Jarvik do not seem to Thomson a good diagnostic character. His research showed that these depressions are basically similar in location in all rhipidistians and are shallow or deep depending upon the size of the mandibular teeth that fit into them when the mouth is closed. If Thomson is correct and the porolepiforms and osteolepiforms are members of the same order rather than two separate ones, Jarvik's hypothesis that urodeles and other tetrapods spring from a dual source is undermined.

Thomson finds still another reason for abandoning Jarvik's version of diphyly: it seems to him that in comparing modern amphibians to rhipidistians directly, Jarvik has used an approach of dubious validity. Jarvik has assumed for the purpose of his argument that urodeles and anurans have retained minute anatomical configurations present in Devonian fishes. To Thomson such conservatism on the part of the amphibians seems unlikely. The similarities that exist, he thinks, are more probably the result of convergence or parallelism than of phylogenetic relationship. Proof of the latter cannot be established without tracing the ancestry of modern amphibians back through the forms which preceded them in Paleozoic and early Mesozoic times. Jarvik made no attempt to find among these earlier animals, the labyrinthodonts and lepospondyls, any types which showed structures intermediate between osteolepiforms and anurans or between porolepiforms and urodeles. Without such evidence, Thomson believes, Jarvik's theory of amphibian origins cannot stand. He thinks that neither Jarvik nor any one else can tell how the rhipidistians gave rise to the tetrapods until many more of the transitional forms are known.

Fossil Record and History of the Early Amphibians

Paleontologists hunting for material that would fill out the record of amphibian evolution have difficulty for two reasons. First, the small size and delicacy of many of the extinct forms make it certain that their preservation was a rare occurrence. Second, the continental deposits that contain amphibian fossils are accessible in only a few places in the world. Sediments laid down in swamps or in streams of the late Devonian and early Carboniferous years, when the amphibians underwent their initial radiation, have been explored in no more than a handful of isolated locations in the United States and Europe. Workers are aware that their picture of the ancient amphibian community is extrapolated from the small and perhaps not entirely representative sample of fossil forms they have gathered from these sites, but they feel, nevertheless, that they understand the broad trends in the history of these early tetrapods. However the group arose, its diversification was rapid: from what must have been a limited base in Devonian times, numerous tribes of labyrinthodonts and the highly differentiated lepospondyls evolved before the Carboniferous period was far advanced. Although certain types of labyrinthodonts exhibited the stout body and strong legs that characterized the ichthyostegids, many paralleled the lepospondyls in developing the elongated body and weak appendages that apparently facilitated swimming in swampland waterways. Other labyrinthodonts grew large, broad-headed, and flat, like the horned nectridean lepospondyls, and lay on the bottom using their wide jaws as deadly fish traps. Adaptation to continuous underwater living was accompanied by selection for neotenic forms-species retaining the external gills and partially cartilaginous skeletons that fitted them as larvae for their aquatic environment. The changes in body form that assured the early amphibians success in the swamps also assured their eventual failure to survive. As the marshy lowlands diminished in the Permian period, the animals that could live nowhere else began to die out. The extremely specialized little lepospondyls disappeared first. The labyrinthodonts, which were larger, staved off extinction longer, but by the end of the Triassic period they are also absent from the fossil record.

Watson's Analysis of Labyrinthodont Evolution

The longer history and more numerous remains of the labyrinthodonts have made it possible to study their interrelationships in greater detail than those of the lepospondyls. Systematic analysis of these amphibians began as long ago as 1842, when Owen first called attention to the complex infolding of the dentin that distinguished their teeth from those of other tetrapods. Although he coined the term "labyrinthodont" to describe them, the animals with these teeth continued for another 75 years to be lumped with all pre-Jurassic amphibians as "stegocephalians." When, in the second decade of the twentieth century, Watson undertook to investigate the labyrinthodonts specifically, he found that fossils belonging to more than a hundred genera of these animals had accumulated in collectors' closets. After sorting those available to him according to, age and reviewing their structure comparatively, Watson was able to outline the course of labyrinthodont evolution. The earliest members of the group had had a relatively high, rounded skull with a well-ossified braincase. Characteristically, the ventral basioccipital and lateral exoccipital bones at the rear of the braincase contributed to a single condyle, through which the skull articulated with the vertebral column. The braincase was movably articulated also with the epipterygoid ossification of the upper jaw. Palate as well as jaw was suspended flexibly, because the pterygoid bones could move upon the medial parasphenoid, which adhered to the underside of the cranial floor. In these primitive labyrinthodonts, the pterygoid elements were broad and filled the central region of the palate, leaving minimal spaces, or interpterygoid vacuities, between them on either side of the parasphenoid. In more advanced labyrinthodonts Watson noted that the skull became depressed and less extensively ossified. The basioccipital bone withdrew slightly from the condyle, allowing enlarged exoccipitals to bear most of the stress of the cranio-vertebral joint. The palate lost its mobility through the development of a firm attachment between the pterygoid and parasphenoid bones, where a movable articulation had existed before, and wide interpterygoid vacuities appeared as the two pterygoid elements spread away from the knife-like process of the parasphenoid which lay in the midline. During the final phase of labyrinthodont evolution in the Triassic period, the trends established earlier persisted. Although superficial differences distinguished one family of late labyrinthodonts from another, Watson perceived that in all the forms he examined the skull was extremely flattened, the palate lightly built and broadly connected to the parasphenoid, and the condyle effectively paired because of the complete absence of the basioccipital from the articular surface.

Watson based his account of the evolution of the labyrinthodonts largely upon changes in the structure of the skull, because in amphibians, as in other aquatic vertebrates, the skull was the part of the skeleton most frequently preserved. The collections to which he had access included some postcranial remains, however, and he studied them also for evidence of progressive modification. His observations convinced him that labyrinthodonts could be categorized as primitive or advanced on the basis of their vertebrae as well as their skull. The few vertebrae that were known from the Carboniferous period had centra composed of double rings: in each segment both the intercentrum and pleurocentrum encircled the notochord and constricted it. These vertebrae, described as embolomerous, Watson accepted as being primary for the labyrinthodonts since they were associated with the most ancient members of the group. In postcranial material of Permian age, Watson found that the typical vertebra was rhachitomous: neither part of the centrum formed a complete ring; the intercentrum, wedge-shaped, clasped the notochord from below, and the smaller pleurocentrum, consisting of separate left and right halves, braced the notochord dorsally. As the Permian labyrinthodonts gave way to the extreme forms of the Triassic period, the pleurocentra dwindled to cartilaginous elements and then disappeared entirely. The flat-headed amphibians at the end of the labyrinthodont line had stereospondylous vertebrae, in which the notochord was encased by expanded intercentra articulating firmly with one another. The sequence of vertebral types led Watson to conclude that the primitive labyrinthodonts, embolomeres, had given rise to the rhachitomes and that these forms had eventually produced the ungainly, doomed stereospondyls.

Romer's Modification of Watson's Evolutionary Scheme

By fitting the hitherto uncorrelated labyrinthodont fossils into an evolutionary sequence, Watson reshaped paleontologists' understanding of a major group of the ancient amphibians. Subsequent research bore out his contention that the stereospondyls were a terminal branch of the family tree, but his idea that all primitive labyrinthodonts were embolomeres did not prove true. As Carboniferous labyrinthodonts became better known, it was evident that rhachitomous forms existed among them. Their characteristic vertebrae were not found at first, but skulls were unearthed that resembled those of Permian labyrin thodonts with rhachitomous centra. By 1947, Romer decided that he could not justify the maintenance of a separate order for the genera represented b these remains on the basis of Wat scheme son's supposition that, being of Carboniferous age, they could not belong to the Rhachitomi. Such a category, the order Phyllospondyli, had been invented earlier for tiny, gilled amphibians in some ways reminiscent of the rhachitomes and then used to house the increasing number of Carboniferous forms that looked more like rhachitomes than like proper embolomeres. It occurred to Romer that all the "phyllospondyls" were really rhachitomous labyrinthodonts; the gilled individuals, called branchiosaurs, were larvae, and the larger, early specimens were adults antecedent to the rhachitomes of the Permian period. Even before the discovery of their typical vertebrae in Carboniferous deposits confirmed the antiquity of the rhachitomes, Romer's hypothesis that they, not embolomeres, constituted the original labyrinthodont stock was strengthened by evidence from another source: Jarvik's studies showed that ichthyostegids, the oldest labyrinthodonts of all, had had rhachitomous vertebrae. Since the structure of their centra could be traced to the ancestral rhipidistians, it seemed certain that the rhachitomous condition was the primitive one in labyrinthodonts and that embolomerous vertebrae were an early, but secondary, development.

The Temnospondyls

Romer revised Watson's classification to reflect the division of the Carboniferous labyrinthodonts into two lines. The known rhachitomes of that period he assigned to the order Temnospondyli, a group which also included the later rhachitomous and stereospondylous animals. The embolomeres he placed near the base of the order Anthracosauria, whose members split away from the ancestors of the temnospondyls and eventually gave rise to reptiles. If the fossil record presents an accurate picture, the temnospondyls were the more numerous and progressive of the two kinds of amphibians. In each period they produced populations highly adapted for the narrow niche which they occupied, and when environmental upheavals extinguished these overspecialized tribes, they managed until the end of the Triassic to replenish their kind. The hardiest temnospondyls exhibited the same evolutionary

(a)

(b)

Fig. 3.7. Restorations of two Lower Permian temnospondylous labyrinthodonts. (a) Eryops, (b) Cacops.

tendencies as the more vulnerable members of the group but did not express them to such an extreme degree. They seem not to have lost the robust body and supportive legs that enabled them to leave the water. *Erops* and its relatives, representatives of the early central temnospondyl stock, apparently remained sufficiently terrestrial to survive the droughts that stranded species whose legs had become too small to bear them. The descendants of the edopoid amphibians, whose skulls had changed in the way Watson described, were creatures like *Eryops*, strong and stocky animals that probably spent as much time out of the water as in it. Several lines of temnospondyls, derived either from eryopoids or their immediate predecessors, radiated landward early in the Permian period. They lost all trace of lateral lines in the skull roof, grew well-ossified appendages, and in one group, the dissorophids, even developed a row or two of protective armor plates down the back. By Permian time, however, the opportunity amphibians once had to live on land unchallenged had passed. The dissorophids and their allies surely had to compete with the emerging reptiles, and defeat in that encounter rather than grievous structural or physiological limitations probably brought about their extinction.

Despite the impermanence of the swamps, the temnospondyls were forced to take refuge in them. There they diversified, producing a number of families whose interrelationships paleontologists are still trying to determine. Tracing the evolution of the various groups is difficult because, in addition to the traits that distinguished them, temnospondyls of different kinds developed similar adaptive features. Since it is often impossible to discriminate between likenesses due to phylogenetic connection and those due to parallelism, hereditary lines are hard to define. The eryopoid descendants that survived beyond the Permian period showed gradual enlargement of the intercentra: paleontologists believe that they have identified correctly the sequence of forms leading from the rhinesuchoids, Permian temnospondyls with small pleurocentra, to the Triassic capitosaurs, great flat-headed animals in which intercentra alone enclose the notochord. Several Triassic temnospondyls are known, however, which do not fitt into this lineage. The trematosaurs, for instance, exhibited a parallel reduction of the pleurocentra, immobilization of the palate, and pairing of the occipital condyles, but they had high, often long-snouted skulls that were constructed somewhat differently from those of the rhinesuchoid animals. Romer suggested their derivation from the archegosaurs, early Permian rhachitomes that had evolved the long snout and jaws presumably as an adaptation for hunting fishes. Contemporary with the trematosaurs

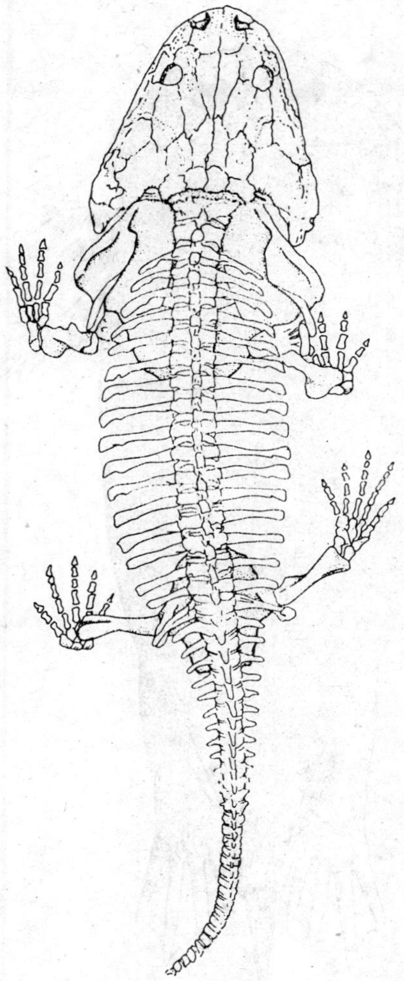

Fig. 3.8. Skeleton of Metoposaurus (=Buethneria).

and rhinesuchoids were temnospondyls, with eyes far forward in the skull, whose origins are also not well understood. Watson sought to ally these forms, the brachyopids, with a much older stock, the trimerorhachoids, relatives of the eryopoids that had developed a short face as early as Carboniferous times. Romer objected to Watson's scheme because *Trimerorhachis* and its kin, despite their short faces, had skulls that were longer than those of the brachyopids and unlike

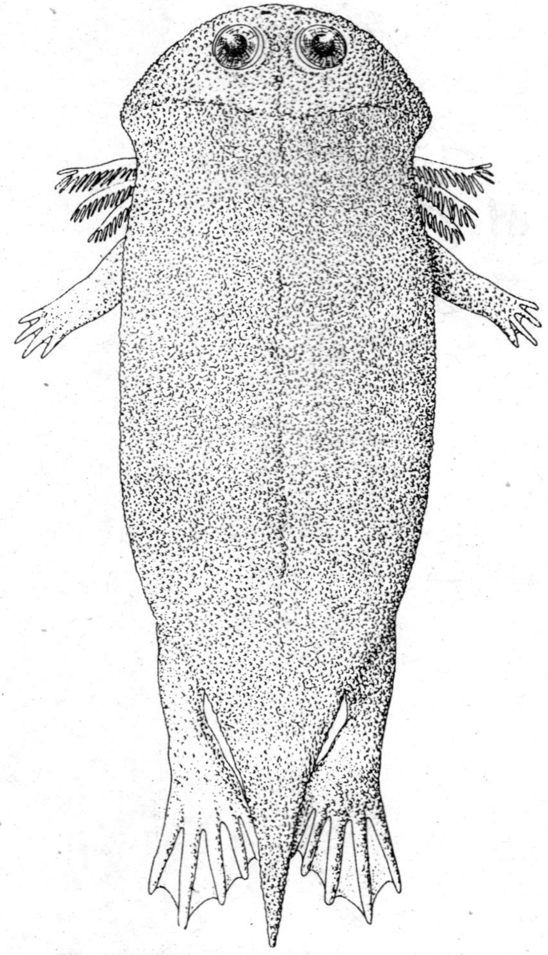

Fig. 3.9. Restoration of Gerrothorax, a brachyopid.

them in the arrangement of the bones of the postorbital region. He pointed out, too, that millions of years elapsed between the time when the trimerorhachoids flourished and the brachyopids appeared. He explained the similarities between the two groups as another example of parallelism: in both lines the animals retained the underdeveloped snout of the immature branchiosaur stage as part of their adaptation for aquatic living. Two kinds of late Triassic stereospondyls, descended perhaps from the brachyopids, were also short-faced. That larval traits were maintained in these forms and in the earlier ones is borne out by the discovery of specimens with branchial arches and external gills. The short face is further evidence of the often repeated lapse into neoteny, it seems, rather than a clue that would help paleontologists untangle the family history of the swamp-bound temnospondyls.

Rarity of Fossils of Early Temnospondyls

Although relationships among the late temnospondyls are not always clear, enough fossils are known to elucidate the nature of the group as a whole. By contrast, the early Carboniferous animals are shadowy creatures whose remains have eluded collectors almost completely. The rare rhachitomes that have been discovered in Lower Carboniferous deposits are all of questionable relationship to the main stock. *Otocratia*, known from a single skull roof found in Scotland, resembled the ichthyostegids in lacking intertemporal bones and having nostrils very near the edge of the upper jaw. Certain temnospondyls of the first part of the Carboniferous period from Scotland, Nova Scotia, and Linton, Ohio, had oddly elongated orbits that mark them as divergent forms. Called loxommids, the animals with the peculiarly shaped orbits have been described as rhachitomes which underwent at an early date the flattening of the head and the adaptation for bottom living that occurred again and again in temnospondyl history. For a while, paleontologists suspected that the remains of an amphibian taken from Lower Carboniferous strata at Greer, West Virginia, were those of a more generalized type. Romer, who studied the specimen, finally

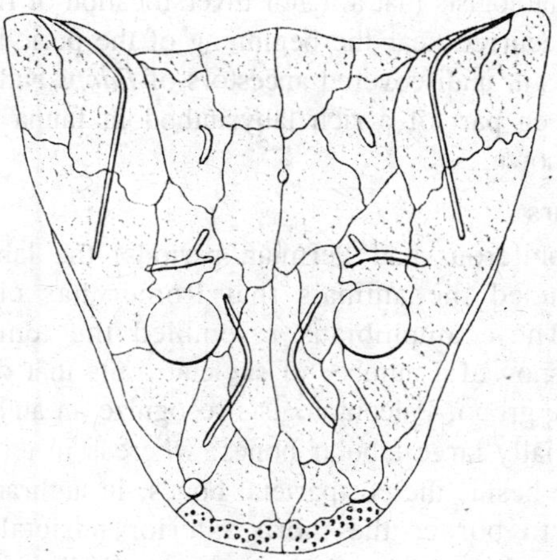

Fig. 3.10. Skull of early temnospondylous labyrinthodonts—Greererpeton, a colosteid.

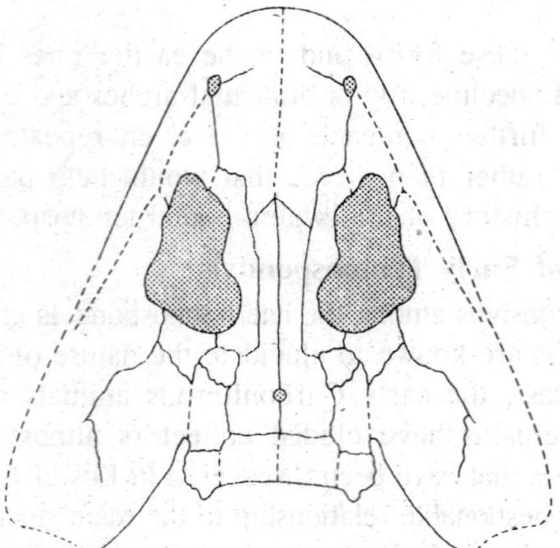

Fig. 3.11. Skull of early temnospondylous labyrinthodonts—Loxomma, a loxommid.

concluded that it did not represent the longsought ancestor of the later temnospondyls. *Greererpeton*, as he named the form, had orbits set far forward, shallow otic notches, and no intertemporal bones, characteristics which rendered it unsuitable as a precursor for rhachitomes like *Edops*. The structure of the skull of *Greererpeton* led Romer to ally it to *Otocratia* and to two later Carboniferous amphibians, *Colosteus* and *Erpetosaurus*. The group stemmed, possibly, from the old ichthyostegid stock and represented a line evolving independently of the contemporary temnospondyls. The lengthening list of divergent families which existed during the first half of the Carboniferous makes it evident to paleontologists that a major diversification of rhachitomous amphibians had occurred near the beginning of the period or perhaps even earlier. The still undiscovered ancestors of *Edops* and later forms had apparently been part of a rich labyrinthodont fauna that is only beginning to be known.

The Anthracosaurs

In the Carboniferous and Permian periods, the labyrinthodont assemblage included the animals that Romer has classified as anthracosaurs. These amphibians resembled the temnospondyls superficially but showed a number of skeletal traits that distinguished them as a separate group. Paleontologists recognize an anthracosaurian skull by its especially large tabular bones: whereas in temnospondyls these elements lie beside the postparietal bones, in anthracosaurs they extend forward to border the more anterior parietals. In early anthracosaurs the otic notch at the rear of the skull was slit-shaped instead of scalloped or ovoid, as in other labyrinthodonts. In front of the notch, the union of cheek and roof bones was so loose that the two parts of the skull almost always separated after death. Although the form of the otic notch changed in later anthracosaurs and the skull roof and cheek eventually became firmly united, the skull remained generally more primitive in the anthracosaurian amphibians than it did in the rhachitomes and stereospondyls. With few exceptions, the anthracosaurian skull never broadened and flattened. It retained the single occipital condyle and even the intertemporal bone that disappeared in numerous temnospondyls. The palate kept its movable articulation with the braincase and did not develop large medial vacuities at the expense of the pterygoid bones.

The anthracosaurs were less conservative in the structure of the vertebral column. They differed from their rhachitomous ancestors in evolving first enlarged, then fused pleurocentra that encircled the notochord. On the basis of centrum construction, two lines of anthracosaurs have been discerned: in one, the Carboniferous embolomeres, the intercentrum grew ringshaped also, so that pleurocentrum and intercentrum contributed equally to the strength of the column; in the other, known from Permian representatives called seymouriamorphs, the intercentra gradually diminished in size until the pleurocentra were left as the principal supporting elements. The two lines apparently diverged early in the Carboniferous period, for primitive members of both are already present in the deposits at Greer. *Proterogyrinus* seems well on its way toward the development of equal-sized, ring-form pleurocentra and intercentra, there being only a small area in the middorsal region of each element that remained unossified. *Mauchchunkia*, on the other hand, had wedge-shaped intercentra beneath the notochord and pleurocentra either in the form of rings around it or completely ossified disks.

Anthracosaurs Ancestral to the Reptiles

It is impossible to tell from the fossil evidence why the anthracosaurs did not radiate as widely as the temnospondyls. Either the rhachitomes and stereospondyls were inherently more adaptable, or, by virtue of superior design, they curbed the spread of the anthracosaurs in every niche to which both had access. Whatever the reason, the anthracosaurs were a smaller and less enduring group than the other labyrinthodonts: the embolomeres became extinct at the end of the Carboniferous period, and the seymouriamorphs disappeared by the end of the Permian. The relative failure of the anthracosaurs as amphibians is paradoxical because the characteristics which distinguished the seymouriamorphs survived in the reptiles and played a part in their success. For a time, paleontologists speculated that the stout-legged seymouriamorphs might have been the first reptiles, but the discovery of a branchiosaur larva of a member of the group and of adult forms with typical amphibian modifications for aquatic life led them to keep the seymouriamorphs in the amphibian fold. Finding true reptiles of Carboniferous age indicates that the first members of the new class must have sprung from an older part of the anthracosaur line, perhaps from relatives of the Upper Carboniferous *Gephyrostegus* (=*Diplovertebron*), a form with small intercentra and ring-shaped pleurocentra. If, as Hotton believes, the Lower Carboniferous *Mauchchunkia* is morphologically intermediate between *Gephyrostegus* and the Upper Devonian *Ichthyostega*, the branch of the anthracosaurs leading to reptiles may have separated very early from the rest of the assemblage. If that was the case, *Seymouria* and its Permian relatives were the very late and somewhat specialized remnants of a stock that had long ago given rise to more promising offspring.

Origin of Modern Amphibians is Obscure

The origin of reptiles from archaic amphibians is better understood than the emergence of the frogs and toads, the salamanders and newts, and the caecilians that constitute the modern members of the class Amphibia. The caecilians, legless, worm-like animals which burrow in the soil of the tropics, are represented by only one specimen in the fossil record. The living forms have been set apart in the order Apoda (Gymnophiona) but are often thought of as having at least a connection with the salamanders and newts of the order Urodela (Caudata). Those animals and the frogs and toads which are members of the order Anura (Salientia) have been found as fossils, but known extinct forms are essentially modern in structure and give no hint of the older amphibians from which they have descended. Most of the urodelan and anuran fossils come from rocks of the Cenozoic era, which began only 65 million years ago. The specimens can usually be assigned to families that have living representatives or even to extant genera. Remains from the Mesozoic era are much more sparse: less than a dozen frogs and about the same number of urodeles have been described. Only one of these forms, the frog *Triadobatrachus* (=*Protobatrachus*) lived in the Triassic period. The others are of Jurassic and Cretaceous age and already exhibit the skeletal modifications that distinguish the members of the modern orders from the labyrinthodonts and

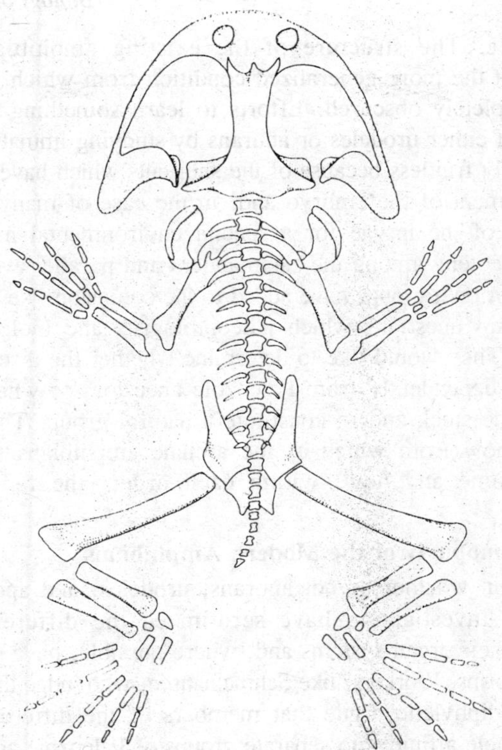

Fig. 3.12. A restoration of the skeleton of Triadobatrachus (=Protobatrachus), a Lower Triassic amphibian thought to be a primitive frog.

lepospondyls. Even *Triadobatrachus* had an essentially frog-like skull. Its iliac bones jutted forward, as they do in typical anurans, and the vertebral column had shortened, suggesting the reduction that was to characterize the later frogs. Although the two bones in the lower leg were not fused, as they are in modern forms, the hind limbs were larger than the front ones, an indication that the animal may have been able to jump. *Triadobatrachus* was a contemporary of the late temnospondyls but does not seem to have been closely related to any of them.

The lack of fossil specimens intermediate between anurans or urodeles and the older amphibians has forced paleontologists and students of the living animals to base their speculations about the evolution of the group upon evidence from the anatomy and embryology of modern species. This approach has presented difficulties that have so far proved insurmountable. The structure of the existing amphibians is so specialized that the more generalized condition from which it derived is almost completely obscured. Efforts to learn something about the early history of either urodeles or anurans by studying immature forms have been nearly fruitless because of the shortcuts which have appeared in the development of the embryo and, in the case of many anurans, the adaptation of the larvae for a special environmental niche. The frequency of neoteny among the salamanders and parallel evolution in different lines of both groups have added to the confusion. Nevertheless, there are certain questions which paleontologists and biologists still try to answer. They would like to determine whether the three modern orders arose independently from Paleozoic ancestors or whether they sprang from one stock and so constitute a natural group. They would also like to know from which of the archaic amphibians the living forms have come and how, within each order, the families are interrelated.

Diphyly or Monophyly of the Modern Amphibians

To ascertain whether or not anurans, urodeles, and apodans are monophyletic, investigators have scrutinized the differences and similarities in their organ systems and, where possible, have compared them to fossil forms. Workers, like Schmalhausen, who judge the modern animals to be diphyletic argue that members of the different orders show unmistakable affinities to separate groups of Paleozoic amphibians and that these affinities are more significant than mutual similarities which could be due to parallelism. The advocates of monophyly take the opposite view: they contend that the large number of peculiar characteristics which anurans, urodeles, and apodans share is strong evidence of their common ancestry. They question the conclusive resemblance of these forms to unlike Paleozoic types and minimize the importance of the structural differences which set the three groups apart. Parsons and Williams, who have declared themselves for monophyly, attribute the salient differences among the modern amphibians to their specialization for different modes of life. While the urodeles have maintained the long, muscular body and tail that characterize animals which undulate the body as they move, the anurans have turned to a peculiar form of locomotion dependent upon great enlargement of the hind legs. The caecilians evolved secondarily the solidly ossified skull and limbless trunk that are advantageous in burrowing. Parsons and Williams point out that despite the radical modification of the skeleton associated with their diverse habits, urodeles, anurans, and caecilians have in common certain traits that are not found in any other vertebrates. For instance, the crown and base of each tooth are separated by a narrow band of uncalcified dentin or fibrous tissue, a condition closely paralleled only in one isolated family of teleost fishes. In the middle ear, the end of the stapes implanted against the fenestra ovalis has, either next to it or fused with it, a bit of bone derived from the embryonic otic capsule. In the sensory portion of the ear, there is a patch of receptor cells so peculiar to the amphibians in its location that it is called the papilla amphibiorum. Other parts of the sensory and nervous systems of modern amphibians contain similar structures as well. Eyes, except in the caecilians, where they are extremely reduced, contain unusual light-sensitive elements known as green rods. The brain in all three types seems simplified in the same manner, the cerebellum being strikingly reduced even in the highly active frogs and toads. Anatomists acknowledge, also, that no other vertebrates have fat bodies anterior to the gonads for storage of reserve nutriment or glands in the skin like the mucous glands and poison glands of the amphibians.

The Lissamphibia and the "Protolissamphibian"

Convinced that the unique structures shared by modern amphibians are too numerous to attribute to coincidental in dependent evolution, Parsons and Williams regard the living animals as a natural group, which they have called the Lissamphibia. In an effort to locate among the archaic amphibians the group that might have served as the common stem of this assemblage, they have listed the traits that the ancestral animals should have had. The "protolissamphibian" that emerges from their description, however, does not correspond closely to any known fossil type. No specimen presently in paleontologists' collections has the additional middle ear bone, the peculiar teeth, and the extra bones in the lower jaw that Parsons and Williams assumed would have appeared in the precursor of the lissamphibians. Only one form, a lower Permian rhachitome, which Bolt described in 1969 and named *Doleserpeton*, seems possibly a representative of the protolissamphibian stock. A small upland animal with strong legs and a compact body, *Doleserpeton* had teeth like those of modern amphibians and showed no peculiar characteristics in the skull or postcranial skeleton that would have debarred it from the ancestry of the extant animals. The vertebrae of *Doleserpeton* still possessed both intercentra and pleurocentra, but one of the elements (Bolt presumes it is the intercentrum) is much reduced. Its disappearance in the descendants of

Doleserpeton would have left a single, cylindrical centrum in each vertebra, like that in the lissamphibian structure.

Classifying the Modern Amphibians

In Schmalhausen's opinion the growing menace of the reptiles on land and the secondarily aquatic reptiles and stereospondyls in the water was important in shaping the course of the evolution of the modern amphibians. The early urodeles and anurans which did not go underground permanently, like the apodans, survived only because they were able to hide or live where the other animals could not go. Schmalhausen believed that the Permian and Triassic forebears of the modern forms migrated into streams too shallow for the ponderous temnospondyls and to uplands too cold for the sun-loving reptiles to tolerate. Their confinement to such areas throughout the Mesozoic era he regarded as borne out by their virtual absence from the fossil record, for highland animals are rarely preserved. When their tetrapod enemies disappeared from the waters downstream during the Cretaceous period, a number of anurans and urodeles returned to the habitat of their ancestors, but many species, including the most primitive frogs, remained in mountainous regions. As additional evidence for this theory, Schmalhausen pointed out that the tadpole form of the larva, universal among anurans, would have evolved only as an adaptation for existence in swift currents of upland brooks: the streamlined shape and sucker-like organs of attachment of the newly hatched animal enabled it to feed on algae without being battered by the water or swept away. The less radical modification of the urodele larvae Schmalhausen attributed to their always having remained on the bottom to feed, where currents were less strong.

Determining Relationships among Anurans

Whether or not Schmalhausen's speculations are correct, it does seem that diversification within the anuran and urodelan groups took place where conditions were not conducive to fossilization. Since forms transitional between primitive types and more advanced ones are not available, investigators have had difficulty in constructing a phylogenetic scheme to underpin the classification of the modern animals. Like students of lissamphibian ancestry, biologists interested in the subsequent evolution of the anurans and urodeles have had to rely on analysis of the characters of the living species. In categorizing the frogs and toads, Noble took as his guide the shape of the vertebral centra. He decided that the amphicoelous form, in which both ends were concave as in fishes, was more primitive than either opisthocoelous centra,

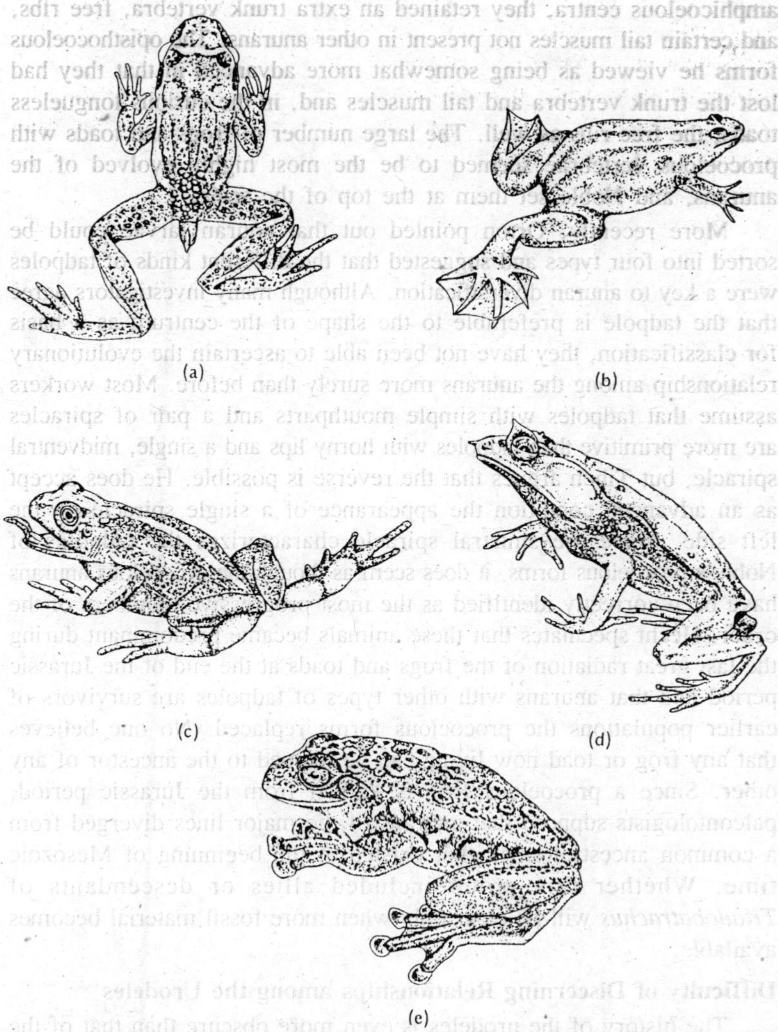

Fig. 3.13. Recent anurans. (a) Ascaphus, an ascaphid. (b) Xenopus, an aglossid. (c) Nectophrynoides, a bufonid. (d) Megophrys, a pelobatid. (e) Polypedates, a tree frog.

with a concavity only at the back, or procoelous ones, in which the anterior surface was depressed. He regarded bell toads (ascaphids) of New Zealand and the northwestern United States as representing the lowest level of anuran differentiation, because besides having amphicoelous centra, they retained an extra trunk vertebra, free ribs, and certain tail muscles not present in other anurans. The opisthocoelous forms he viewed as being somewhat more advanced in that they had lost the trunk vertebra and tail muscles and, in the curious tongueless toads, the free ribs as well. The large number of frogs and toads with procoelous vertebrae seemed to be the most highly evolved of the anurans, and Noble set them at the top of the scale.

More recently, Orton pointed out that anuran larvae could be sorted into four types and suggested that the different kinds of tadpoles were a key to anuran diversification. Although many investigators agree that the tadpole is preferable to the shape of the centrum as a basis for classification, they have not been able to ascertain the evolutionary relationship among the anurans more surely than before. Most workers assume that tadpoles with simple mouthparts and a pair of spiracles are more primitive than tadpoles with horny lips and a single, midventral spiracle, but Tihen argues that the reverse is possible. He does accept as an advanced condition the appearance of a single spiracle on the left side. Since the sinistral spiracle characterizes the tadpoles of Noble's procoelous forms, it does seem as though the procoelous anurans have been correctly identified as the most progressive members of the order. Hecht speculates that these animals became predominant during the last great radiation of the frogs and toads at the end of the Jurassic period and that anurans with other types of tadpoles are survivors of earlier populations the procoelous forms replaced. No one believes that any frog or toad now living can be likened to the ancestor of any other. Since a procoelous frog is known from the Jurassic period, paleontologists suppose that anurans of the major lines diverged from a common ancestral stock that existed at the beginning of Mesozoic time. Whether this group included allies or descendants of *Triadobatrachus* will be clear only when more fossil material becomes available.

Difficulty of Discerning Relationships among the Urodeles

The history of the urodeles is even more obscure than that of the frogs. Biologists recognize the living newts and ships among the salamanders as belonging to eight families but can say little or nothing about their interrelationship. Although it was suspected that animals with external fertilization were less advanced than those in which fertilization is internal, the supposition that families whose members fertilize their eggs externally were phylogenetically distant from the rest ran counter to the anatomical evidence. The tendency toward

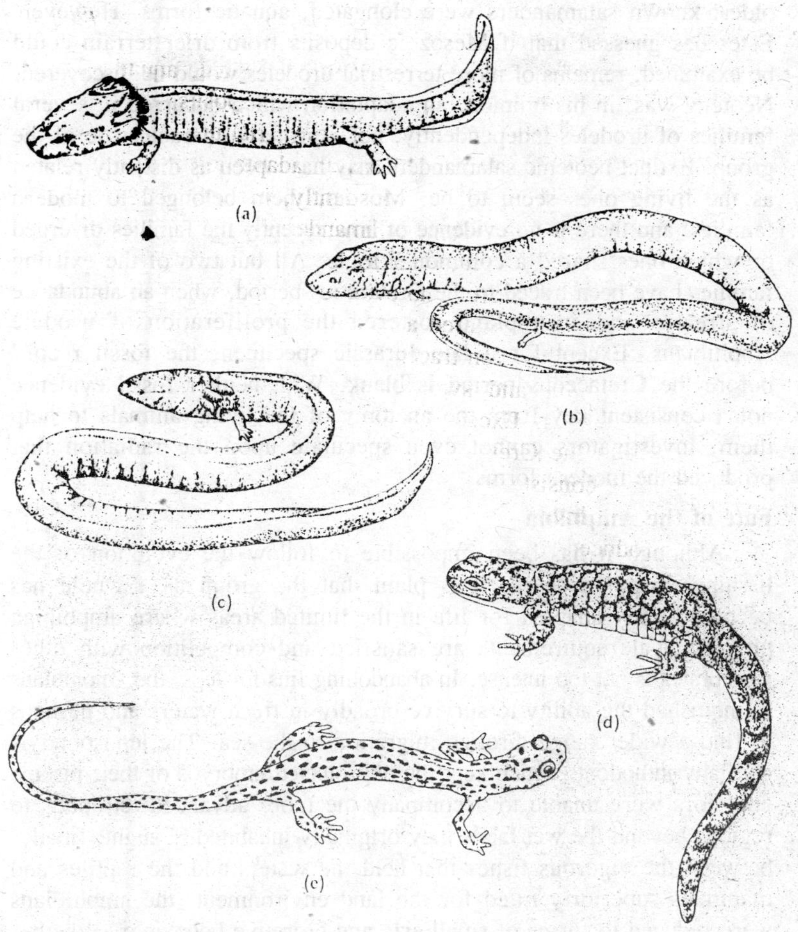

Fig. 3.14. Recent urodeles. (a) Necturus, a proteid. (b) Amphiuma, an amphiumid. (c) Siren, a sirenid. (d) Ambystoma, an ambystomid. (e) Eurycea, a plethodontid.

neoteny was also found unreliable as a guide to relationship. The sirenid salamanders, or "mud eels," of the United States retain larval gills and are permanently aquatic like the European *Proteus* and American "mud puppy" *Necturus*, but the unique specializations of the sirenids preclude their being allied closely to the latter animals. The "congo eel" *Amphiuma*, another salamander which retains gills and gill clefts, appears related to the gill-less salamandrids and the gill-less and lungless plethodontids on the basis of other characters. The idea that neoteny might be a primitive trait was put forth at first because the oldest known salamanders were elongated, aquatic forms. However, Estes has guessed that if Mesozoic deposits from drier terrain could be examined, remains of more terrestrial urodeles would be discovered. Neoteny was, in his opinion, an adaptation which occurred in several families of urodeles independently, not a primary characteristic of the group. Extinct neotenic salamanders may have been as distantly related as the living ones seem to be. Most of them belonged to modern families, and there is no evidence of how recently the families diverged or which ones shared a common lineage. All but two of the existing families have been traced to the Cretaceous period, when an abundance of streams and swampland fostered the proliferation of urodele amphibians. Except for a late Jurassic specimen, the fossil record before the Cretaceous period is blank. With neither fossil evidence not a consistent key from the anatomy of the living animals to help them, investigators cannot even speculate upon the radiation that produced the modern forms.

Fate of the Amphibia

Although it has been impossible to follow the evolution of the lissamphibians in detail, it is plain that the group as a whole has become highly adapted for life in the limited areas where amphibian physiological requirements are satisfied and competition with other vertebrates is not too intense. In abandoning fins for legs, the amphibians relinquished the ability to survive broadly in fresh waters and perhaps to find a wider range through migration to the sea. The lcpospondyls and labyrinthodonts which kept the unprotected embryos of their piscine ancestors were unable to accompany the more advanced tetrapods to regions beyond the wet lands they originally inhabited. Caught, finally, between the vigorous fishes that held the waters and the reptiles and mammals superiorly fitted for the land environment, the amphibians were reduced to forms of small size and secretive behaviour. Whether the lissamphibians represent animals on their way toward extinction or forms that will hold their place as long as any vertebrates is another question that no biologist would attempt to answer.

4

CLASSIFICATION OF AMPHIBIA

Biologists have discovered, described, and given names to about 1.5 million of the many millions of species of plants, animals, and microorganisms that exist at present. They have also named many species that formerly lived on Earth but are now extinct. The role of classification is to provide a unique name for each of these organisms and put them in a hierarchy of increasingly inclusive taxa (groups) based on their evolutionary relationships, so that they can be unequivocally recognized and associated with other species having a common ancestry.

Systematics and Taxonomy

The science of discovering the diversity and revolutionary relationships of organisms is called systematics, and aspects of systematics involved with naming and classifying organisms form the subdiscipline of taxonomy. The basic taxon (group) is the species, and ideally ail higher or more inclusive taxa comprise an ancestral species and all its descendants. Determination of this relationship is based on all members of a group sharing one or more derived features (evolutionary novelties). For example, all frogs and toads have two of the ankle bones greatly elongated to form a fourth segment in the hind leg (all other limbed amphibians and reptiles have three segments in the hind leg). This condition provides greater leverage for leaping and serves to distinguish the group, the order Anura, in which frogs and toads are placed.

Names and Classification

Scientists use Latinized names for groups of organisms in order to provide universality and stability and to avoid the necessity of translating the name into many different languages. No matter what one's native language, the Latinized scientific name is immediately associated with the same group of organisms.

The fundamental category in all classifications is the species. Species are populations of organisms that share one or more similarities not found in related species. Species form closed genetic systems, although some closely related species may occasionally hybridize. The name of species is composed of two words: the first is the name of its genus (the generic name) and indicates its closest relationships; the second name (the specific name) is uniquely used for that species within a genus. The generic name always begins with a capital letter, and the specific name is always written in the lower case, and both names are always printed in italics; for example, the name of the Gaboon viper of tropical Africa is written *Bitis gabonica*. Species names are frequently based on a distinctive feature, or may refer to the species' place of origin, or may honor a person by thing a Latinized form of their name.

In some instances when populations of anyone species are separated geographically and have relatively consistent differences, they may be recognized as subspecies and then have a third name: for example, *Lampropeltis triangulum triangulum* for the eastern milk snake and *Lampropeltis triangulum gentilis* for the plains milk snake of North America.

Scientists also use a standard set of terms for each level in the hierarchical arrangement of taxa and a single Latin name (always capitalized) for each different taxon. The most inclusive category to

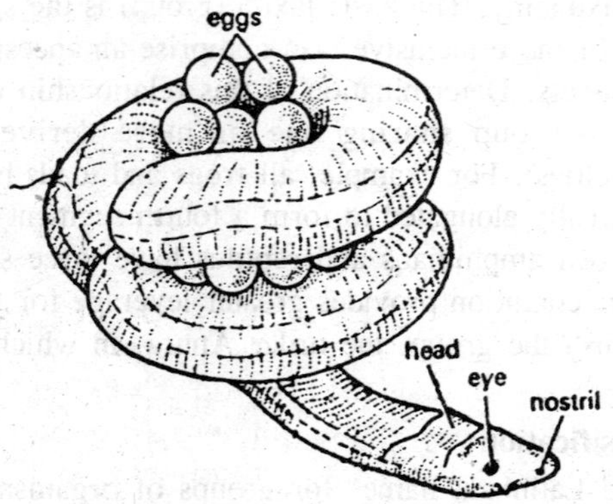

Fig. 4.1. Ichthyophis. Female coiling around the eggs.

which amphibians and reptiles belong is the kingdom Animalia. This taxon contains all organisms that are multicellular, obtain their energy by ingesting food from the surroundings, and usually develop from the coming together of two cells, a large egg and a small sperm.

The kingdom Animalia is composed of 35 phyla, each phylum representing a distinctive major biological organization and lifestyle. Ampbibians and reptiles belong to the phylum Chordata, which consists of animals that have a single dorsal nerve cord; a cartilaginous notochord (at least embryonically) that lies below the nerve cord and extends the length of the body; and gill arches and gill pouches (which are present only embryonically in reptiles, birds, and mammals). In other chordates, including larval amphibians, the gill arches support functional gills and the gill pouches in the throat region open to the outside.

Three subphyla are recognised within the Chordata; amphibians and reptiles are located to the subphylum Craniata. Animals belonging to this group have chambered hearts, semicircular canals in the inner ear, skeletal gill supports, and a protective cartilaginous or bony cranium around the brain. Most members of this group also have segmentally arranged skeletal elements (vertebrae) surrounding and protecting the spinal cord.

The next division in the hierarchy is the class Amphibians and reptiles are placed in different classes, the Amphibia and Reptilia, respectively. Classes are further divided into orders; there are three living orders of amphibians and four of reptiles. Orders are divided into families, families into genera, and each genus comprises one or more species. The classification adopted in this book also makes use of the suborder, which is a category more inclusive than a family but less inclusive than an order.

These hierarchical categories form a set of taxa defined by shared similarities. This means that species within a genus are more similar to one another than to species in other genera within the same family; genera in the same family are more similar to one another than to genera in other families; and so on. It also means that, progressing up the hierarchy, fewer and fewer features are shared by the species included in all order, a class, a phylum, or a kingdom. The hierarchy thus reflects in reverse the branching pattern of the evolution of life, because the decrease in similarity correlates with time of divergence—that is, the greater the similarity the more recent the divergence.

Vernacular and Coined Names

Amphibians and reptiles are known by vernacular or common names that vary from language to language, country to country, and even within the same country. In some cases, the same name may be used for a number of similar appearing species. In others, one name is used for entirely different creatures; for example, in different pans of the United States the name "gopher" is applied to a turtle, a ground squirrel, and a subterranean rodent called a pocket gopher. Because of the potential for confusion these names are nor used in formal, scientific identification. Species names that had never been used by local people for amphibians and reptiles have in recent years been coined for use in popular works or field guides, especially in English. These names, alas, lack even the dignity of being truly vernacular, as they are inventions of their authors, duplications of scientific names, and usually uninformative when read by someone who has a different native language.

Species

There are about 4,950 species of living amphibians and about 7,400 species of living reptiles. Although some of the larger and more spectacular representatives—such as the "hairy frog", the leatherback turtle, and the Komodo dragon—are very distinctive, many closely related forms are characterized by subtle differences. External form and structure, size and body proportions, and colouration are usually the basis for separating allied species, with differences in the number, shape, and structure of scales being emphasized in reptiles. Recent studies of chromosomes have been valuable in distinguishing between cryptic species—those whose external features differ only slightly from their closest relatives.

Because many kinds of amphibians and reptiles are small and secretive, occur very locally, or are inhabitants of unexplored regions of the Earth (especially tropical rainforests), previously undescribed species are discovered each year. During the past decade about 80 new species of amphibians (mostly frogs and toads) and 25 species of reptiles (mostly lizards) have been described each year. Sometimes, as our knowledge of a particular group increases, forms previously thought to be distinctive are shown to belong to a single species. These factors, plus the cases in which conflicting evidence for its interpretation) leads different scientists to disagree on the status of named forms, reflect the differing numbers of species recognized by one or another authority. It is safe to say, however, that new species of amphibians and reptiles will continue to be discovered at these rates well into the future. It should be noted that while amphibians and reptiles are often regarded as minor remnants of an earlier evolutionary radiation, the number of species of living amphibians exceeds that of mammals, and there are many more species of living reptiles than of mammals.

Genera

A genus contains one or more speices that share at least one derived, or advanced, feature. The calssification of amphibians into different genera is usually based oninternal features, especially those of skeleton, because their smooth bodies possess relatively few significant external differences. Skeletal morphology is also used in reptiles, but scalation (especially the scales on the head) frequently provides a basis for generic groupings. Occasionally a single, very distinctive species may be classified in a genus of its own because or one or more unique features; for example, most American and Eurasian treefrogs are placed in the genus *Hyla*, but the spiny-headed or crowned treefrog is placed in a separate genus. *Anotheca* as *Anotheca spinosa*. Frequently, however, genera originally thought to contain only a single species have newly discovered living or fossil species added to them

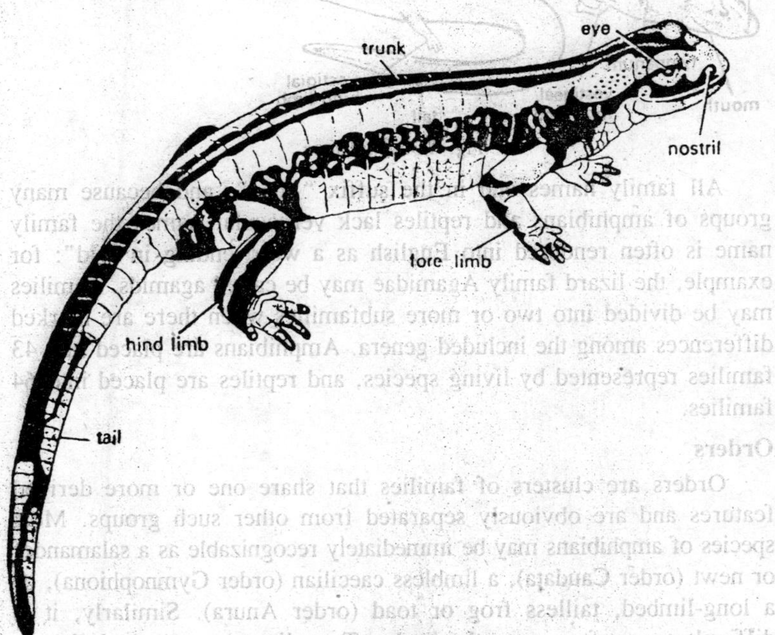

Fig. 4.2. Salamandra maculosa.

as the result of additional research. Currently we recognize about 430 genera of living amphibians and 898 genera of living reptiles.

Families

Genera are clustered into different families on the basis of one or more shared features of internal morphology. Frequently it is difficult for the layperson to readily identify which family a species belongs to, from only its external features; even a specialist may have a problem with family placement of unfamiliar amphibians, for external resemblances are deceiving. A few families of amphibians have only a single very distinctive genus containing one or a few species. In theory such families should be regarded as sole relicts of a larger group of species, the majority of which have become extinct.

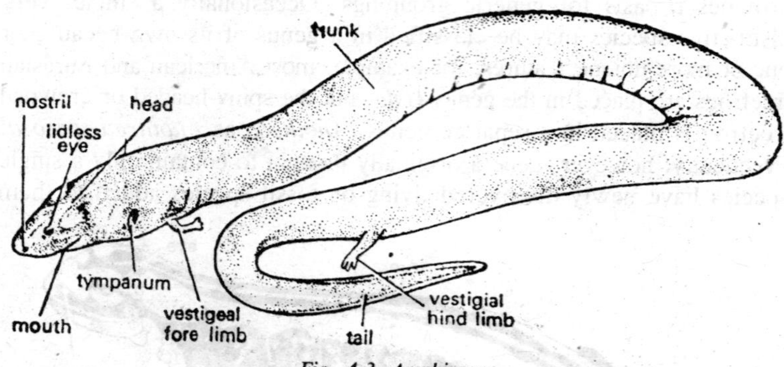

Fig. 4.3. Amphiuma.

All family names end in the suffix "-idea", and because many groups of amphibians and reptiles lack vernacular names the family name is often rendered into English as a word ending in "-id": for example, the lizard family Agamidae may be called agamids. Families may be divided into two or more subfamilies when there are marked differences among the included genera. Amphibians are placed into 43 families represented by living species, and reptiles are placed into 64 families.

Orders

Orders are clusters of families that share one or more derived features and are obviously separated from other such groups. Most species of amphibians may be immediately recognizable as a salamander or newt (order Caudata), a limbless caecilian (order Gymnophiona), or a long-limbed, tailless frog or toad (order Anura). Similarly, it is difficult to confuse a turtle (order Testudinata), a typical lizard,

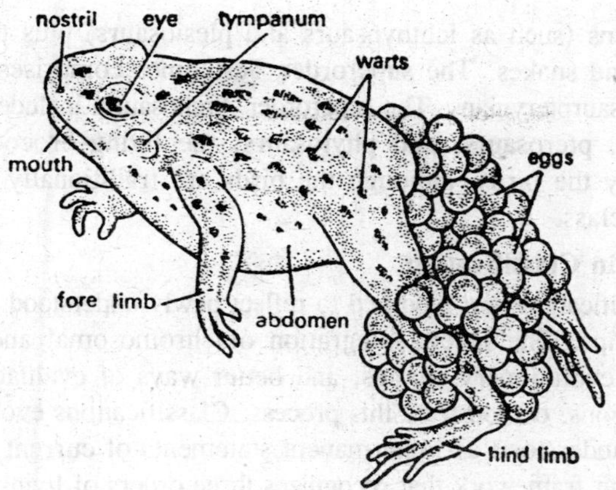

Fig. 4.4. Alytes.

amphisbaeman, or snake (order Squamata), or a crocodilian (order Crocodilia) with one another. The tuatara of New Zealand superficially resemble lizards but are unlike any lizard in obvious details of scalation and numerous internal features. Tuatara are the sole living representatives of the order Rhynchocephalia, but a number of extinct rhynchocephalians are known from the Mesozoic, 245 to 65 million years ago.

There are also many extinct orders of amphibians and reptiles with no representatives alive today.

Higher Classification

Scientists agree that modern amphibians are related to one another and in turn are related to a cluster of ancient amphibians that date back to the upper Devonian, about 380 million years ago, but became extinct by the end of the Triassic, 200 million years ago. The difficulty of interpreting these relationships and the status of other fossil "amphibians"—including the lineage most closely allied to early reptiles—leaves the higher classification of all so-called "amphibians" in a state of flux. At this time, it seems best not to use the conflicting classification schemes, except to not that the three living orders are usually regarded as forming the subclass Lissamphibia.

Reptiles, on the other hand, form two distinct subclasses based primarily on differences in skull and jaw characteristics. The subclass Parareptilia includes several extinct lines. The living reptiles and many extinct groups are now recognized as part of the subclass Eureptilia and its three superorders. The superorder Lepidosauria contains many fossil forms (such as ichthyosaurs and plesiosaurs) plus the tuataras, lizards, and snakes. The superorder Testudines comprises turtles and the fossil saurpterygians. The superorder Archosauria includes the extinct dinosaurs, pterosaurs, and phytosaurs, the living crocodilians, and technically the birds, although the birds are traditionally placed in a separate class.

Changes in Classification

Classifications are modified to reflect newly understood evolutionary relationships. The recent integration of chromosomal and molecular data into evolutionary studies, and better ways of evaluating existing classifications, demonstrate this process. Classifications evolve and they must be understood as impermanent statements of current knowledge. The present framework that recognizes three orders of living amphibians and four orders of living reptiles seems firmly established, but some revision in the number and positioning of families, especially for those with larger numbers of species, is to be expected as relationships become clearer.

5

HABITATS AND ADAPTATIONS

Amphibians and reptiles occur throughout much of the world, even in the Himalayas and the Andes above 4,500 meters (15,000 feet). They are present on all continents except Antarctica and some reptiles can even be found on tiny, remote islands. They live in rainforests, woodlands, savannas, grasslands, deserts, and scrub; they occur on the ground, underground, in rock crevices and under debris, up trees, and in marshes, swamps, lakes, streams, ponds, and the sea. Some even take to the air briefly.

Where Amphibians Live

Only Antarctica, extreme northern Europe, Asia, and North America, and most oceanic islands do not have native amphibians Caecilians are pantropical, but are absent from Madagascar, New Guinea, and Australia. Salamanders are widely distributed in temperate regions of the Northern Hemisphere, although one family, the Plethodontidae, is most diverse in tropical Central and northern South America. In contrast, frogs are found nearly worldwide, with the greatest diversity of species in the tropics—nearly the same number or species (about 90) occur at single sites in the upper Amazon Basin in Ecuador as occur in all of the United States.

Relatively few species or amphibians are aquatic (water-dwelling) as adults, although most of them have aquatic larval stages (for example, tadpoles in frogs). Most salamanders and caecilians, and about half of the species of frogs, are terrestrial (ground-dwelling). Some of these, especially the caecilians, spend most of their time below ground. A few salamanders and many frogs are arboreal (tree-dwelling).

Aquatic Amphibians

Four families of salamanders in North America (Proteidae, Cryptobranchidae, Amphiumidae, and Sirenidae) are strictly aquatic, as are two cryptobranchids in China and Japan (the giant salamanders), and one proteid in southeastern Europe (the olm). These, and some North American plethodontid salamanders that live in subterranean waters, are neotenic; that is, they never complete metamorphosis from the larval stage; they include some of the largest salamanders, which breathe by means of gills and retain larval tail fins. The sirens (family Sirenidae) and congo eels (family Amphiumidae) are limbless, or nearly so, and swim in a serpentine manner. In contrast, the mudpuppies and olm (Proteidae) and the giant salamanders and hellbenders (Cryptobranchidae) have depressed bodies and normal limbs; they are more sedentary and crawl about on the bottoms of lakes or rivers.

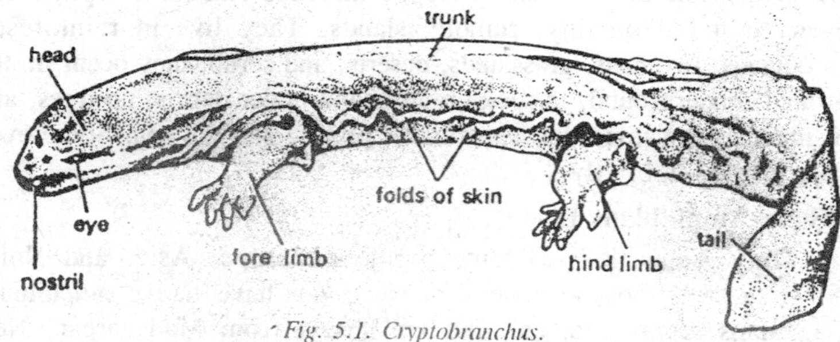

Fig. 5.1. Cryptobranchus.

Most of the so-called pond frogs of the genus *Rana* have powerful hind limbs and fully webbed feet, they are well adapted for leaping from land into water, where they are excellent swimmers. Practically all frogs are capable of swimming, but members of two families, in particular, are highly specialized swimmers. The paradox frog and its

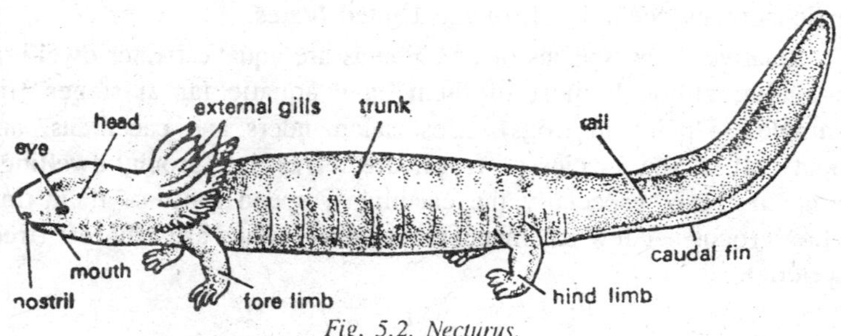

Fig. 5.2. Necturus.

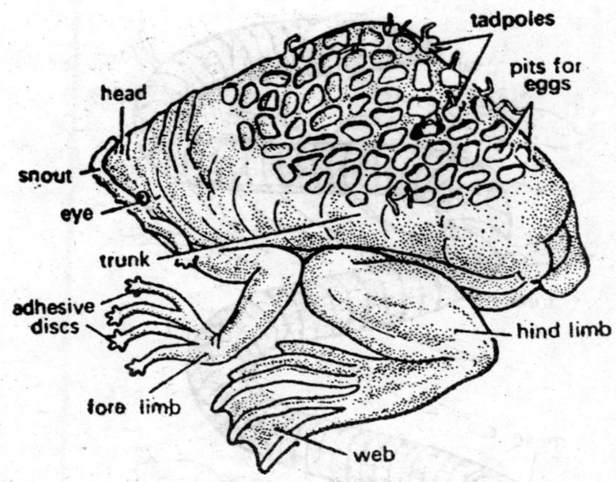

Fig. 5.3. Pipa pipa (female).

allies (family Pseudidae) in South America have extremely powerful hind limbs, huge fully webbed feet, and dorsally protruding eyes: the frogs are active near the surface of ponds. The clawed frogs of Africa and the Surinam toad and its relatives in South America (family Pipidae) also have fully webbed feet, but they have depressed bodies and relatively small eyes. They rest and feed on the bottoms of ponds and come to the surface periodically to gulp air. If ponds dry up, these frogs burrow into the mud and remain there until the ponds fill with water again.

One subfamily of caecilians (Typhlonectinae) in South America is strictly aquatic. These eel-like amphibians live in lakes and rivers.

Burrowing Amphibians

While few salamanders are known to burrow in mud or soft soils, the caecilians are the most adept burrowing amphibians. These slender, worm-like, limbless animals have muscular bodies and compact skulls. Their eyes are covered by skin or bone, sot hey are blind: but with a protusible, sensory tentacle lying in a groove or cavity on the head, caecilians are capable of detecting environmental cues and prey (mostly earthworms) by chemosensory means. Caecilians propel themselves through the soil by concertina locomotion, in which part of the body remains in static contact with the earth, and the adjacent parts of the body are pushed or pulled forward at the same time.

Frogs of diverse groups have a large, keratinized, spade-like tubercle on the inner edge of each hind foot. Spadefoot toads (family Pelobatidae) in North America, rain frogs (Microhylidae) in Africa, sand frogs (Ranidae) in Africa, the Central American burrowing frog

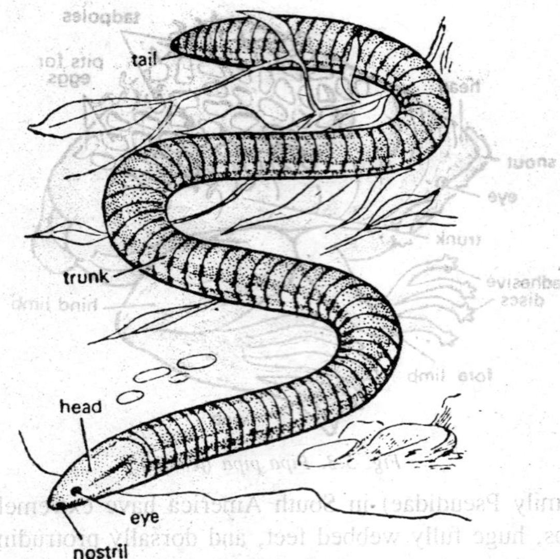

Fig. 5.4. Ichthyophis (male).

(Rhinophrynidae), and the burrowing frog and Holy Cross toad (Myobatrachidae) in Australia, all use lateral scooping motions of the

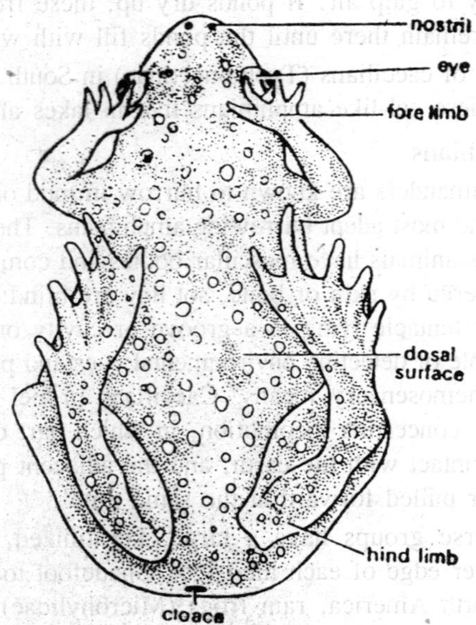

Fig. 5.5. Bombinator.

feet to bury themselves quickly in soft dirt or mud. A few frogs burrow head-first and have calloused pads on their snouts. African shovel-nosed frogs (Hemisotidae) have pointed snouts and burrow by vertical motions of their heads. The Australian turtle frog and the sandhill frog (Myobatrachidae) have blunt snouts and dig with their forefeet.

Frogs may bury themselves to seek daytime retreats or more importantly for long periods of aestivation (in a torpid condition) during dry seasons, which may last for many months. During these times, frogs are subject to desiccation, so before aestivating they fill their urinary bladder with water. Some frogs, such as the African bullfrog (Ranidae), the water-holding frog (Hylidae) in Australia, and horned frogs and their allies (Leptodactylidae) in South America, reduce the risk of desiccation even further by making a cocoon. Once they are deep in their burrows the frogs shed the outer layer of their skin and secrete mucus, which together with the shed skin hardens to form a cocoon around the frog.

Arboreal Amphibians

The terminal segment of each finger and toe of many treefrogs (Hylidae, Centrolenidae, Hyperoliidae, Rhacophoridae, and a few Microhylidae) is expanded into a specialized toe pad, by means of which these frogs are able to adhere to vertical surfaces. Many treefrogs also have slender bodies and long limbs, modifications that allow them to leap from one leaf or branch to another and to hold onto their perches. In one group of hylids (*Phyllomedusa* species) in the American tropics, the innermost fingers and toes are elongated and opposable in

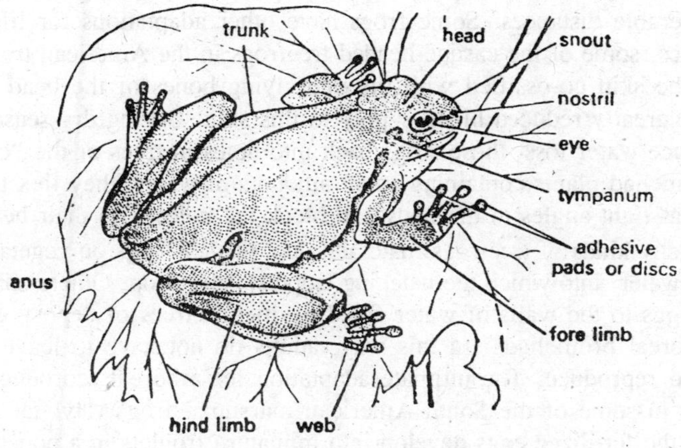

Fig. 5.6. Hyla.

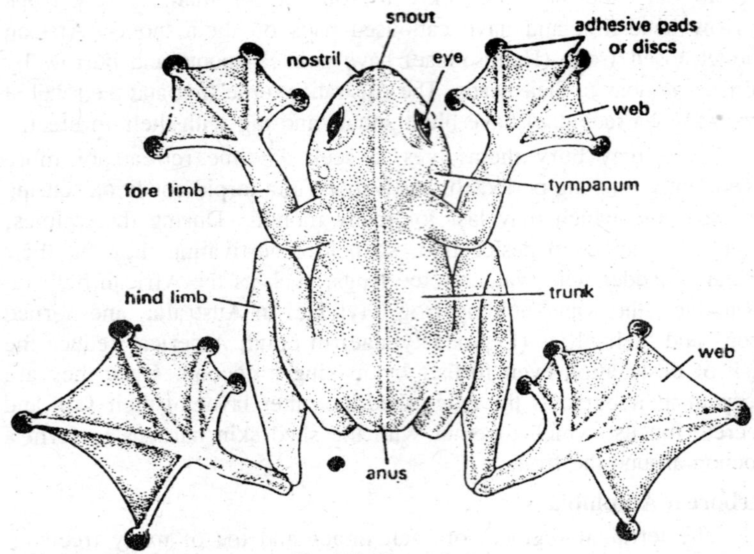

Fig. 5.7. Rhacophorus.

the outer digits. These frogs walk about in trees by grasping branches. A few treefrogs (some *Agalychnis* and *Hyla* species in Central America and some *Rhacophorus* species in southeastern Asia) not only have huge, fully webbed hands and feet but also fringes of skin along the limbs. These modifications combine to provide an extensive surface area when the limbs are partially extended and the fingers and toes are spread apart; after leaping from high perches these frogs can glide considerable distances. Some frogs have other adaptations for life in the trees, some of the casque-headed treefrogs in the American tropics have the skin co-ossified with the underlying bones of the head and have a greatly reduced blood supply to this skin. During dry seasons, to reduce water loss, these frogs back into holes in trees or the "cup" of bromeliad plants containing small amounts of water; they flex their heads at right angles to the body and block the holes with their heads.

Many kinds of treefrogs mate and deposit their eggs on vegetation above water, into which the hatching tadpoles later drop. Others adhere their eggs to the walls of water filled cavities in trees or deposit eggs in arboreal bromeliads. In this way, adults do not have to leave the trees to reproduce. The ultimate adaptation for arboreal reproduction occurs in some of the South American marsupial frogs (Hylidae), in which the fertilized eggs develop into miniature froglets in a pouch on the mother's back. Thus, it is possible for generations to pass without any individuals leaving the trees.

6

Amphibian Behaviour

Reptiles and amphibians are the only land vertebrates unable to control their body temperature by internal, physiological means. By and large, this means that unless they regulate their body temperature behaviourally, it will simply rise and fall with the temperature of their immediate surroundings. This places enormous constraints on every aspect of their life. Hatching, growing, feeding, reproducing, escaping from predators, and even just being active depends on attaining and maintaining a body temperature that allows them to function normally. Consequently much of the behaviour of reptiles and amphibians is aimed at temperature regulation, and this requirement underlies much of the complex and often bizarre behaviour of individual species.

Because amphibians have permeable skin and, as a rule, are unable to regulate their body temperature by physiological means, many aspects of their behaviour are dictated by the environment. Most amphibians are nocturnal and are active only with environmental conditions are sufficiently moist to prevent their bodies losing too much water by evaporation.

However, some amphibians exhibit water-conserving behaviour and even behaviour that regulates their body temperature. For example, diurnal (day-active) frogs living in cool climates often bask in the sun and thereby raise their body temperature before foraging for food. Some of them periodically enter water to acquire moisture and to lower their temperature. The North American bullfrog regulates its temperature by changing its position in relation to the rays of the sun. Other postural changes, evaporative cooling by mucous secretions, and periodic rewetting of the skin, all serve to reduce or stabilize the bullfrog's body temperature. Conversely, tiger salamanders, living in the Rocky Mountains of North America, and tadpoles in many parts of the world increase their body temperature by moving into shallow water on sunny days. However, almost nothing is known about the behaviour of the secretive, burrowing caecilians.

Feeding Behaviour

Most amphibians simply sit and wait for prey to come within reach. A few heavy-bodied frogs, such as South American horned frogs, lure agile prey by waving the long toes on their hind feet. Other frogs, especially the small poison frogs in tropical America, actively forage for small insects by day and find their prey by sight. A Central American burrowing frog (genus *Rhinophrynus*) apparently locates subterranean termite tunnels by their odor, then breaks open the tunnel, inserts its nose, and laps up the termites.

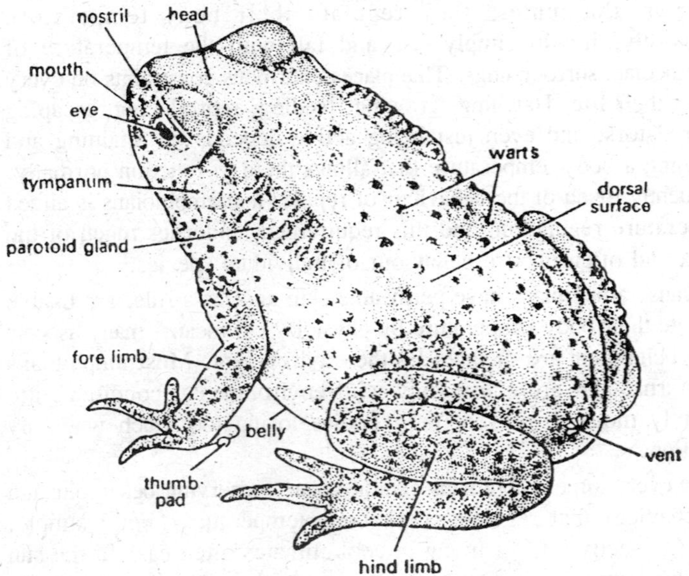

Fig. 6.1. Bufomelanostictus.

Defensive Behaviour

Faced with potential predators, many amphibians feign death, present an enlarged image or the least-palatable part of the body to the predator, confuse the predator by changing the characteristic shape of the body, or change their colour as a warning. In some cases they may even attack the predator. Salamanders that have concentrations of poison glands on their tail may lash it at predators; others have many poison glands on the back of the head and butt their head at the predator. Some Asiatic salamandrids are capable of protruding the tips of their ribs through the poison glands on their flanks, greatly increasing the chances of a predator encountering unpalatable secretions. Many salamanders exhibit caudal autotomy—that is, they can break off a portion of their tail—so if a predator grasps the tail, it will break off and continue to wriggle, thereby distracting the assailant while the salamander escapes.

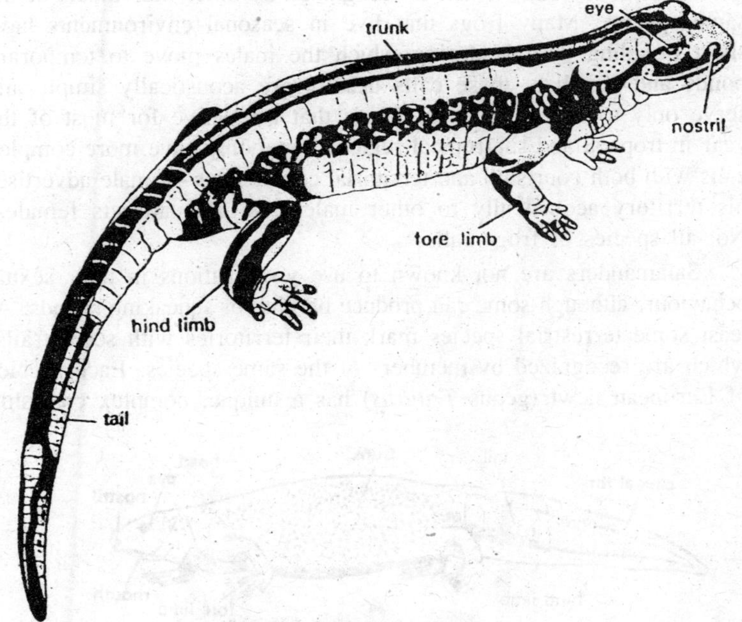

Fig. 6.2. Salamandra maculosa.

Feigning death is widespread among frogs: some fold the limbs tightly against the body and lie motionless on their backs; others assume a rigid posture with the limbs outstretched. Toads commonly innate their lungs, thereby puffing up the body and presenting a larger image to the predator, at the same time lifting their body off the ground and sometimes tilting it toward the predator. Some South American leptodactylid frogs (*Physalaemus* and *Pleurodema*) have large glands in the groin resembling brightly coloured "eyespots", so that when a frog lowers its head and lifts its pelvic region, it looks like the head of a larger animal and presumably frightens off the would-be predator. When faced with danger, some African and South American treefrogs open their mouth (often displaying a brightly coloured tongue) in an apparently threatening pose. Some large frogs actually leap and snap at predators. Many kinds of frogs emit a loud scream when disturbed or grasped by a predator; the noise may serve not only to frighten the predator but also warn other frogs of danger.

Amphibian Reproductive Behaviour

The most obvious reproductive behaviour of amphibians is the sound made by frogs, especially the mating call made by males. Each species has a distinctive call, which is recognized by other individuals of the same species. Many frogs that live in seasonal environments have short breeding periods during which the males move to temporary ponds and vocalize, these calls usually are acoustically simple and serve only to attract females. Frogs that are active for most of the year in tropical and subtropical regions commonly have more complex calls with both courtship and territorial components—a male advertises his territory acoustically to other males and also attracts females. Not all species of frogs call.

Salamanders are not known to use vocalizations in their sexual behaviour, although some can produce barking or squeaking sounds. At least some terrestrial species mark their territories with scent trails, which are recognized by members of the same species. Each species of European newt (genus *Triturus*) has a unique, complex courtship

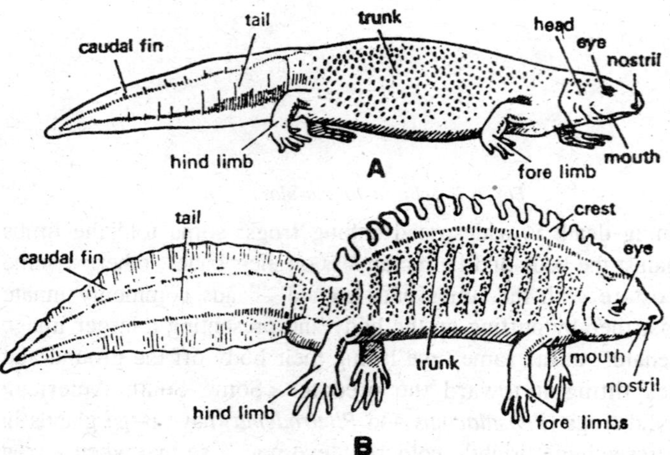

Fig. 6.3. Triton. A—Female. B—Male.

behaviour, in which the male must elicit a response from the female before he can initiate the next phase of courtship. Courtship takes place in water, and during the mating season males develop bright colours and elaborate tail fins. The male positions himself in front of the female and initiates a series of tail movements, stimulating her visually, by touch, and by smell. New World lungless salamanders engage in a tail-straddling walk in the water or on land. The male initiates courtship by rubbing the female's head with his chin, secreting chemicals from glands on the chin which stimulate her. He then moves forward and the female straddles his tail, pressing her chin to the base of his tail. If the female does not follow closely, the male may swing around and violently slap her head with his chin. Courtship terminates successfully when the female follows the male until he deposits a gelatinous capsule with a cap of sperm (spermatophore), which she then picks up with the lips of her cloaca.

The only frogs that have elaborate courtship rituals are some of the aquatic tongueless frogs. In the South American genus *Pipa*, the male grasps the female around the waist while she performs a series of somersaults during which eggs are expelled; the male then fertilizes the eggs and sweeps them onto her back with his feet, where they implant and develop.

Care of Eggs and Tadpoles

Most amphibians that deposit their eggs in water immediately abandon them. In species that deposit their eggs on land, however, it is usual for a parent to stay with the eggs, preventing marauders from stealing them. Many frogs take care of their young in more active

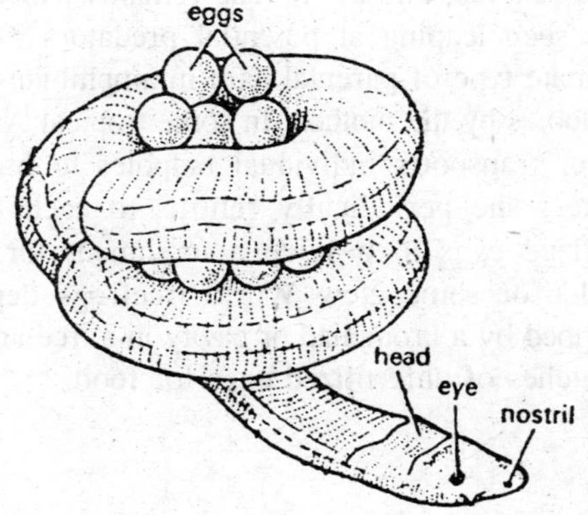

Fig. 6.4. Ichthyophis. Female coiling around the eggs.

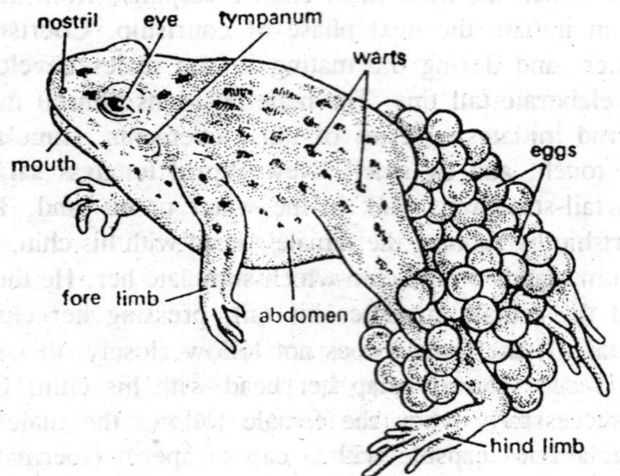

Fig. 6.5. Alytes.

ways. For example, females of the South American marsupial frogs transport eggs on their backs or in a dorsal pouch. Males of Darwin's frog in Chile pick up hatching tadpoles from egg clutches on the ground and carry them in a vocal sac until they complete their development. Males of the hip-pocket frog in Australia collect tadpoles and transport them in pockets in their flanks. Females of the Australian gastric-brooding frog swallow fertilized eggs; the eggs hatch, and the tadpoles complete their development in the stomach, which has to turn off its digestive functions during brooding.

Observers have seen the parents of some South American pond frogs (genus *Leptodactylus*) protect their free-living tadpoles; the tadpoles move in schools, and the female remains with them. Females have even been seen leaping at potential predators of the tadpoles. Perhaps the ultimate type of parental care in amphibians is the feeding of free-living tadpoles by the mother. In some tropical American poison frogs, the mother transports individual tadpoles to bromeliad plants which hold water; she periodically returns to each bromeliad and deposits unfertilized eggs to provide nourishment for each tadpole. Similarly, females or some New World treefrogs deposit fertilized eggs in water cupped by a bromeliad or cavity in a tree and subsequently provide other batches of unfertilized eggs for food.

7

Ecology of Amphibians

Amphibians Living in Caves

Caves of some sort are found in almost every state in the union. But southwestern Pennsylvania, West Virginia, western Virginia, southern Ohio, southern Indiana, Kentucky, Tennessee, southern Missouri, northwestern Arkansas, and eastern Oklahoma abound in caves because this is a limestone region where conditions are ideal for cave formation. Caves contain three different ecological niches, one gradually merging into the next. First, there is the mouth of the cave, fairly open and with almost the same amount of light and roughly the same temperature as is found just outside the cave. Gradually, as one penetrates farther, there is less light until one reaches the twilight region, where the light is dim, just as it is at dusk.

As one descends still farther, one reaches a zone where there is no light at all–a zone of eternal blackness, darker than night. Here the temperature remains almost constant the year round. Caves would appear to be ideal homes for amphibians, and so they are for certain salamanders. We find very few frogs in caves, and none in the region described above. Why this should be is a mystery as yet unsolved. Many salamanders are found in the mouths of caves. Among them are the Green Salamander, *Aneides aeneus*, mentioned earlier, the Kentucky Spring Salamander, *Grinophilus porphyriticus duryi*, Wehrle's Salamander, *Plethodon wehrlei*, and the Slimy Salamander, *Plethodon glutinosus glutinosus*.

Farther inside the cave, in the twilight zone, we find that the numbers fall off sharply. In this niche we find only the Dark-sided Salamander, *Eurycea longicauda melanopleura*, and the Cave

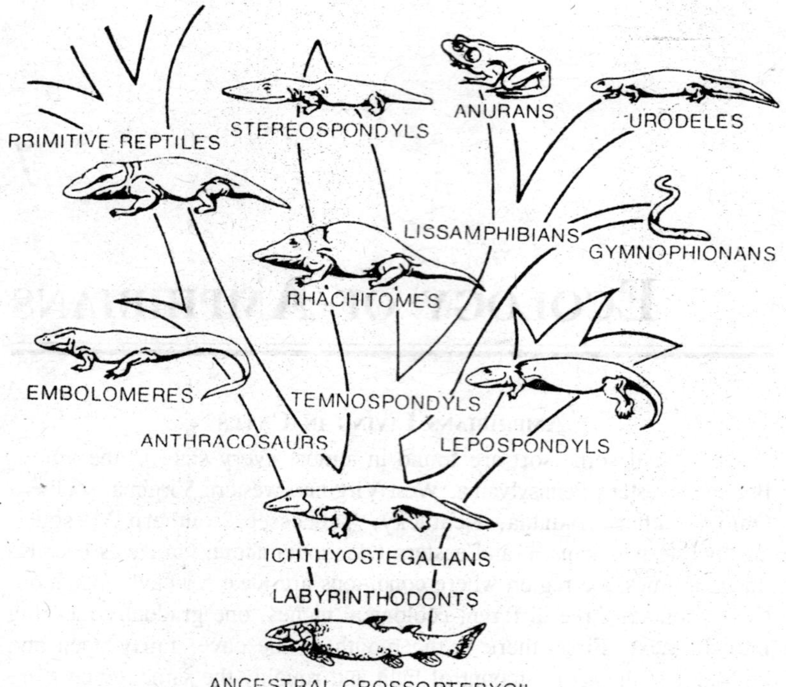

Fig. 7.1. A "family tree" of the Amphibia.

Salamander, *Eurycea lucifurga*. As we look over this list, we notice that without exception, all these salamander are of the family Plethodontidae. All have large eyes, the better to see with in the falling light of the cave. But these salamanders cannot really be considered true cave dwellers. The ones found inside the cave mouth are as often or oftener found outside it. The two cave salamanders of the twilight region are also found frequently outside, though during certain months they are evidently more numerous in the cave.

Dr. Oliver tells of investigations carried on concerning the seasonal occurrence of *Eurycea lucifuga* in a cave Tennessee. The cave was visited once a month. No salamander at all were found in March, June, November, and December. From 1 to 10 were found in January, April, May, July, and October. From 10 to 20 salamanders were found in February. It is impossible to draw any conclusions at all from these figures. If, on the same day, the investigator had also made a search outside the cave for this salamander, and recorded all he found outside each month, together with temperature and humidity readings inside and outside the cave (he may well have done these things, but they are not reported by Dr. Oliver, and we have not found the original article), we should be better informed and might learn many things.

The true cave dwellers of the world may be named on the fingers of one hand, and we are fortunate in having three of them in this country. They are the Olm, *Proteus anguinus*, found in the eastern Alps; the Grotto Salamander, *Typhlotriton spelaeus*; the Texas Blind Salamander, *Typhlomolge rathbuni*; and the Georgia Blind Salamander *Haideotriton wallacei*. All are salamanders; all are blind when adult; all are white or very pale in colour; all but one are neotenous; all but one are plethodontids; all live in the deepest recess of the cave; and all are rare and extremely restricted in their range.

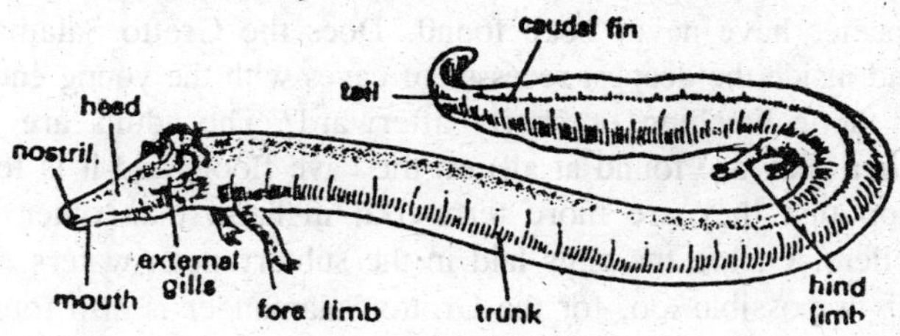

Fig. 7.2. Proteus.

The Olm, Proteus anguinus, belongs to the family Proteidae and is of the same family as, but of a genus different from, our native mudpuppies. Proteus is neotenous, as are the mudpuppies, but, unlike them, he has lost his colour and his eyes–or rather the eye do not develop sufficiently to enable him to see. When Proteus is exposed to light, its skin becomes pigmented. This pigmentation is retained even if the specimen is subsequently returned to darkness. This fact, as innocuous as it may appear, led to one scientist's suicide. A fascinating account of how the arguments over Proteus culminated in Dr. Paul Kammerer's suicide is contained in the first chapter of Willy Ley's Salamanders and Other Wonders.

Normally Proteus produces one or two young ovoviviparously. But in captivity, if the temperature of the aquarium water rises above 50° F., eggs may be laid. One or two young may hatch from these eggs; however, to date, any such young have not survived for more than a few days. In nature, of course, the water temperature remains constant, and young born alive are the rule. The Olm makes his home in the subterranean waters of the eastern Alps, and is rarely found outside

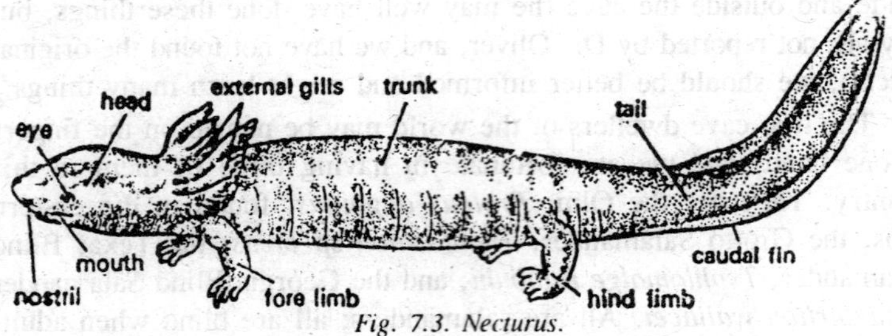

Fig. 7.3. Necturus.

them. The Grotto Salamander, *Typhlotriton spelaeus*, is found in limestone caves of the Ozark Plateau of Missouri, Arkansas, Kansas, and Oklahoma. Nothing is known of the breeding habits, and the eggs of this species have never been found. Does the Grotto Salamander lay on land inside the deepest recesses of caves with the young entering the water upon hatching or shorty afterward? The adults are often found (when they are found at all) on the cave floor, and it is felt by Dr. Bishop that they are more terrestrial in habit than other blind cave dwellers. Or are the eggs laid in the subterranean waters of the cave? This is possible too, for the Grotto Salamander is also found in these underground streams. Or is it possible that the adults leave the caves and go to streams just outside its mouth and deposit their eggs there? This may seem farfetched, but it is not so ridiculous as it sounds and it cannot be ruled out as possibility. Fir it is strange but true that the larvae of the Grotto Salamander are most frequently found in streams outside the cave, though a few larval specimens have also been captured in subterranean waters. Do the larvae hatch from the egg in the cave and then leave it to spend their larval life outside? If this be so, why do they leave the cave at all? And since it is known that their larval life is frequently spent in surface streams, why do they, when adult, leave the world of light and retreat to the complete darkness of the cave?

One answer to the last question lies in the adult Grotto Salamander's requirements. Inside the deepest recesses of the cave, the temperature remains constant year round, varying only by a degree or two. In addition, this air has a higher oxygen and moisture content than that of surface air. It is known that the Grotto Salamander cannot tolerate high temperatures. Perhaps, too, it cannot tolerate fluctuations in temperature either, and in order to survive it retreats to the only place in this region which meets its requirement sin these respects–

namely, the cave. The larval Grotto Salamander is grayish-white above, with dirty white legs and belly. It has well-developed eyes that see as well as the eyes of an other salamander larva. When the time for metamorphosis approaches, the larva enters the cave and transforms there in utter darkness. As it transforms, or soon after, the eyelids close and grow together; the rods, cones, and eye muscles degenerate until the adult can no longer see. Of course, in the complete darkness of its environment eyes are of no use at all, but it seems curious that blindness is not an inherited characteristic in this species.

Larval Grotto Salamanders do not lose their sight during metamorphosis if they are kept in a lighted place. What does the adult Grotto Salamander look like? It is small, averaging about 3½ inches long, though larger and smaller specimens have been found. The head and body are long and slender, as are the legs. It is a pale flesh colour with small orange spots scattered on the top of the tail. The eyes show as small dark spots, and are completely or almost completely covered by the fused eyelids. The Grotto Salamander is the only non-neotenous blind cave dwellers known. This summarizes what is known about this rare species. And since so very little is known, many questions occur to us–questions that will have to wait to be answered until some persistent scientist has done more research on these salamanders. Rarer and more restricted in locality is the Texas Blind Salamander, *Typhlomolge rathbuni*. It has been found only in the area around San Marcos, Texas, about thirty miles southwest of Austin.

In the Grotto Salamander, we have seen a species that has not, as yet, become completely adapted to cave life, for the larval period is spent at least in part outside. But the Texas Blind Salamander is a true cave dweller, for neither adult nor larvae have ever been found except in deep wells and the subterranean waters of caves. Nothing is known of the breeding habits in nature, and the eggs have never been found, though one captive specimen did lay eggs in the laboratory. Dr. Noble stated that thyroid "may be absent" from this salamander. Whether this phrase means that it is absent from some specimens and not from others, or whether it means, as is probable, that sufficient data is lacking to make a positive statement, is not entirely clear to us. Be that as it may.

The Texas Blind Salamander is neotenous with permanent gills. The larvae resemble the adults closely in form, but they are, of course, of a smaller size. The average length of the adult Texas Blind Salamander is 3¾ inches, though individuals over 5 inches in length have been found. The general colour of these animals is white, but parts of the head, body, and tail are suffused with iridescent shades of pale pink, blue, and lavender, which must give this rare creature a beautiful but strange appearance. The legs are very long–much longer than those of the Grotto Salamander and slender, with long fingers and toes. Probably its most outstanding feature is its head, which is very broad in front of the gills and gently tapers to a long snout. The outline of the head viewed from the top resembles the outline of a dig's head. In a side view, however, this salamander's head looks not at all doglike, for from this angle it may be seen that the snout is very greatly flattened (herpetologists term it "depressed"). The eyes show as small black spots, but they are covered with skin. The eyes are completely useless, for not only are there no eye muscles or rods and cones, but even the lens is nonexistent.

Our remaining salamander is the Georgia Blind Salamander, Haideotriton wallacei. It is the rarest of them all, for only one specimen has ever been found and described. This individual was an adult female that came from a well 200 feet deep in Albany, Georgia, a town about 150 miles south of Atlanta. It is small salamander, 3 inches in total length, neotenous, with permanent red gills. The head is broad, the snout equally broad and very gently rounded. The viscera and larger blood vessels may be seen easily through the thin white skin of the body and legs. There are no eyes at all, nor are there any dark eye spots that might indicate where they once were. These are the blind cave dwellers of the amphibian world.

Before leaving them altogether, let us consider for a moment the possible reasons for their rarity. Caves make ideal homes for many salamanders. They are primarily nocturnal creatures, and the cave with its darkness, even temperature and humidity, and high oxygen content would seem to be ideally suited to their needs. Why, then, are there so few of them? It may be because relatively few amphibians are able to tolerate the high concentrations of limestone found in many caves. It may be that most species require some light in order to survive. Perhaps they need more temperature variation than is present in most caves. And most important of all, perhaps there is an insufficient food supply to support great numbers of salamanders in the cave. We may wonder about all these things, but until we learn more about amphibians we cannot know why more of them do not make the darkest inner recesses of a cave their home.

It seems surprising that those salamanders who are true cave dwellers have not become highly adapted to the darkness, as have stead of the huge eyes and phosphorescent bodies of many deep-sea fishes, the cave-dwelling salamander have only pigmentless bodies and blindness. Nor have any extrasensory organs been developed on these cave creatures, though it is felt that sensory perception is greatly heightened. And since no green plants grow where there is no light, animals who live in the darkness are, of necessity, carnivores. The deep-sea fishes have evolved all kinds of devices for more efficient food getting, ranging from large teeth to elaborate "lures" to entice their prey within grabbing distance. But the blind cave-dwelling amphibians have none of these, and are the same as noncave salamanders. Perhaps we are too impatient. Deep-sea fish were in existence millions of years before there were any amphibians or any caves. Is it possible that the blind cave dwellers may become as highly adapted in time? No one can know, for in nature all things are possible, but it is interesting, we think, to speculate.

Terrestrial Amphibians

Just as there are many types of ponds, streams, ditches and swamps, so are there many types of forests and fields. There are forests that consist of deciduous trees, and there are those whose stand is principally coniferous. There are the immense rain forest of the Northwest and the tropics and the drier jungles. There are the forests made up of large trees of tremendous antiquity, and thickets made up of new trees and saplings. The same variety is present in fields. There are plowed fields and fields under cultivation. There are hayfields and land which is used for pasturage. We could go on almost infinitely, describing the type of soil, the moisture content, and so on, and we should not be straying from our subject in doing so either. Mineral content, moisture content, amount and type of vegetation–all these are pertinent factors in the make-up of an ecological niche or habitat.

Forests and fields have one broad characteristic common to them both. They are both more or less dry land. The forest dwellers, as a general rule, prefer the darker, damper atmosphere, where cover such as rocks, leaves, leaf mold, and logs is always readily available. The field dweller can tolerate the more open situations found in most fields. Those who are least susceptible to desiccation are often found in plowed fields. Let us examine a few of these dry-land dwellers; first, the more moisture-loving woods dwellers, then the grass-field dwellers, and last the dwellers of plowed field and area sparsely vegetated.

Of all the dry-land amphibians, none is more moisture-loving than the Spotted Salamander. *Ambystoma maculatum*, found throughout most of the northeastern half of the United States, Florida and southern Georgia excepted. Herpetologists tell us that these salamander live in the woods, hiding by day under logs and rocks during rainy weather. We are told that during drier weather, they make shallow burrows underneath the surface. It is fortunate for us that the scientists give us this information, for few of us would be able to find it out for ourselves. Though we have looked long and persistently in wet weather and dry, making lengthy excavations to explore tunnels with openings under logs and rocks, we have never been fortunate enough to find a Spotted Salamander outside the breeding season. This might not be so extraordinary if this creature were small and slim-bodied, or inconspicuously marked. But it is not. It is a large, stout-bodied animal, almost 7 inches long, black in colour with a row of yellow or orange spots on either side. It is a very striking animal, and one couldn't possibly overlook it, especially when one is searching for it.

Neither is the Spotted Salamander rare; indeed, it is among our most common salamanders, as anyone can prove for himself by going to the ponds at night in the early spring. Though the Spotted Salamander may be seen in ponds found in the open, it is the woodlands ponds that they prefer and where they are most numerous. Here they may be counted by the hundreds during the evening; and a little later their egg masses serve as proof that one has not counted all that were present. The males are a little larger than the females, and have a protuberant vent. So swollen is it that one can sometimes distinguish the sexes while they are swimming. The nuptial dance consists of both sexes swimming over and around one another, the males nudging each and the females. "Dance" is a good description for these courtship activities, but one must not interpret the dance too literally. There is nothing organized or stylized about it. Occasionally the dance is carried on in a very limited area; there are many participants and the water fairly boils with their activities.

More often, there are scattered groups of fifteen or twenty salamanders seemingly swimming in haphazard fashion. But as you watch, you will see that they do indeed twine in among one another, over and below, then swim off only to return in ones and tows to nudge one another again. Usually this is done in a leisurely manner–unhurried caresses in passing one another by. Every so often a salamander will surface, gulp a breath of air, and return to his courtship activities. We might mention that Spotted Salamanders appear to be a different colour at night than when seen in daylight. As you look at them by flashlight, they have slate-gray bodies with light gray spots. The following morning one finds the pond bottom littered with the white gelatinous bases of the male spermatophores; they show up clearly against the dark leaves of the bottom. Not a salamander is to be seen. It is as if they had vanished into air. Where have they gone? We do not know. We do not find them under logs or stones by the ponds. Nor have we found them under logs and stones on the pond bottom, though it is possible that they may be there. Do they leave the pond and burrow in during the day? This seems a not unlikely possibility.

The breeding season extends over a week or two in any given locality. The Spotted Salamander are not like some of the burrowers that we shall describe later, whose breeding activities are completed in one or two nights. The breeding season is in March or April. The eggs are laid several days after impregnation in a roundish mass attached to submerged vegetation. They may be deposited in several small masses or one large one. The total complement ranges from 25 to 250 eggs, with larger, older females laying the larger number. Geneally, we find that most large eggs masses contain approximately 100 to 125 eggs. The egg itself is dark brown above, whitish below. The envelopes that surround the egg may be clear or they may be light green. The green colouring is caused by algae that penetrate the capsule and grow there. Surprisingly enough, the alga does the egg no harm; indeed, algae and eggs are mutually beneficial. The algae give off oxygen as a by-product of photosynthesis. This additional oxygen in the egg capsule makes larger embryos, a lower mortality rate, and a shorter incubation period. The algae benefit from the carbon dioxide given off by the embryo as a waste product of respiration. When does the alga enter the egg capsule? At any particular stage in the egg's development? Why does one find some eggs with algae, while others in close proximity have none? These are interesting questions that may someday be answered by herpetologists.

The mass of jelly may be clear or it may be milky white. We find both kinds in any given pond. What governs the colour? Does it depend on the female who lays the mass, or are environmental conditions responsible? The incubation period is between 1 month and 2 months, depending on the temperature. The larva is a light sandy or greenish yellow colour with darker spots scattered over the back. Balancers are present at hatching, as are the buds of the forelegs. The dorsal fin is high and extends up as far as the forelegs. Transformation occurs after a larval period of 2 to 4 months. The adult colouration complete with spots is usually assumed within a week of metamorphosis. Of all our local salamanders, the Spotted Salamander are our favourites. The plump body, the head with its large dark eyes and bluntly rounded snout, and the docile, gentle manner make this animal very engaging. How aggravating it is not to be able to find him outside the breeding season! What we should do is to keep them in a terrarium. They evidently live to a considerable age when properly cared for. Dr. Oliver tells us that the maximum age recorded for a captive is 25 years.

The Slimy Salamander, *Plethodon glutinosus glutinosus*, is another stout-bodied woodland salamander. It averages slightly smaller than the Spotted salamander, being about 6 inches long, though larger individuals have been found. It belongs to the Plethodontidae, as is evidenced by its nasolabial grooves. Its head is different from that of the Blunt-mouthed, or Mole, salamander, the snout being more pointed. Its range is about the same as the Spotted Salamander's, but extends a little farther south, down into half of Florida, and not so far north. It is entirely absent from the New England states.

The Slimy Salamander has a black back with numerous scattered white flecks. The belly is slate-gray. This salamander get its popular name from the copious sticky secretions produced by its mucous glands. Do these exudates help keep this salamander's skin moist under adverse conditions? Probably. All Lungless Salamanders are easily killed by desiccation, but this is more marked in the Slimy Salamander than in some of the other woodland salamanders, such as the Red-backed Salamander, for instance. The Slimy Salamander is found under surface materials in the woods, occasionally in caves or in rock crevices, and very often, especially during dry periods, it burrows into the soil.

Very little is known about the breeding habits of the Slimy Salamander. By piecing together scattered accounts and evidence gathered through dissection, it seems probable that the mating season is in the fall, in the North, while the laying season is in the early spring. However, this may not hold true in the South, for eggs attended by a female were found in a cave in Arkansas in August. Eggs and mother have been found in West Virginia in June. The eggs, about 15 to 18 in number, are laid in a small mass. Dr. Bishop feels that in the North the eggs may be found in the hiberation sites deep in rock crevices. It will be no simple matter to find them, but their discovery and precise observations on the larvae that hatch would indeed be a nice feathers in an amateur herpetologist's cap.

Far more common and less secretive is the Red-backed Salamander, *Plethodon cinereus cinereus*, found in the Northeast as far south as North Carolina and west to Indiana and Wisconsin. This salamander prefers coniferous woods, and we find them more frequently in the vicinity of evergreen trees.

The Red-backed Salamander has two colour phases. All of them have a bluish-black or charcoal-gray belly liberally speckled with bluish-white flecks. Perhaps we should say almost all of them have mottled bellies. Once we found a Red-back whose belly was dark gray with only a few widely scattered white specks. The red-backed phase has a broad band of red or orange extending from the head back to the tail. The sides are black. The lead-backed phase is a uniform dark gray or black on its back. Some individuals are intermediate between these two phases, with the dorsal stripe of a faded gray interrupted in places by the darker gray of the sides.

In some localities Red-backed Salamanders may predominate; in others, Lead-backed Salamander are more frequently found; in still a third place, they may be found in approximately even numbers. Dr. Conant tells us that at high elevation Lead-backs may be scarce or absent altogether. But whether it is a Red-backed Salamander or a

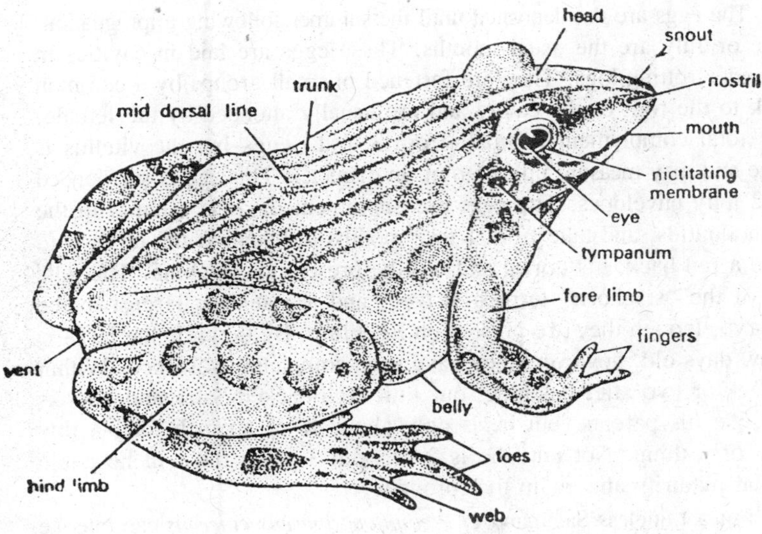

Fig. 7.4. Bombinator.

Lead-backed Salamander, they are the same subspecies, *Plethodon cinereus cinereus*. A few days ago we listened to a discussion of industrial melanism in the Pepper-and-Salt Moth in England. In areas near large industrial cities, tree trunks and vegetation are blackened from soot. In these regions the black, or melanistic, variety of the Peeper-and-Salt Moth predominates in a 3 to 1 ratio. In regions not polluted with industrial smog, the normal colouration predominates–also in a 3 to 1 ratio. Both the melanistic and the normal colouration are valuable to the animal in the preferred environment. We are not suggesting that industrial melanism is responsible for the two-colour phases of the Red-backed Salamander, but we wonder whether, in situations where one or the other colour phase predominates, there is not some good reason why it does. In other words, in some way not known at present, either phase may perhaps have a definite survival value for its possessor.

The Red-backed Salamander varies in size from 2½ to 3½ inches, and though the Lead-backs appear a little smaller, it is probably their dark colour that gives us this impression. The males are shorter than the females, with a more swollen snout. The premaxillary teeth are enlarged in the male–but you will need a magnifying glass to see this feature. The breeding season is in the fall on land, from late September to late November.

The eggs are not deposited until the summer following impregnation; June or July are the usual months. These eggs are laid in cavities in or under rotting logs. They are fastened in small groups by a common stalk to the roof of the cavity and are usually attended by the female. The total complement is small–from 3 to 13 eggs–but the vitellus is large and may measure up to 5 mm. in diameter. This egg is surrounded by 2 jelly envelopes. The eggs are unpigmented, as is usual with the plethodontids, and take from 1 to 2 months to hatch. If the larva is to have a red back, his dorsal and will be present tat hatching. The gills are of the "staghorn" terrestrial type, and are at their largest in the embryo, though they are present on hatchling. By the time the larva is a few days old, however, they have been completely absorbed. Within a week or two after hatching, our little Red-backed Salamander looks just like his parents, but he is only about ¾ of an inch long–a tiny mite of a thing. Not until he is 2 or even 3 years old will he attain sexual maturity and begin to reproduce.

For a Lungless Salamander, *Plethodon cinereus cinereus* can tolerate amazingly dry conditions. This is all the more surprising since it is a fairly small salamander, for the skin area is proportionately larger in a small animal than in a larger one when the surface is related to weight. We might expect to find that the larger plethodontids are less prone to desiccation, but such is not always the case, as a comparison of the large Slimy Salamander and the small Red-backed Salamander proves.

During the winter months the Red-backed Salamander hibernates underground. The depth to which it burrows varies considerably, and may be as shallow as 3 inches or as deep as 15 inches. Possibly in the more northerly portions of the range they may hibernate even below this maximum recorded depth. In this species hibernation is governed principally by lowering temperatures, and is not dependent on any inherent rhythm. Dr. Oliver gives us some very interesting facts concerning the Red-backed Salamander's hibernation. He reports that one investigator found that he mortality due to hibernation was as high as 57 per cent. How such an exact figure could be determined, we cannot imagine. Is it really true that over half the Red-backed Salamanders die from cold every year? We do not feel that we are being disrespectful if we view this figure with skepticism. Perhaps the mortality rate is this high in certain years on in certain unfavourable localities. But perhaps this high percentage does not hold good for all years in all localities. How was this figure arrived at? Was it done by finding a hibernation site and counting the animals that were dead and those that were alive? Or was it done by marking off a collecting area and making a fall count and a spring count? Either of these methods has its flaws.

In the first case, hibernation sites are difficult to find, and one cannot feel sure that one has not found the ones that are less suitable. The second method also has its disadvantages. Though one does not see hundreds of terrestrial salamander gathered together in a common breeding site, there still exists a certain gathering together of the clan. We find more Red-backed Salamanders clustered together in scattered groups in the fall than at any other time of year. The fall count is therefore apt to be large. However, we don't think that the smaller spring count necessarily indicates a high hibernation mortality rate. Whereas in the fall there are two urges that prompt these salamanders to collect – the breeding season and then hibernation–in the spring neither of these conditions is present. Probably the salamanders emerge in small groups at different times, and scatter. Though scientist can predict to the exact hour when the vernal equinox will occur, they cannot predict with equal accuracy just when each salamander will emerge from hibernation.

Be all this as it may, it seems certain that whatever the percentage, the hibernation mortality of the Red-backed Salamander is high. Though Nature has not provided the Red-backed Salamander with an innate knowledge of when to hibernate, she has partially compensated him in another respect. It has been found that in the winter, the water content in the body is greatly lowered. It is believed that this enables the animal better to withstand low temperatures without freezing.

The prize for the earliest spring breeding frog in the North goes, not to the Spring Peeper, but to the Eastern Wood Frog, *Rana sylvatica sylvatica*. Herpetologists are not in agreement as to whether the Wood Frog should be divided into subspecies or not. Some authorities divide the Wood Frog into three subspecies: *Rana sylvatica sylvatica*, found in the northeastern United States; *Rana sylvatica cantabrigensis*, found in eastern Canada; and *Rana sylvatica latiremis*, found in the western Canada and Alaska. Others feel that only two subspecies are called for: *Rana sylvatica sylvatica* and *Rana Sylvatica cantabrigensis*. Still a third group feels that the colour differences and leg length as related to body length are not sufficiently constant and distinct to warrant subspecific division. The latter group feels that the relatively longer legs of our United States Wood Frog change gradually to the shorter

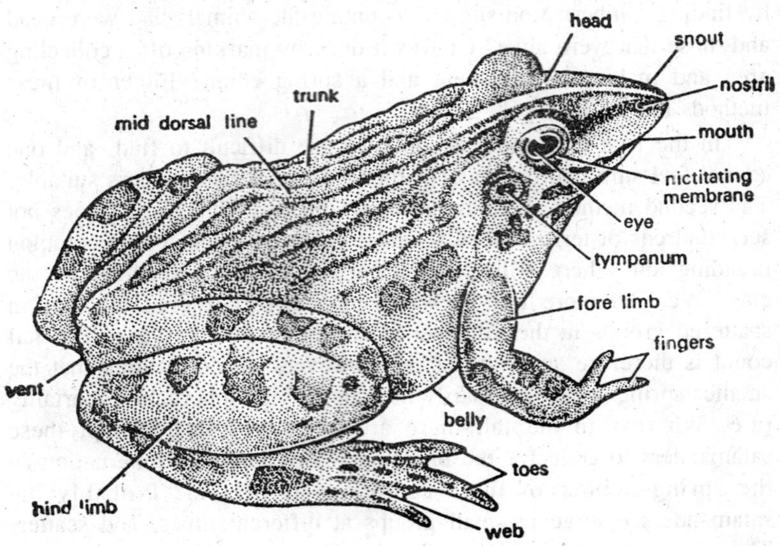

Fig. 7.5. Rana tigrina.

leg of the Northern Wood Frog and that this, therefore, is not an innate characteristic but merely one produced by physical differences in their range. We amateurs are not qualified to judge who is right in this controversy. Let us therefore take the middle course and assume, as a great many herpetologists do, that this species is divided into two subspecies. Perhaps industrious amateurs in northern. Minnesota and Wisconsin, where Rana sylvatica cantabrigensis occurs, will through leg and body measurements of many individuals, be able to shed some light on the confusion.

Our Eastern Wood Frog, *Rana sylvatica sylvatica*, is a small brown frog of from 2 to 3 inches long. In the cool spring water of the ponds where it breeds, the male appears to be dark or reddish-brown, unmarked except for the light line along the upper jaw and the light-coloured dorsolateral folds. But after the breeding season, in the warmer weather, its colour lightens considerably and the dark mask in back of the eye and dark lines from eye to snout stand out like dark brown velvet against the light tan of the rest of the body. You will notice, if you look closely at the photograph of the Wood Frog in his summer dress that the lower part of the iris is considerably darker than the upper part and blends into the mask. This indicates the purpose of the mask. It breaks up the conspicuous eye on an animal which is otherwise inconspicuous against the leaves of the forest floor. The mask, therefore, is an excellent camouflaging mark. Sometimes the Eastern Wood Frog has barred legs and a few dark spots scattered over the back. The belly is white except where the hind legs join the body, where it is a light greenish colour.

The Eastern Wood Frog is, as we said, a very early breeder. AS a general rule, they do not breed in the North before the middle of March or early April, but farther south they may start their breeding activities in February. Water temperature has been found to be a reliable indicator of when a number of frogs will lay their eggs. The average water temperature is 50°F. when the Eastern Wood Frog lays its eggs.

The males may be distinguished from the females by their toe webs, which are convex, while those of the female are concave. The vocal sacs of the males are internal, and show only as generalized throat swelling when expanded. The males lie on the surface of the water or may call under water. The voice usually resembles the quacking of ducks, and many hundreds all call at the same time. The call does not carry well, and is inaudible a very short distance away.

Sex recognition is almost entirely by trial embrace. All nearby moving objects are seized; those which croak or whose bodies have not sufficient girth are released promptly.

The breeding period in any given pond is short; the peak of the breeding activities is over in 2 or 3 days. When you hear hundreds of "ducks" quacking in a pond where there are no ducks to be seen, you may be sure you have come upon Wood Frogs breeding. But though you ears assure you the frogs are there, your eyes may not convince you. We have found that during the height of breeding activities in woodland ponds, the frogs are almost oblivious of our presence, even in the daytime. But where large breeding groups have chosen a pond in a more open situation, they seem to be more alert to danger and the greater number of them choose to call under water rather than while floating on its surface. However, though the Wood Frog is hard to see in his dark spring dress, once you have spotted him on the pond bottom he is quite easy to scoop up with net or hand. He flattens himself out on the bottom and depends on protective colouration and absolute immobility to save him. As your hand slowly descends, he is not so apt to dart away to another spot as are many other frogs.

The eggs are laid in a roundish mass attached to submerged vegetation. There are from 2,000 to 3,000 black eggs in a mass. Each eggs is surrounded by 2 jelly envelopes; only the outer one is distinct. The vitellus is small and averages under 2 mm. in size. Because these frogs are early breeders, they choose the warmest parts of the pond in which to lay their eggs. This often means that they lay close to shore in shallow water. Twenty-five or more bunches of eggs are often found laid so near to one another that soon after the jelly has swelled, it is impossible to tell exactly how many masses have been deposited, for each merges into the others around it.

Laid in such shallow water, the eggs are very vulnerable to desiccation and freezing. The shallow water of ponds often evaporates rapidly in the warmer days of spring. Many of the uppermost eggs are left high and dry, and these, naturally, do not hatch. However, the same high temperatures that cause rapid evaporation of the pond water also hasten the development of the egg. Dr. Oliver gives us the precise hatching time at different temperatures. At a water temperature of 50° F., eggs hatch in 275 hours, or a little over 11 days, whereas the eggs hatch in 3 to 4 days at a temperature of 68°F. Eggs that are near the surface may be killed by freezing temperatures. But even though the top of an eggs mass may be frozen, the eggs of the lower part will develop normally. The eggs and embryos of the Eastern Wood Frog can tolerate a considerable temperature variations–a 38° F. variation, to be precise–and as long as water temperatures do not fall below 37°F. or rise above 75°F., these eggs will hatch satisfactorily. Moreover, for some reason, eggs develop faster when temperature variations occur; the shortest incubation period is found in those eggs which are initially subjected to a low temperature and which finish their development at a higher temperature.

The tadpole of the Eastern Wood Frog is greenish olive in colour, with a high tail crest. It is wholly carnivorous in diet. We often see groups of these tadpoles all nibbling on the freshly killed body of some unfortunate casualty of the breeding season. We see them nibbling at other tadpoles' tails. Is this because food is scarce, or have they, as do so many animals, some way of knowing that the one they attack is unhealthy and not long for this world? Or is it plain cannibalism? Of course, their diet also consists of the minute insects and entomostracans found in the pond.

The Wood Frog tadpoles have a larval period of 1½ to 2½ months. As they transform, they leave their natal ponds wearing the light summer dress that makes them so inconspicuous in the forest. Outside the breeding season, our best collecting results have been in wooded spots near our river. We have found Wood Frogs in the forest near small trickling springs, but not so often as we do near the river. Perhaps it is the younger ones, always more gregarious and less elusive than any older adult amphibian, that prefer the damper, more open areas around the river. Though the Wood Frog is common, he is hard to find outside the breeding season. His alertness and shyness are all the more noticeable because this behaviour is so unlike his behaviour in the spring.

True land frog that he is, the Eastern Wood Frog hibernates under the soil of the forest floor. But as one might expect from a frog that can breed in such cold weather, the Wood Frog is also able to tolerate colder weather in the fall than many other species can. The latest date that we have found a Wood Frog was September 29th. However, since we do not know the home range of any particular individuals, we have no way of knowing the precise hibernation date. It seems quite likely that these frogs may be found abroad in October also, and even later in the more southerly portions of their range.

The Oak Toad, *Bufo quercicus*, is our smallest native toad, Its maximum size does not exceed 1¼ inches, and it is for this reason

Fig. 7.6. Bufo.

that it is frequently mistaken for a young Southern Toad. The Oak Toad is found along the coasts of our southern states from North Carolina south to Florida and west to Mississippi.

The Oak Toad varies in colour from almost black to a light gray. There is always a light-coloured stripe extending from the tip of the snout to the vent along the middle of the back. In addition, there are, in the lighter-coloured individuals, for or five pairs of black spots on either side of the vertebral streak. Many of the tubercles are red, and those on the black-barred arms and legs are frequently spiny. The palms of the hands and soles of the feet are orange.

When one looks closely at the Oak Toad, the first thing that be comes apparent is the extreme shortness of the head in relation to the rest of the body. The parotoid glands are oval and elongate, as a general rule, though an occasional specimen may have parotoids that are roughly triangular in outline. If we take a specimen in hand, and look at his foot, we find that the outer-sole tubercle os small, the inner one large and equipped with a cutting edge.

The breeding season is an extended one from April to September. However, these small toads do not go to the breeding ponds until warm rains occur. Unlike many toads, the vocal sac of the male Oak Toad is kidney-shaped when expanded. This characteristic indicates that this toad is closely related to the large Great Plains Toad, *Bufu cognatus*, and the Texas Toad, *Bufo compactilis*, whose males also possess kidney-shaped vocal sacs. The voice of the calling Oak Toad is most unfroglike, and resembles very closely the peeping ob baby chicks. Only the lower part of the male's throat is discoloured, while the rest of the throat and underparts are grayish white. Like most male toads, the Oak Toad male has dark excrescences on the first, second, and third fingers of the hands.

Though almost all true toads lay their eggs in strings of jelly, the Oak Toad, perhaps because of its minute size, proves an exception to this generality. We find female Oak Toads laying their black eggs in short bands of 3/8 of an inch or less in length with 4 to 8 eggs in each bar. The vitellus does not exceed 1 mm. in size. These bars may be scattered haphazardly close together, fastened to water vegetation, or they may be deposited so that they resemble the spokes of a bicycle without the wheel rim. In more precise scientific language, these bars may radiate out from a common focus. The total egg complement of 600 to 750 eggs is small for a toad.

The incubation period for these eggs is not known. Although the Wrights describe the tadpole as being small, grayish in colour, with an upper tail crest marked with black, and a purplish-coloured belly, they include no specific measurements of tadpole sizes just before transformation, nor do they include this species in their tadpole identification key. From this, we conclude that less is known about Oak Toad tadpole than about many other species of tadpoles.

This small toad is a resident of the sandy pie barrens of the South. Its breeding habits and the cutting edge on the innersole tubercle of its foot might lead one to guess that this was a fossorial toad. Actually, this is far from the case. Most toads are primarily crepuscular or nocturnal animals when fully adult. Though the Oak Toad is no real exception, curiously enough these little fellows are found abroad during the day more frequently than most toads. When they do hide during the day, they may be found under logs or in very shallow burrows whose opening in shielded by vegetation. This type of burrow is quite different from those dug by the true burrowing amphibians, as we shall see on the next chapter.

Doubtless, more scientific arguments have been fought over the Leopard Frog, *Rana pipiens*, than over any other species found in this country. The battle has been going on for years, and still continues. The question is over the subspecific division. Wright and Wright devote about forty pages of absorbing discussion to *Rana pipiens*, its various colour forms and mutations, and give their viewpoint on each of the subspecific divisions proposed. From their summary of the species and the controversy over it we can see most clearly the enormous pains the conscientious herpetologist takes to ensure that a given subspecies be a valid one. He must collect hundreds of specimens in the field throughout the animal's range. He must observe many individuals in nature, as well as preserved specimens. He must examine eggs and

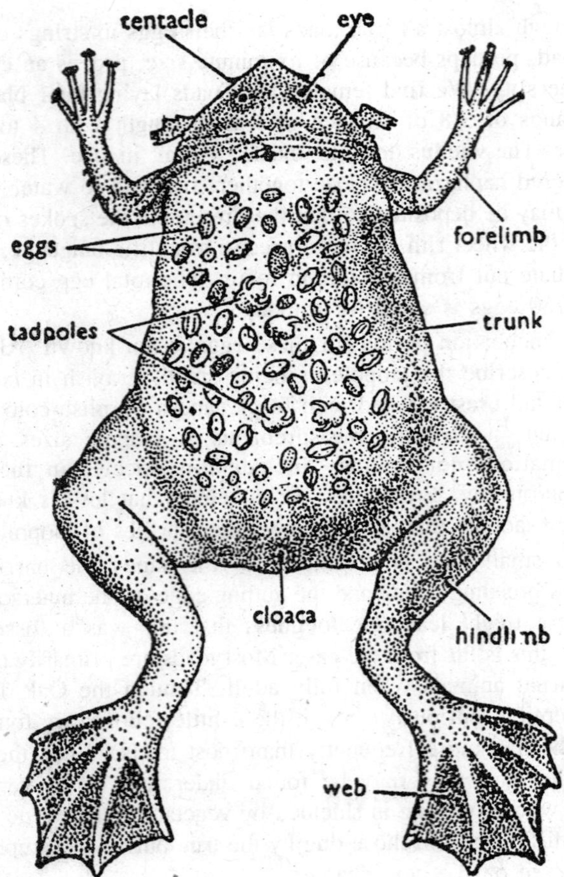

Fig. 7.7. Pipa americana.

tadpoles. In the case of *Rana pipiens burins*, once advocated as a subspecies, it is now fairly well established that burins is merely a colour phase of *Rana pipiens pipiens*, and not a valid subspecies at all. This was finally settled by embryological and genetic study. *Burnsi's eggs* were structurally indistinguishable from those of *pipiens pipiens*. It was found that these eggs differed by one colour gene only. So when one finds a brown unspotted frog in Minnesota or the borders of states surrounding it, one must correctly call it *Rana pipiens pipiens* (*burnsi* mutant).Such is the exactitude of science.

We are in no position to enter into this fascinating argument. Let us therefore stay clear of it as much as possible and consider the

Leopard Frog as one species, bearing in mind of course, that colour variations and slight differences in body measurements do exist. Some variety of *Rana pipiens* is found throughout the United States and Canada. The western states of California, Oregon, and Washington have this species only on their eastern borders, however.

The Leopard Frog is a slender, smooth-skinned frog with long legs. The dorsal ground colour is brown, green,, or tan. The prominent dorsolateral folds are light in colour. Between these ridges are two or three rows of irregularly placed white-bordered round spots. The sides are also spotted with white-bordered black, spots. The legs are barred with light-bordered black bars. The underparts are glistening white with no orange on the underside of legs or groin. Occasionally the underparts of the legs may be a light yellow, and, in Arizona only they may be orange. The italicized words are most important, for we have another frog the Pickerel Frog, which we shall discuss in a moment, that resembles the Leopard Frog and is distinguished from it by the differences in the italicized details only.

The Leopard Frog is a beautiful frog. Fortunately, it is among our most common frogs, so we can enjoy its beauty as often as we like. Though it hibernates in the water, it emerges from hibernation early and may begin breeding as early as March 15th. One may generally hear scattered calls of solitary males as much as three weeks before the breeding season really begins. Not until the water attains a minimum temperature of 41° F. does not the Leopard Frog start laying its eggs.

Unlike many salientia, both the male and female have a voice though during the breeding season we presume that the females are normally silent. The vocal sacs of the male swell out on either side between the angle of the mouth and the shoulder. The call is difficult to describe so that one who has never heard it can imagine it. The start of the call sounds somewhat like a woodpecker tapping on a dead tree, and this sound is followed by several throaty croaks. The frogs call while in the water either when floating on its surface or when submerged. Sex recognition is by voice; when other males are clasped, they croak in protest and are usually released immediately.

The egg complement of between 3500 and 4,500 is laid in a mass attached to aquatic vegetation. The vitellus is black above, white below and is about 1.6 mm. in diameter. It is surrounded by 2 envelopes. The incubation period varies with the temperature of 68°F., the eggs will hatch in 4 days. But in nature we are more apt to find that the eggs have a longer incubation period, especially in the North. It usually ranges from 2 weeks to a month.

The mature tadpole has a brownish body with translucent tail crests finely speckled with black. The belly is an iridescent cream colour, with the intestine showing through the skin. The tadpole attains a mature size of about 3¼ inches before transforming, after a larval stage of 2 to 2½ months.

Thousands of Leopard froglets may be found basking on the shores of their natal ponds in July and August. In all probability you will not even notice them until you approach closely and hear the splashes as they jump into the water. By all means catch one so that you can look at it closely. Note the long slender body and legs with very narrow waist. See the pointed snout. The feet are only partially webbed. These features indicate that it is a land frog of extreme agility. The ground colour of the upper parts is light tan, or greenish depending on the individual. The white-outlined dark-brown spots stand out clearly against the lighter background on a warm day early in July. It is hard to believe that this frog is so difficult to see in his natural surroundings.

We do not know how long he stays near the banks of the pond where he was born. Unfortunately, we do not live near enough to a pond where Leopard Frogs breed in great profusion to be able to make the daily observations necessary to establish this. It is not too long, but whether "too long" means several days or several weeks we can not say. Be that as it may, during the latter part of July and throughout August and most of September we begin to find the current year's crop of Leopard Frogs hopping around the field and untrimmed edges of the lawn.

We rarely if ever spot this small creature before it spots us and jumps to elude us. Its hop is long and low, and generally it jumps three or four times, each jump being in a slightly different direction. Of all our frogs, the Leopard Frog is the longest jumper in terms of distance jumped. Its prodigious leaps exceed even those of the Green Frog, and Dr. Oliver tells us that a leap of a little over 44 inches has been recorded for this frog.

The Leopard Frog is a frog of grasslands, preferring the damper parts. You are not likely to find him jumping around in a cornfield or wheatfield. You may find him among your untrimmed lawn borders, or, in the early morning, when the grass is dew-drenched, even out on the lawn itself. As the Leopard Frogs grow to the adult size of from 2 to 4 inches, they are more wary and stick to the longer grass. We do not find adults with anything like the frequency that we find the young.

As the temperature begin getting colder in the fall, the Leopard Frogs head for the ponds to hibernate. In the warm days of fall they may be seen swimming about, but as the water grows colder they remain on the bottom and finally dig into the detritus to spend the winter. They must dig deep enough so that the body temperature does not fall below 31° F. Below this, the Leopard Frog will die, for the coordinating centers of the nervous system cannot function.

How does the Leopard Frog fare in hot weather? Can he, begin a dry-land frog, withstand great heat? Surprisingly enough, we find that he cannot. The maximum temperature he can survive internally is only a fraction over 64°F. This helps to explain why we occasionally find desiccated Leopard Frogs on the roads or on a plowed field in summer. They got too hot while crossing it, and died. This also explains why the Leopard Frog prefers long grass and moist spots.

Twin brother to the Leopard Frog is the Pickerel Frog, *Rana palustris*, found in the eastern half of this country as far west as Texas. It is absent from Florida and the southern portions of Sough Carolina Georgia, Alabama, Mississippi, and Louisiana.

Like the Leopard Frog, the Pickerel Frog is a spotted frog. However, its spots are squarish with dark borders, arranged regularly in two rows between the dorsolateral folds with an occasional third spot in between. The sides are also regularly spotted. The underparts are white except for the undersurfaces of the hind legs and groin, which is orange or dark yellow. Probably the most outstanding characteristics of this frog is its odor, which once smelled is unmistakable. Unfortunately, the young are more difficult to separate from young Leopard Frogs. The undersides of the young Pickerel Frog's legs are not orange, but yellow or flesh-coloured, and the acrid Pickerel Frog odor is not present. To add to the confusion, the spots of young Leopard Frogs are not always light-bordered. But with a little practice, one can usually tell the difference by the way in which the spots are arranged. These, regularly arranged in rows in the Pickerel Frog, cover a larger area than do the spots of the Leopard Frog, in which there is a greater proportion of ground colour. The Pickerel Frog nearly always has a ground colour of tan or brown, and rarely if ever exhibits the green colour sometimes found in Leopard Frogs.

The skin secretion of the Pickerel Frog produce the animal's characteristic odor. They are extremely toxic to other animals—even

Fig. 7.8. Rhacophorus.

to amphibians of a different species. But we shall discuss the protective properties of this frog's skin secretions in the chapter on defense. What we should like to ask here is why the young do not have the adult odor. Is it because their skin secretions are not toxic? No, it is not. We have seen a very hungry Garter Snake mouth a young Pickerel Frog but refuse to eat it. However, from the snake's behaviour, we believe that the skin secretions of the young are not so potent as those of the fully matured adult. Though the snake refused to eat this frog, it did repeatedly bit it, and showed no sign of discomfort after doing so. This is not the case when a snake makes a mistake and seizes an adult Pickerel Frog.

The Pickerel Frog is a slightly smaller frog than the Leopard Frog, being between 2 and 3 inches long at maturity. It emerges from its hibernating retreat in the pond a little later than the preceeding species, and begins breeding when the water temperature is 57° F. or above. Unlike the Leopard Frog, it has no large external vocal sacs and its snoring call produces only a small swelling between the ear and arm on the sides of the head. The males are equipped with the enlarged thumb pads of most ranids.

The eggs are brown above, yellow below, and are a fraction of a millimeter smaller than those of the Leopard Frog. The total complement of 2,000 to 3,000 is laid in a mass of jelly, attached to submerged vegetation in shallow water. After an incubation period a little longer than that of the Leopard Frog, the Pickerel Frog eggs hatch.

The mature tadpole is about 3 inches in length, with a cream-coloured iridescent belly. The intestines are visible through the skin.

The tail crests are not transparent and are heavily washed or blotched with purple or black. After a larval period of between about an inch long. These little fellows are even more beautiful than the Leopard Frogs. You will find, if you examine them closely that all of them have a very metallic luster, making their ground colour appear bronze, gold, or copper, depending on the light. These frogs are living gems; few people who have ever seen closely would dispute it.

Just as the Leopard Frog is hard to see against a natural background or grass, so too is the Pickerel Frog. The photograph showing this camouflage was taken on the lawn. If the subject had been posed in meadow grass, you would have seen nothing but grass in the picture.

The Pickerel Frog seems to head for its hibernation fairly late in the fall, and not before the first frost. We have found them on land as October 6th, though at this time of year they are easier to find in the river, where many of them probably hibernate, even though it is too swift a stream in which to breed in the spring. However, the ones we find there are immature adults, never full-grown frogs. Is this because the adults prefer the quieter ponds or is it merely that we do not see them because they are more secretive than the young?

The North American continent is not the only continent where spotted meadow frogs are found. In Africa there is the Spotted Grass Frog *Rana grayi*, who resembles our Leopard Frog in markings and our Green Frog in body build. This frog has a very unusual habit that makes it worthy of brief mention. Though it often lays in water in the manner of most Ranidae (all except one other), it also sometimes lays its large eggs in depressions in the earth near a pond. The outer egg capsule is clear, hard, and glassy so that no dirt adheres to it. The eggs may be washed into the water by rains and may hatch there, or they may hatch on land. In the latter case, the tadpoles wriggle the short distance necessary to reach the water. Dr. Rose, who gives us these interesting facts about Rana grayi, wonders whether there may not be two subspecies of this frog–one that lays on land and one that does not. Or is Rana grayi one species with individuals laying on land or in the water, depending on weather conditions or other external factors? As yet, no taxonomist has gathered enough specimens of this frog to determine whether or not it should be split into two subspecies.

The Marbled Salamander, *Ambystoma opacum*, belongs to the Mole Salamander family. Its breeding habits, however, are quite different from those of all the other members of the family. This salamander is found in the eastern part of the country as far west as Texas and as far north ad Connecticut. It is absent from the southern part of the Florida.

Its habitat is much drier than that preferred by most ambystomids. It is found on sandy or gravelly hillsides. There is one spot near our home that should be ideal for the Marbles Salamander, but we have never succeeded in finding one. Since we live in the northerly portions of its range, they are, we feel sure, not at all common here. Although herpetologists tell us that they are less moisture loving than many others of the family, it seems probable that they are quite secretive, as are so many other Mole Salamanders.

The Marbled Salamander is stout-bodied with a blunt snout. It is among the smaller members of the family, being only about 3 to 4 inches in length. The ground colour is black, dark brown, or dark gray, with the belly of a little lighter hue. This dark ground colour is interrupted by bands of white or light gray. In some individuals the bands are narrow. In others, they are so extensive that the creature appears to be white with a series of large black blotches down the center of the back.

The males may be distinguished from the females by their smaller size, protuberant vent, and by their colouring. The male's bands are white, while those of the female are gray. Dr. Bishop's handbook gives us, among others, three photographs of adult. Marbled Salamanders. Two of these are females with narrow gray bands. The remaining photograph is that of a male who appears white with black botches. Naturally, one cannot draw any conclusions at all from only three photographs. But it did occur to us to wonder whether perhaps all males may not be more extensively marked with white than the females. To prove this, however, would require extensive collecting of both sexes throughout the range area. Possibly there is nothing at all that would warrant such investigation.

Unlike other ambystomids, the Marbled Salamander breeds in the fall, from September to December, depending on the temperature. Naturally, it breeds earlier in the North, later in the South. The courtship is on land. The eggs, between 50 and 225 in number, are pigmented. The 2.7 mm. vitellus is surrounded by 2 envelopes. These eggs are deposited singly in a depression on the ground, under leaves, bark, stones, and so on. The spot chosen is one that is likely to be flooded by fall remains with the eggs, but for just how long has not been determined. If the rains come soon after laying, and flood the eggs, she leaves them. But suppose a month or two elapses before the rains come? Will she stay with her eggs all that time? What is the maximum length of time this species will brood? The eggs themselves have a very variable incubation period of from 13 to 207 whose hatching is depending on rain to wash them into the water or flood them. They are evidently unable to hatch on land.

After an aquatic larval period of 5 to 6 months, they transform, and leave the water. At this stage they are from 2½ to 3 inches in size, dark brown or black with light specks liberally sprinkled over the ground colour. We do not know how soon they develop their adult marbling. As a matter of fact, the precise number of days or weeks necessary for any species of newly transformed amphibian to attain mature adult colouration is known for very few species.

The Giant Toad, *Bufo marinus*, was for some years considered to be the largest toad in the world. It still ranks among the biggest, for specimens of 9 inches have been found in British Guiana. This very large animal never attains such an immense size in this country – 7 inches is its maximum growth here. It is found in a very small section of southwestern Texas in Zapata, Starr, and Hidalgo countries along the Rio Grande. Recently, it has become established in southern Florida also. It becomes more common and larger in Mexico and South America.

The Giant Toad is found around houses, gardens, fields and water tanks. During the day they generally hide under logs and rocks or in shallow burrows. In colour, this toad is brown, reddish, green, or black on the dorsum, with enormous triangular pitted parotoid glands. Underneath, it is a pale yellow or dirty white. Its voice is described as a low trill; when they are singing in chorus in the tropics the noise is evidently deafening.

Depending upon where it lives, the Giant Toad breeds at different seasons. The Wrights tell us that they breed from February to July in Bermuda, in Trinidad from August to October, and in British Guiana possibly the year round except for February and October. In certain areas this toad may lay twice a year, though this has never been proved as a definite fact, nor do we know in what regions it may breed biannually. In the United States the breeding season depends upon rains, and the Giant Toad may breed any time between March and September.

Like most toads, Bufo marinus lays its eggs in strings of jelly, but I have never been able to find a description of the size of the vitellus, number of envelopes, or the total complement. The eggs may hatch in less than 3 days. The tadpoles resemble those of the American Toad, and transform after a larval period of about 1½ months. Most amphibians cannot tolerate salt water. The eggs and tadpoles of the Giant Toad, however, actually develop faster, with a lower mortality rate, in slightly saline water (15 per cent salt water). We shall have much more to tell about the Giant Toad in later chapters when we examine the subjects of defense and economic importance.

Another dweller of plowed fields and gardens is the American Toad, *Bufo americanus americanus*, found throughout most of the eastern half of the United States. In Florida and parts of North Carolina, South Carolina, Georgia, Alabama, and Mississippi, it is replaced by the Southern Toad, *Bufo terrestris*, of similar appearance and habits. Within its range, the American Toad is likely to be confused only with Fowler's Toad, *Bufo woodhousei fowleri*. Differentiation between the American Toad and Fowler's Toad is not always easy because the two sometimes hybridize, and the offspring thus show some of the characteristics of both species.

Typically, the American Toad has a short, very broad body, and measures from 2 to 4 inches in length. Fowler's Toad is shorter–from 2 to 3 inches–and is more slender in body build, with longer legs. Both are some shade of brown, reddish, drab olive, or gray above. This ground colour is punctuated by dark spots. In the American Toad, there are only one or two large warts in each of the dark spots. In Fowler's Toad three or more warts are enclosed by the dark spots. The belly of the American Toad is dirty white with scattered dark spots. That of Flower's Toad is of the same colour with but one black spot on the chest.

The two species of toads are also differentiated by structural differences. The most outstanding are in the shape of the parotoid glands and in the cranial or ridges on the head. These crests are not folds of skin but are made by bony ridges of the head underneath the skin. If you look at the diagram of the toad and the photograph of the toad's back, you will notice that in both, the parotoid glands are kidney-shaped. This is characteristic of the American Toad. The parotoids of Fowler's Toad are oval. Notice the two ridges between the eyes, known as the supraorbital or interobital crests; the ridges in back of each eye, known as the postorbital crests; and (in the photograph only) the short ridges that join the postorbital crests to the parotoid glands. These last are known as the preparotoid longitudinal crests. In the American Toad the postorbital crest is either separated from the parotoid

Fig. 7.9. Hyla.

gland, as in the diagram, or connected to it by a preparotoid longitudinal crest, as in the photograph. In Fowler's Toad, however, there is no preparotoid longitudinal crest, and the postorbital crest ouches the parotoid gland horizontally. And while we are talking about identification by crests, we might add that the supraorbital crests of the Southern Toad are thickened and clublike where they end at the back of the head. The cranial crests of toads provide a valuable means of identification of many species of toads, which is the reason we discussed these rather technical details here. They are of value, however, mostly in adult identification, for immature toads do not always show the clear crest development manifested in the adults.

The American Toad emerges from hibernation in March, April or May, depending on the temperature. It breeds later than the Wood, Leopard, and Pickerel Frogs, for, unlike them, it breeds in temporary rain puddles or in very shallow water. Sometimes, the eggs are ruined by freezing, but generally the toad does not start breeding until the danger of freezing water is past.

The song of the toad is a high musical trill. It is of almost unbelievable sweetness, whether heard close at hand or from a distance, in chorus or solo. The breeding male seems slim and alert-quite unlike his fat, dusty midsummer self. As he sits in or on the edge of his puddle, he takes air into his mouth and forces it into his vocal sacs. Pushing himself very erect with his forelegs, his throat expands so that it forms a grayish white bubble, flecked with charcoal-grey. The sweet trill emerges. At the end of the call, as the bubble reduces, the male toad looks amazingly like a child drawing a large bubble of bubble gum into his mouth. I have never heard toads trilling in the morning. But they are often heard from noon far into the night in humid or rainy warm weather.

Male toads, at the height of the breeding season, are very eager. They quite frequently grab rocks or other males in their anxiety to find a mate. When a male is grasped by another male, he struggles and gives a series of soft rapid chirps. This is generally sufficient to inform the other male the he has made a mistake. The more alert the male, the better his chances of securing a mate are, for Dr. Oliver tells us that it has been found that males far outnumber the females at the breeding ponds. We wonder whether there is a reasonably constant ratio of this imbalance between the sexes. In our mudhole we found the ratio of males to females to be about 5 to 1. But since the mudhole is artificial and not an ideal breeding habitat anyway, we cannot assume that this ratio would hold true in other, more natural breeding ponds.

Sooner or later a large female, swollen with unlaid eggs, will approach and nudge a singing male. Instantly he embraces her, and the pair are ready to begin to lay the eggs. The female finds a spot to her liking. The male lowers his legs so that they are between the female's and places his feet on or near her anus. Not until his feet are in this position will the female express any eggs. If you watch the breeding pair closely, you can see one side of the female's body expand, contract, and return to normal position, followed by the same movements on the opposite side. As these contractions occur, the eggs are expelled in two strings of jelly. The male contracts his stomach, and you can see him jerk slightly as he forces out his sperm to fertilize the emerging eggs. After a few eggs are expelled, the couple surface to breather (the two string of jelly trailing from the female's cloaca), and one can see their nostrils rapidly "blick" open and shut. Then they submerge and more of the eggs are laid. As the eggs emerge, they look like black-and-gray beads.

If one returns to look at the eggs an hour or so after the pair has finished laying and separated, one will find that the eggs in their inner compartments have turned and are now so oriented that the black part is facing up, the grayish-white side down. The jelly has swelled and shows clearly. The strings lie partially coiled, twined over and through vegetation and on the pond bottom. The eggs themselves are very small, with a vitellus measuring between 1 and 1.4 mm. The total complement may be as high as 20,000 or as low as 4,000. In a day or two the coils have straightened, and the eggs mass is difficult to see, so discoloured is it by the mud of the pond.

In 3 to 12 days, the eggs are all hatched, and the tiny black tadpole begins his life in the puddle. But whether in the puddle or in the home aquarium, the toad tadpole is a cute little morsel. His upper parts are a velvety black–though close examination in a strong light may show that some tadpoles are not quite so black as they first seemed, but are a very dark brown instead, with tiny flecks of greenish bronze. The tail crests are a translucent milky white, and the tail itself is rather short with a rounded tip. When viewed from above, the body is vaguely triangular in outline–the tip of the triangle being the snout, with the base just above the tail insertion.

The behaviour of toad tadpoles is quite different from that of most frog tadpoles. They are extremely gregarious, and group together in bunches. We do not notice this in the aquarium, but in the puddle it is a quite frequent occurrence. Why do they swarm in this manner? Do toad tadpoles do so naturally? I don't think the latter is true, for we have never seen this swarming together in the aquarium. In captivity, the little fellows rise to the surface and swim round and round the bowl with rapid lashings of their small tails. We do not see swarming so frequently in larger ponds where ample room for development exists and where the tadpoles need for fear desiccation because of water evaporation. It has been suggested by herpetologists that this swarming of toad tadpoles may in some may hasten transformation. It is an interesting theory, but there is no proof that it is true (and none that it isn't).

The larval period in nature is a little less than 2 months. In the home aquarium, at an average room temperature of from 68°F. to 70° F., we find the larval period to be 34 to 40 days. This short tadpole stage makes these tadpoles ideal specimens for children to observe. In those species of amphibians that have a larval period of 2 to 3 months, the changes are slower and the children's interest is apt to fade away before the final transformation occurs. But with toad tadpoles changes occur rapidly, and thus the imagination and enthusiasm remain to the end.

The newly hatched toadlet is tiny–often less ¼ inch long. He is no larger than a fly, and far more engaging. Now that he has assumed his adult form, he is a dusty dark-brown colour, with fine stipplings of bronze still remaining. For a while this tiny fellow will stay near the water, but as the days pass he will become increasingly a dry-land creature. A week or two after transformation, the toadlets may be found hopping all over the lawn and fields. But you need watchful eyes to see them, so well do they blend in with their surround and so small are they. Unlike the adults, the toadlets are markedly diurnal.

If a warm rain occurs soon after transformation, you may be fortunate enough to see a mass migration away from the natal ponds and puddles. Thousands of little toadlets may be seen at this time, and this undoubtedly where the ancients got the idea that toads were rained down from the sky.

We see more adult toads during the months of July and August than at any other time of the year. They have taken up their summer residence in and around our lawns and gardens. Though they hide by day under rocks and logs or in the earth, we can see them in the evenings or at night. According to Dr. Oliver, their nonbreeding home range rarely exceeds an area 100 feet square. Where ideal conditions exist, it may be much less. The adult toad in the photographs is a dear fried of ours. She lives in one of our puppy yards, where she is not annoyed by our cats and the dogs–even the puppies pay little attention to her. By day she sleeps under the puppies' bench, and is always at home when a photographic model is needed. By night she can be found abroad in the puppies' yard–an area 20 by 15 feet. We say her nearly every evening last summer, and have never found her in any of the adjacent yards. Will she be there next summer, or will she change her residence? We can hardly wait to find out if she will return to us and her home range after breeding next spring.

Toads, unlike many amphibians, do not depend solely on lowered temperatures to start hibernating. Out toad disappeared on September 8th, exactly ten days before the first frost occurred. This was not unusual, for toads usually do burrow in before the first frost. How much before it? Is our observation of ten days an average figure? How can a toad "know" that frost is soon coming? Do the shortening days and lengthening nights inform it that it is time to dig in for the winter? We do not know much about the factors that may influence hibernation time, but undoubtedly it will be found that in many species hibernation is caused by a number of complex, interrelated phenomena.

We like all amphibians for their own unique qualities, but we must admit that of them all we are fondest of the toad. How it is that this gentle, placid creature with a jewel of an eye is regarded by many as the personification of ugliness we shall never be able to understand. Of all our amphibians, the toad is the easiest for most people to observe, for he is of moderate size and spends much of his active life in and around gardens. Perhaps you toll will come to love this quiet, useful animal if you get to know the individuals around your home.

FRESHWATER AMPHIBIANS

There are many different kinds of streams. Some are small, with muddy bottoms, and flow very slowly through woods or fields. The inhabitants of this type of stream are likely to be much the same ones as those found in ponds or ditches. There are brooks with a gravelly bottom that are roaring torrents of white water during periods of maximum rain or snow, but they slow up to become trickling brooks in the summer. There are mountain streams that start high in the mountains and consist of very cold cascading water year round. Then there are the rivers, large and small; but we shall use the term to designate those bodies of running water usually of sufficient depth to permit boating of some sort even in periods of drought.

Rivers and streams, whatever their type, have one important feature common to them all. All have running water, constantly changing. This water is better oxygenated than the still water of ponds and lakes. Therefore, it should not surprise us to find that many notenous and Lungless Salamanders live in rivers and streams. They have become admirably adapted to swift-water dwelling; you will not usually find them or their larvae in the ponds of forest or field. What adaptations have been made to enable these amphibians successfully to make their homes in the rivers and streams? Differences between pond dwellers and stream dwellers are marked in egg, larva, and adult. The eggs of stream-dwelling salamanders are large and unpigmented.

An unpigmented egg is not, as the term implies, transparent, nor are all of them white. Many unpigmented eggs are white; some are yellow; all are very light in colour. The vegetal pole is not visible in these unpigmented eggs. Unpigmented eggs are usually larger than pigmented eggs, and in general they are deposited singly or in a number of small clusters and are hidden rather than exposed. Most stream dwellers either deposit their eggs under stones or else select spots where quiet pools are formed; a few scoop out a nest so that the eggs will have a better protected place in which to develop. It can easily be seen that the unprotected clump of eggs laid by many ambystomids would be easily washed away and dashed against rocks by swiftly running water. Therefore, it was necessary for the stream dwellers to modify the primitive form of the egg mass before laying could take place in swiftly running water. In this country we have, properly speaking, only one true mountain-brook-type frog. Because the Tailed Frog is a unique animal in more ways than one. Africa and especially Asia have several true mountain-stream frogs.

Frogs of the African genus *Heleophryne* produce large unpigmented eggs, few in number. This much has been ascertained by dissection. Whether they are laid on land, as Dr. Rose suggests, or laid in the brook is not known, as the eggs have never been found in nature. We know a little more about the Asian mountain-brook forms largely owing to the work of Dr. Liu. *Staurois chunganensis* attaches its unpigmented large eggs to the underside of rocks in the mountain streams where it lives. The eggs of other mountain-brook dwellers are not known. We see that the true mountain-stream frogs, whose eggs are known, produce unpigmented large eggs, and these are concealed under rocks so that they are protected from the onslaught of rushing water.

What advantages a large unpigmented egg has over a small unpigmented one is hard to imagine. Stream-dwelling salamander larvae have short gills, and these, as a general rule, are not very bushy, though there are exceptions. Their bodies are more elongated, and the tail crests are reduced in size. Body crests so characteristic of the pond-dwelling larva are absent altogether. The stream-dwelling tadpole also has similar modifications in body shape, tail crests, and in the form of the mouth. The typical cascade tadpole body is shaped like a canoe paddle and is much flattened. The tail muscles arc thick and powerful, the fail tip rounded and blunt, the tail crests low and thick.

All of these adaptations make the stream-dwelling tadpoles very powerful swimmers. Because the flattened body and low tail crests offer the least resistance to the current, the animal is unlikely to be caught by the rush of water and dashed to his death on a nearby rock. A useful adjunct to these is the mouth shape of the true stream tadpole. Those species found in the fastest waters have large suctorial lips with which they adhere to the rocks while resting. In addition, many tadpoles of the Asian genus Staurois have a large adhesive disk on the belly, formed by the lower lip and a semicircular ridge of skin on the belly. Mention should also be made of another curious mouth-part adaptation found on the tadpoles of *Megophrys minor* of western China and *Phyllomedusa guttata* of Brazil. These tadpoles are surface feeders and have what are known as funnel or umbrella mouths. The umbrella mouth is located on the tip of the snout, not, as with other tadpoles, on the underside. The lips are large, and the lower one may be folded up so that it completely covers the mouth opening. When these tadpoles are feeding, they face upstream and extend the lower lip.

The oncoming water pushes food particles into the funnel thus formed. Then the mouth parts (there is no horny beak) either accept the food into the mouth or reject it. The umbrella-mouth tadpoles have the true stream-dweller's body and tail. However, they do not live in as swiftly running water as those tadpoles with suctorial mouths. Nor are the adults aquatic, as are the adults of the true cascade-type tadpoles. Stream-dwelling adult amphibians often have lungs that are much reduced in size. In water, the lungs act as hydrostatic organs providing buoyancy, which is undesirable from the standpoint of a stream-dwelling amphibian. And since the water of streams is well oxygenated, the animal usually has no trouble in satisfying its oxygen requirements with reduced lungs. Among the salamanders, we find many plethodontids, or Lungless Salamanders, living in rivers or streams. The Plethodontidae, our largest family of salamanders, are either terrestrial or stream dwelling. Those that are terrestrial often lay on land, but those that return to the water to lay their eggs always return to streams.

The outstanding exception is the Four-toed Salamander, *Hemidactylium scutatum*, a terrestrial plethodontid who lays in wet spots on the edge of ponds or swamps and whose larvae possess the long bushy gills and body crests of the pond-type larva. *Astylosternus robustus*, the Hairy Frog of Africa, whom we mentioned earlier, is believed to be a mountain-brook dweller. His reduced lungs are adequate for his needs except during the breeding season, when the metabolic rate of the males is much increased. To accommodate the greater oxygen requirement at that time, the males develop "hairs" that act, as we pointed out, much as the skin and external gills do. But the Hairy Frog does not look very much like other true stream-dwelling frogs as typified by *Staurois* and *Rana margaretae* of Asia. These adults have a rather flat slender body with long legs. The toes are fully webbed, and all digits are equipped with large, well-developed toe disks. In addition, there is generally a sexual dimorphism in size–the males being much smaller than the females. This feature is of considerable advantage to a couple in amplexus in running water for the water thus passes over and around them rather than sweeping them with it. These then are some of the adaptations made by the amphibians to enable them to live in running water. Let us now look at the detailed lives of a new of these stream dwellers.

The Greater Siren, *Siren lacertina*, is a large eel-like neotenous salamander found throughout Florida and along the coasts of Virginia, North Carolina, South Carolina, Georgia, and Alabama. It does not confine itself to swiftly running streams, though it is found there frequently. It may also be found in ditches, ponds, swamps, and lakes. The Greater Siren attains a maximum size of 3 feet, and may be as big around as a woman's writs. It has small gills and front legs only. Although the animal with its long streamlined shape appears to be all tail, close examination shows that it is really all body, the head and tail combined consisting of only about 1/3 of the total length. It colouring is inconspicuous. The back and sides are gray, very faintly marked with yellow; the belly is blue, marked with muddy yellow. Little is known about the breeding habits of the Greater Siren.

Anatomical evidence indicates that the eggs are fertilized externally, but no one has ever seen Siren breeding. The eggs and larvae are rarely found in nature. Siren has laid eggs in captivity. These were brown above, white below, surrounded by 3 jelly envelopes. They were laid singly and, including all envelopes, were about 9 mm. in size. But herpetologists do not jump to the conclusion that the eggs need necessarily be laid singly in nature just because they were so deposited in the laboratory. Captive specimens do not always lay in their characteristic manner. Although the adults of the Greater Siren have long been known to science, the eggs and larvae are seldom found in nature even by herpetologists who are looking for them. They are probably well hidden in water weeds or in mud.

Much much more needs to be learned about this species before we have a reasonably complete story of its life. The family Sirenidae consists of nine subspecies within two genera: *Siren*, with four fingers, and *Pseudobranchus*, with but three. The eggs and larvae of the Dwarf Siren, *Siren intermedia nettingi*, have been found in nature and described. These eggs were found in April, deposited in aquatic nest, and were in a mass. Eggs and larvae are also known for a few other species. The eel-like salamander *Amphiuma* means *tridactylum* is another neotenous stream dweller. In contrast to the Sirenidae, the Amphiumidae lack gills as adults and possess both forelegs and hind legs. The legs are so small that they may easily be overlooked. Therefore, the Amphiumidae are frequently mistaken for eels. These diminutive limbs have but three toes, instead of the more usual four in front and five behind. Furthermore, the toes are greatly reduced in size, the first and third being, as a general rule, mere buds.

Unlike most salamander toes, there are no bones in Amphiuma's toes. They are merely supported by cartilage. This incidentally, is true of the toes of the Sirenidae also. The Three-toed Amphiuma is a very long-bodied animal, and specimens measuring 40 inches are not unusual. It is found in parts of Alabama, Mississippi, Louisiana, and Arkansas. The other subspecies, *Amphiuma means means*, the Two-toed Amphiuma, has a slightly broader distribution, being found along the coasts of Virginia, North Carolina, South Carolina, Georgia, Alabama, Mississippi, and throughout Florida. In colour the Three-toed Amphiuma is inconspicuous, and is brown on the back and upper sides. The lower sides ad belly are gray. It is surprising to find that such a thoroughly aquatic salamander leaves the water and excavates a shallow nest beneath a log in which she lays her eggs. Evidently, the eggs are generally laid in the fall, though specimens have been found brooding eggs in winter.

The female does not always burrow in mud, but quite frequently chooses soil that is fairly dry and that may be some distance from water. The eggs are deposited in rosary-like strings and are large–about 9 mm. The female coils around them during an incubation period of from 1 to 2 months. How long do the hatchling larvae remain in the nest before going to the water? No one knows, but if one lived in the same part of the country as the Three-toed Amphiuma, one could have a great deal of fun in trying to find out. The Proteidae, or Mudpuppies, have the broadest distribution of any of the neotenic salamanders found in this country. Because a list of states in which they are found would be lengthy, we shall merely say that Mudpuppies are found in parts of every state in the eastern half of the country, New Hampshire and Maine excepted, but the area of greatest concentration is in the Middle West. Mudpuppies seem to prefer larger swift-running rivers, and most prefer clear water.

The Mudpuppy, *Necturus maculosus maculosus*, has the widest distribution of all the species, and may be found not only in the clear running waters of large rivers but also in the more sluggish and muddy water of canals and drainage ditches. It is the largest of our native mudpuppies, and may attain a maximum length of 1½ feet. The average specimen falls far short of this, however, and is about 1 foot. Its legs are short and strongly made, with four toes on each of the four digits. The Mudpuppy has gills that are large and bushy and red in colour. It also has lungs when adult. The Mudpuppy has a reddish brown back with black spots scattered over it. The underparts are light gray and may or may not be marked with darker spots.

The mating season of the Mudpuppy is in the fall months of September, October, and November. There has been but a single recorded instance of its courtship activities. The male swam around the female and crawled over her tail and between her legs. The female remained quite passive during this display. After the female has seized the spermatophore, she does not immediately lay her eggs. Implantation is delayed, and the eggs are not deposited until the following May. The female excavates a nest under a log or stone with its entrance facing downstream.

We almost always find stream nests with entrances so oriented, for if the entrance faced upstream the eggs might be damaged by the force of the oncoming waters. In this nest the female Mudpuppy deposits from 20 to 150 eggs, attaching each egg to the roof of the nest. These eggs are nonpigmented, yellow in colour, about 5 mm. in diameter, and surrounded by 3 jelly envelopes. The female remains with the eggs until they hatch I month to 2 months after laying. She may even remain with the larvae until they leave the nest. The larvae possess both fore and hind limbs upon hatching; however, the toes are not developed until later. The larval Mudpuppy does not attain sexual it is about 5 years old at a length of 8 inches. The Mudpuppy's food consists principally, not of insects, as with most amphibians, but of crayfish. Water weeds are also consumed, with insects and fish coming as third in preference.

The most unusual neotenic salamander in appearance is undoubtedly the Hellbender, *Cryptobeanchus alleganiensis*, and its very close relative, *Megalobatrachus maximus*, the Giant Salamander of Japan and western China. The Hellbender is very large, averaging about 11/2 feet, but females have been found that were nearly 21/2 feet long. The body is broad and flat; the legs are short and stout with four toes on the forelegs, five on the hind. The skin is thick but highly vascularized; it is also loose, and on either side it falls into deep wrinkles. In colour, the Hellbender varies from light brown or rusty red to almost black with scattered darker spots, Underneath, it is of a similar bot lighter colour. The Hellbender is found in the swifter waters of large rivers in parts of New Jersey, Pennsylvania, West Virginia, Indiana, Ohio, Kentucky, and North Carolina. It is rarely seen, for it is not only completely aquatic but also thoroughly nocturnal, hiding by day under rocks is the water and coming out only at night to forage for its food. The breeding season is in the late summer and early fall-the end of August through September.

The male excavates a nest under a large rock and lies there awaiting the arrival of the females. When one appears, he pays little or no attention to her until her eggs start to emerge from her cloaca.

The sight of the eggs stimulates the male to activity, and he seizes them and fertilizes them externally as they are laid. The egg is unpigmented yellow in colour, and about 6 mm. in size. However, the 2 envelopes with which it os surrounded make the whole much larger-about 3/4 of an inch in diameter. After the female has finished laying, she departs, leaving the male to guard the eggs. The egg complement of several females, each laying form 300 to 400 eggs, may be entrusted to his care. The larvae hatch on 2 or 3 months and have both fore and hind legs. The gills, which are small are retained for retained for a year and a half. The Giant Salamander of Japan and western China is very similar to our Hellbender. Surprisingly enough, though, it seems to prefer smaller streams, despite its immense size of 4 feet and girth roughly the sane size as a woman's thigh, and its weight of 60 pounds.

All of these neotenous salamanders have established records for longevity in captivity. The prize must go to *Megalobatrachus maximus*, a specimen of which lived in the Amsterdam Zoo for 52 years. The Hellbender is second with 29 years. Third is the Amphiuma, or Congo Eel, with 27 years, and fourth is the Greater Siren with 26 years. We might expect the Mudpuppy to cone next on the list, but unfortunately I have been unable to find any record of its life span in captivity. However, it dose not really matter, for life spans in captivity cannot be taken as reliable indicators of life spans in the wild. These, then are some of the stream dwellers. All are neotenous and most are difficult to observe. Let us now turn to the more typical and common salamanders-the plethodontid stream dwellers. One of the commonest species of stream-dwelling salamanders is the Northern Two-lined Salamander, *Eurycea dislineata dislineata*. Its range extends from Virginia north to Maine, from the Atlantic Coast on the east to Indiana on the west. It is a slender-bodied salamander of about 3 inches when adult. But this average measurement may be shorter on the more southern portions of its range or longer in the more northern parts. The upper parts are yellowish tan in colour with small black spots scattered over the back; on many individuals the spots form a line down the middle of the back. The tan back is separated from the sides by two black lines-one each side—that commence behind the eyes and extend back to the tip of the tail. The underside is yellow.

As with all plethodontids, nasolabial grooves are present. These are very tiny grooves running from the base of the nostril to the upper lip. They serve to free the nostrils and grooves. In order to see these nasolabial grooves on a salamander as small as the Northern Two-lined Salamander, a magnifying glass is required. The difference between the males and females is not, as in so many salamanders, in the character of the vent, but in the character of the snout. During the breeding season the males snout is apt to be somewhat swollen around the nasolabial grooves-and in some individuals small cirri may be formed, These cirri are blunt fingerlike appendages on the edge of the upper lip, and when they are present the nasolabial groove extends over them.

The Southern Two-lined Salamander, *Eurycea dislineata cirrigera*, carries this development of cirri further. The males always have well-developed cirri. Another difference between the sexes in this species is that on the males the teeth on the edge of the upper jaw called premaxillary teeth are frequently elongated and extend forward so that they are exposed. The Southern Two-lines Salamander is similar to the northern variety in looks and habits, but it spends more of its time on land in swampy places, returning to the stream for breeding and egg laying. It is found in the states of North Carolina, South Carolina, Georgia, Alabama, the northern and of Florida, and the southern tip of Mississippi. We cannot call the Northern Two-lined Two-lines Salamander completely aquatic, for though it does live in brooks it is found on their banks or even some distance away quite frequently. I should very much like to know just what conditions or factors prompt them to leave the water.

In the winter, especially in the more northern portions of its range, it is said to hibernate on stream bottoms. We have been amazed, therefore, to find Two-lined Salamander in January and February under rocks on the shore during periods of thaw. On other February days, when the weather seems to be much warmer, not a salamander is to be found. The governing factor in this case may be the water temperature. The Two-lined Salamander is one of those species of amphibians that depend almost entirely on lowering temperatures to warm them that it is time to got into hibernation (or emerge from it). They are not like many amphibians (the toads, for instance), who dig their hibernation burrows long before the real cold weather comes. We suspect that the Northern Two-lined Salamander is not a true hibernator when it makes its home in the swiftly running brook water which, even in subzero weather, forms only a thin coating of ice, not secure enough to support a man's weight. It unquestionably remains on the bottom, but perhaps, though sluggish and inactive during these periods of extreme cold, it is not truly hibernating.

Another argument in support of this theory of mine is that the Two-lined Salamander evidently breeds quite early- in middle or late winter with us. During the spring and summer months, it is not difficult to find this salamander by overturning rocks in shallow waters or in wet spots on the bank. In some years these salamanders are very common in such locations; in other years, in July and August, we find the adults more frequently away from the stream, burrowed under well-decayed logs in the nearby woods. The Northern Two-lined Salamander seems to prefer gravelly streams and rivers. In Connecticut, at any rate, we have never found one in a slow-running muddy-bottomed brook. This may be, of course, because with a swift-running stream at our back door we go to muddy slower-running brooks infrequently.

The courtship of the Two-lined Salamander is a fairly elaborate affair and is carried on in the water. The breeding season is in winter, from January to March or April. The male noses the female and rubs his snout and chin against her face. The secretions from the glands found there arouse her to straddle his tail, facing in the same direction as he. She then places her chin on the base of his tail where another scent gland is located. In this position the two walk off along the stream bottom until the male deposits his spermatophore, which the female retrieves. The female glues her eggs singly to the underside of rocks on the stream bottom. These eggs are unpigmented–as indeed are all plethodontid eggs–white or yellowish in colour and about 3 mm. in size. They are surrounded by 2 jelly envelopes.

Each female lays about 30 eggs, though several females may utilize the same rock. These eggs are not easy to find. We have been actively interested in amphibians for seven years. Each year, almost daily from June through August, we have searched for the eggs, looking in shallow water and deep water, still water and swiftly running water and all the gradations between. We have looked under small stones and rocks so large that had they not been under water we could never have budged them. Nor have we been alone. Our daughter, an ardent herpetologist of eight, has for four years been searching for these eggs, both with us and on her own. We had begun to suspect that we were late in looking for them, for we frequently found what we believed to Two-lined larvae of hatchling size–about ½ inch–in May and June. This year we began egg hunting as soon as the height of the stream waters permitted. On May 27th our daughter came rushing in with the news that she had found them. Sure enough, under a small flat rock no larger than my hand were 20 eggs with pale yellow embryos ready to hatch at any moment. The rock was in shallow quiet water near shore–though when the salamander laid, the rock must have been under water at least a foot deep and perhaps more.

After this momentous we find we continued searching for an hour longer, but we saw no more eggs. Doubtless most of the others had already hatched. So if you have trouble finding the eggs of the Two-lined Salamander, don't give up in disgust, for you are not alone. Keep looking, and one day you will be lucky enough to discover them. The larvae of the Two-lined Salamander are not difficult to find and you will, if you look sharp as you overturn rocks in shallow water, see them in all stages of development, during the spring, summer, and fall. This is because this salamander has a larval period of from 2 to 3 years. The larvae are easier to find in the summer months when the brook waters are lower and shallower, for it is difficult to spot one of these tiny creatures in deep running water. The Northern Two-lined Salamander is a very agile species, and can elude your grasping hand with the greatest of ease just when you are sure you have captured it.

Quite frequently, if the animal's body squirms away, but you have a firm grip on the tail, you will find that you have only the tail, wriggling violently in your hand. The salamander has broken off his tail and departed with little or no inconvenience to himself. The best way, therefore, to catch a Two-lined Salamander is to grab with both hands. Another common Lungless Salamander is the Northern Dusky Salamander, *Desmognathus fuscus fuscus*, whose range is similar to that of Norther Two-lined Salamander, but extends slightly farther South into the northern half of South Carolina, Georgia, and Alabama. Another subspecies, *Desmognathus fuscus ariculatus*, extends along the coasts of the southern states of North Carolina, South Carolina, Georgia, Alabama, Mississippi, Louisiana, and the northern half of Florida.

The Northern Dusky Salamander, *Desmognathus fuscus fuscus*, is not really a stream dweller. It lives on the edges of streams, never straying for from moist soil. It is often found in the water during the summer, but it exhibits its terrestrial habits in its choice of egg-laying site. It hibernates, however, under water in the streams and seems to be more of a true hibernator than does the Two-lined Salamander. We find the Dusky Salamander out of water in midwinter infrequently. The Dusky Salamander is a larger, stouter-bodies species than the Two-lined Salamander. It is between 3½ and 4½ inches in length, the males being somewhat longer than the females, and sometimes attaining a length of 4¾ inches.

Our local Dusky Salamanders are brownish black above. The underparts are pink, heavily speckled with blue-gray. But *Desmognathus fuscus fuscus* is one of those amphibian species that show considerable variation in colour and pattern. Many individuals are light brown or brown above, with two darker, irregularly marginated lines dividing the back from the sides. Unlike the tongue of the Two-lined Salamander, which is of the mushroom type, fastened to the floor of the mouth, the Dusky Salamander's tongue is fastened in the front of the mouth like those of most frogs and toads. The life cycle of the Northern Dusky Salamander is a rather curious one and might be compared to the newt's life cycle in reverse.

After a courtship on land, the female makes a nest under stones or logs in a damp spot near a stream. Here she lays her eggs, about 20 in number. She remains with them, brooding them faithfully until they hatch. The young larva remains on land for about 2 weeks. At the end of that time, it moves down into the water, develops a tail fin and the typical gills of a stream dweller, and remains aquatic for the rest of its larval period. After 8 or 9 months, it transforms, and leaves the water to spend its adult life along the stream margins. You might think that the stouter-bodied Dusky Salamander was less agile than the Two-lined Salamander. And as you lower your hand to capture a specimen, the animal never appears to move. The Two-lined Salamander squirms out from under your descending hand. The Dusky Salamander eludes your grasp altogether by jumping. These are not the only stream-dwelling salamanders of the United States, but they are among the most common.

There are several plethodontids found along the northwestern coast, and also a stream-dwelling ambystomid, but these are limited in their range, and one, at least, is rather rare. But though the West has been neglected by stream-dwelling salamanders, more stream-dwelling Salientia are found there. These frogs do not, however, show the adaptations of a typical mountain stream or cascade frog mentioned earlier. They are not found in waterfalls or rapids, and evidently no adaptations are needed to enable them to survive in the streams they inhabit. The California Yellow-legged Frog, *Rana boylei boylei*, is one of these brook dwellers. It is found along the coasts of southern Oregon and northern and central California. It occasionally inhabits lakes, but above all, this frog prefers streams with a rocky bottom, though it is not unknown in those having a mud bottom. It seems to prefer slower-running brooks, but it never inhabits those that dry up completely in summer. The underparts of many western frogs are red in colour.

The Yellow-legged Frog may be distinguished from all of these by the fact that the underparts of the legs and around the vent are yellow, never red. The rest of the belly is white, and the throat is generally marked with black. Above, it may be black, grayish-brown or greenish, with indistinct spots or blotches. The upper surfaces of the legs and arms are barred with black. There is generally a lighter area on the snout between the two eyes. Unlike many frogs of the family Ranidae whose skin is quite smooth, the skin of the Yellow-legged Frog is warty or tuberculous, and resembles a toad's skin. There are one or two other western frogs whose skin is rough, but not to such an extreme degree as is the skin of the Yellow-legged Frog. The body shape of *Rana boylei boylei* is also rather toadlike, for it is small–about 1½ to 3 inches in length–and very broad. Unlike a toad, its legs are long, and extensively webbed.

This stream dweller is very shy and difficult to catch. It is completely aquatic and is rarely found more than a foot or two away from the banks of streams. It alarmed when basking, it immediately dives into the water, taking refuge under rocks or in the mud. No one has recorded or described the calls of the males, but the breeding season is in April and May. The males may be distinguished from the females by the thumb, which is greatly swollen at the base. The eggs are laid in a submerged mass, with a total complement of about 1,000. The egg, black above, white below, is small–about 2 mm. and is surrounded by 3 distinct envelopes, the largest being 4.5 m. in size. The egg mass is not always found in quiet pools as one might suppose, but frequently in shallow water that flows steadily. Eggs have even been found in swiftly running water. In this connection it would be helpful if herpetologists would measure the speed of the water so that every one would know just what "swiftly running" means. They might, for instance, drop a chip of wood on the water and time the number of seconds it takes to cover a prescribed distance. Then we amateurs could perform the same experiment and thus determine whether our water was running at a speed greater than, less than, or equal to the herpetologist's water at the time of his observation.

The incubation period for the eggs of *Rana boylei boylei* is not known, but one might expect to find it slightly longer than that required for frog eggs laid in the pond. The tadpole is olive green, and transforms after a larval life of about 3½ months. Another stream dweller is the

Canyon Treefrog, *Hyla arenicolor*, found throughout Arizona and New Mexico and in parts of southern and Lower California, western Texas and Mexico. The Canyon Treefrog belongs to the family Hylidae, or Treefrogs. Actually, this is a misnomer, for many treefrogs almost never climb trees.

The Canyon Treefrog is one of these, and is semiaquatic or aquatic in habit, rarely being found very far from the gravel-bottomed clear streams it seems to prefer. In appearance, the Canyon Treefrog is similar to the Eastern Grey Treefrog of the Northeast. It is of moderate size for a hylid, being from 1¼ to 2¼ inches long. It has a granular skin and conspicuous toe disks on all its digits. There is no webbing between the toes. It is brownish or grayish above, with small scattered black spots. The upper parts of the legs and arms are barred with black. The belly and throat are white or light tan; the undersides of the arms, legs, and around the vent are orange or yellow. The authorities seem to be somewhat at odds about the voice of the male Canyon Treefrog. Some say that is a short trill similar to the one made by the Eastern Grey Treefrog, *Hyla versicolor versicolor*. Others describe it variously as goatlike, sheeplike, duck-like. We have not heard this frog calling, so our opinion does not add to the confusion.

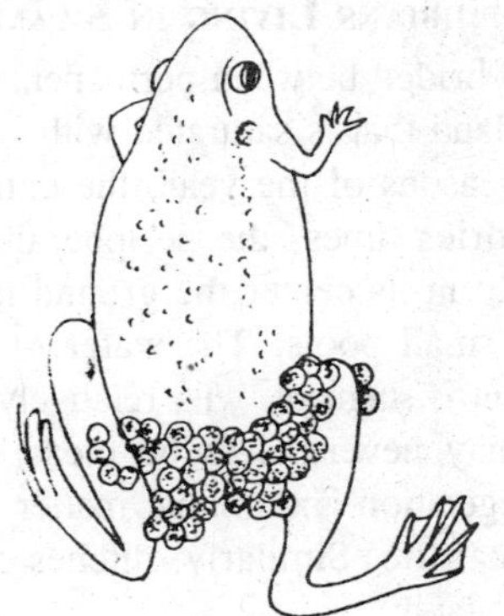

Fig. 7.10. Alytes.

Admittedly, it is difficult to describe sounds verbally. One might suppose, though, that the accounts would be more similar. A duck quacking most certainly does not sound like a sheep bleating or like the trill of a bird. However, the voices of many Salientia sound differently at different times. The trill of the Eastern Grey Treefrog sounds very different when the weather is cool than it does when he is singing alone or in chorus at the height of the breeding season.

Furthermore, the calls of many frogs are ventriloquial, and the same call may sound unalike to two observers, each occupying a separate vantage point. Then too, the location of the frog itself makes a difference. Some species call from under water as well as from above.

The underwater call sounds slightly different from the one voiced in air. All of these factors help to explain why we sometimes find a great disparity in the descriptions of voice by herpetologists. However, the calls of frogs and toads are being recorded. This is a great help to the amateur, who can thus hear the voice of various species separately and in chorus, in different mediums and at different times of year. There are two such records that we know of an the market today. They are listed in the Bibliography. The Canyon Treefrog has a breeding season extending from March to July. The eggs are laid singly, and may be found either floating on the water's surface or attached to leaves at the bottom. The tadpole has an olive-coloured body and tail. Its tail crests are often mottled with orange or red, and the center of the belly is pink. The tadpoles transform in 1½ to 2½ months. These, then, are a few of the stream dwellers. You will find many others if you look for them. You will also find stream visitors, or species that live mostly by its banks rather than in the water.

Amphibians Living in Swamp

Swamps provide a bridge between permanent lakes and ponds and dry land. A swamp is land that is saturated with water but not covered by it. During certain seasons of the year, the entire swamp area may be flooded; in other, drier times, the peripheral areas of the swamp may be quite dry, while in its center the ground may be saturated and riddled with scattered small pools. The water of swamp pools is not usually running water but is stagnant, with relatively low oxygen content. These swamp pools may nevertheless be clear, or they may be so dirty from decaying vegetation and animal matter that they can sustain very few kinds of animal life. Similarly, ditches are a bridge between rivers and streams and land.

Ditches may be large drainage ditches, which are more like canals, with a huge volume of water, or they may be merely roadside ditches no more than a foot or two wide. In ditches, the oxygen content of the water is lower than in river, and the water is more or less stagnant. Of course, it is difficult to generalize about ditches and marshes and the amphibians who live in or around them. Depending upon the season and the rainfall, a swamp or a ditch may become a lake or a stream; in seasons of extreme aridity they may become dry land. Habitat conditions are constantly fluctuating, and as they change, so do the animals who inhabit them. The largest of our native toads is the Colorado River Toad, *Bufo alvarius*. Though the Giant Toad, *Bufo marinus*, attains a larger size in the more southerly portions of its range, the length of the Giant Toads found in the United States does not quite equal the large size of 7 inches often attained by *Bufo alvarius*.

In fact, next to the Bullfrog, the Colorado River Toad is our largest salientian. Most true toads are dry-land dwellers coming to temporary pools only for breeding, but the Colorado River Toad is more aquatic, and lives around cattle-watering troughs, swampy land, ditches, and irrigated fields. It is found in Arizona around the Colorado and Gila rivers to the borders of Nevada and New Mexico respectively, and down into the Mexican state of Sonora. We usually think of swampy areas as being humid, but when we are thinking of southern swamps, not the wet areas of ground found in this part of Arizona. For this is a desert region where the air is dry, and though few amphibians can tolerate such low humidity the Colorado River Toad is well equipped to withstand it. He has a smooth but very thick skin with a few small round warts scattered over his back. The parotoid glands so characteristic of toads are large and slope down sharply toward the arm.

In addition, there are several large warts on the legs - one on the femur, or first segment of the leg, one or more on the tibia, or second segment of the leg. In colour, this toad is greenish with brown or orange warts. The underparts are white. The eyes are large and very beautiful, with the dark pupil and yellow iris veined with red. The males may be distinguished from the females by their heavier arms and by the horny excrescences on top of the fingers. The voice of the male when held in the hand has been variously described as a chirp similar to the note of the American Toad, a cluck like contented chickens, and, when in chorus, loud calls. They breed at night in shallow ponds, during or after rains, from March through August. The eggs, about 8,000 in number, are laid in strings of jelly. They are black or very dark brown above, light below. The vitellus is sometimes round, but just as often it is wedge-shaped, and triangular in outline.

It is interesting to speculate as to how the eggs become wedge shaped. Do some females always lay round eggs and others wedge shaped ones? Or does a female lay wedge-shaped eggs one year and round eggs another? If the latter be true, what governs the shape of the eggs? Has age any bearing on the matter? Going a little further, does one find a preponderance of one egg shape during drier or wetter

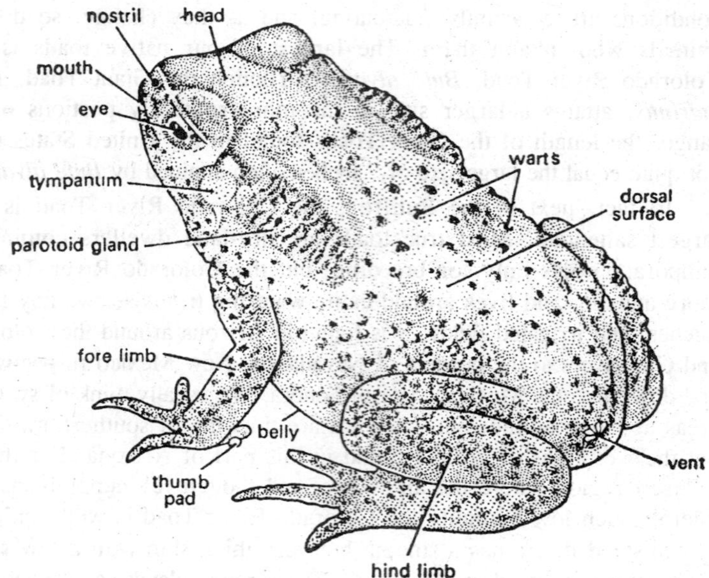

Fig. 7.11. Bufo melanostictus.

periods or earlier or later in the year? Does the shape of the egg influence in any way the length of the incubation period? If so, how? We don't know the answers to these questions, but discovery of the facts is well within the range of the amateur.

The tadpoles of the Colorado River Toad have never been precisely described for science. They have been seen, but evidently a large-enough series have not been collected and studied.

Colorado River Toads are not especially numerous. One is fortunate to find them when they are breeding, for evidently they are rather scarce at other times. They are quite secretive, and seek cover during the day in moist spots under watering troughs. By night they are seen more frequently, but they are never common.

In sharp contrast to our largest toad is the smallest of all our native Salientia, the Little Grass Frog, *Hyla ocularis*, found along the coasts of North Carolina, South Carolina, Georgia, and throughout Florida. One needs sharp eyes to see this tiny little fellow, for he is slender of body and only about ½ inch long when adult. This frog is usually some shade of brown or tan on top, but sometimes specimens that are grayish-green or brownish-red are found. The most constant identification mark is a dark line running from the snout through the eye back to the arm or beyond it, though in some specimens this line may be missing. Sometimes one finds a dark triangle between the eyes and a central dark back stripe and two dark side stripes, but quite often these markings too are absent. The vent is whitish or very light tan with dark speckles that are sometimes scant, sometimes profuse.

For so small a creature, the Little Grass Frog has amazingly long but very slender legs. And though it lives in swampy areas around pine woods and cypress ponds, it toes are nearly webless. The round digital disks characteristic of all the Hylidae are distinct but so tiny as to be of little practical use in climbing. However, this frog does not climb very much or very high, confining its climbing to low bushes. It is more often found on the ground around the swampy edges of ponds and in marshes.

The Little Grass Frog is able to turn its head up, down, or to either side without turning its body. Most frogs are unable to do so. A few, such as the Eastern Grey Treefrog are able to cock their heads slightly, but have not freedom of head movement of the Little Grass Frog.

The males may be distinguished from the females by their "pleated" throats; that is, the skin on the throat is not smooth but falls into wrinkles or folds. The vocal sacs when expanded form a large round throat bubble. The call is loud for so small a creature. It is very high-pitched and quite cricketlike in quality–or it may sound like the peeping of newly hatched chicks. The breeding season is an extended one from January to September, though Dr., Conant states that they may breed the year round in Florida.

About 100 eggs are deposited singly in shallow water. They are brown above, cream below, and are very tiny–about .7 mm. in size. The envelope measurement does not exceed 2 mm.; obviously, these eggs are hard to find. The tadpole is greenish in colour on top, with scattered black spots. The tail is long, and the tail crests are high and extending far up on the body. After a larval life of 1½ to 3/8 of an inch long.

Though the Little Grass Frog seems to prefer swampy ground, they may sometimes be found in quite dry areas. Mrs. Goin tells us that she finds them around her place, not in the wet winter and summer, but in the dry spring and fall. Why does she not find them in wetter weather? It is known that amphibians make long (never exceeding two miles and usually less than half a mile) migrations to and from suitable breeding sites. Dr. Oliver says that these distances are in marked contrast (especially in some species) to their normal everyday travels, which in many species do not exceed a circle with a radius of a few hundred feet. Possibly these facts may help to explain why Mrs. Goin found Hyla ocularis during dry weather. They may have been merely journeying to or from the breeding sites.

It is believed that most amphibians do not breed more than once a year. But when we realize that the Little Grass Frog breeds during any month of the year in Florida, and that Mrs. Goin's captures were in the spring and fall, we wonder whether this species may possibly breed twice a year. A widespread marking of breeding individuals over several years would be necessary to prove or disprove this theory. It would be a good project for some ambitious amateur herpetologist residing in Florida. Another small frog belonging to the family Hylidae is known as the Cricket Frog, genus *Acris*. Some species of Cricket Frog are found throughout the eastern half of the United States, New England, New York, North Dakota, most of Pennsylvania, West Virginia, and South Dakota excluded. All of them are swamp dwellers with similar appearances and life histories.

The Southern Cricket Frog, *Acris gryllus gryllus*, is found in the coastal lowlands of North Carolina and South Carolina and most of Georgia, Alabama, Mississippi, and the western arm of Florida. Elsewhere in Florida, it is superseded by the Florida Cricket Frog. *Acris gryllus dorsalis*, which differs from it principally in thigh pattern. The colour variation of the Southern Cricket Frog is extreme; individuals of black, brown, red, green, or gray may be found. The markings are variable also, and may be dark or red. Often there is a triangle between the eyes, with the apex on the snout, and another triangle just behind this, with the apex turned toward the rear. There may be, in addition, a central back stripe and a side stripe. The legs are barred with dark. There is also a light line running from the eye to the arm.

Probably the most reliable identification mark is the thigh pattern, which is, unfortunately, not visible when the frog is at rest. In order to see it, you must catch your frog and straighten out one hind leg. On the rear of the femur, you will find a dark evenly edged stripe on the Southern Cricket Frog. In addition, the upper part of the thigh and around the anus is sprinkled with small tubercles. In the Florida Cricket Frog there are no anal turbercles, and there are two stripes. In the Northern Cricket Frog, *Acris gryllus crepitans*, anal warts are present, together with one irregular edged dark stripe. The Northern Cricket Frog also has more extensive webbing on its toes.

The Southern Cricket Frog is a small frog from ¾ to 1¼ inches in length. It has a slender body, long legs, and feet that are more completely webbed than those of the Little Grass Frog. Its finger and toe disks are small and are not used for climbing. It is more like the true frogs in appearance than are other hylids. The Southern Cricket Frog spends most of its life in swamps and bogs in fairly open situations. When alarmed, this little frog shows its aquatic affinities by heading straight for water, and with a leap dives down into the muck to hide.

While the female's throat is a light tan, that of the male is a much darker brownish yellow, sprinkled with black dots. This marked throat discolouration is a reliable indicator of sex. The vocal sacs expand into a round throat bubble when the male is calling. Miss Dickerson tells us that the voice may be imitated by hitting two marbles or pebbles together several times with increasing rapidity. I have tried this. Neither the marbles nor the pebbles sound very much like the recording of the Southern Cricket Frog's voice, but of the two the pebbles make the better approximation. The breeding season is from February to October in periods of maximum rainfall.

The eggs number about 250 and are laid singly, either attached to water weeds or on the bottom in the shallow water of temporary pools. The eggs are a little larger than those of the Little Grass Frog but not enough larger to make them readily visible. The tadpole is olive in colour with a black tail tip. As with so many of the hylids, the tail crest is high and the body deep. The spiracle is found on the left side. It is not just an opening in the body in this species, but is equipped with a tube that stands out at an angle from the body, with the opening at its end. The larval period is an extended one of 2 to 3 months or even longer.

The Southern Cricket Frog is the commonest frog within its range. It may be found in every month of the year, though naturally it is less frequently collected during the colder months of December, January, and February. Does this frog hibernate or not? Some say that it does hibernate on land. Others say that it does not but that during the colder months of the year it goes to the larger bodies of water so that it may take refuge there in the advent of freezing weather. Which of these opposite schools of thought is right? We don't know, but perhaps they both are for their own locality.

The Southern Cricket Frogs are active creatures; moreover, they are found both during the day and at night. Owing to the small size, however, they are easily overlooked or mistaken for grasshoppers. Their long legs make them wonderful jumpers. They vie with the Leopard Frog for the maximum distance jumped by our native frogs. But if the distance jumped be related to body length, they are far and away the best jumpers we have, for they can jump between 35 and 40 times the length of their own bodies. Imagine, for comparison, how far a six-foot man should be able to jump in order to compete–240 feet! The world record for the standing jump is only a little over 12 feet. The world record for the running broad jump, made by Jessse Owens in 1935, is 26'8¼. Let us assume that Owens is 6 feet tall. You may work out his relative jumping ability accurately if you like. I am content to say merely that it is a little over four times his height. However, frogs make standing broad jumps, and here we find that an exceptional man can jump only about twice his height. Of course, frogs and men are built quite differently, but, this comparison demonstrates the disparity in their relative jumping abilities clearly.

Familiar to most people living in the eastern half of the United States, southern Florida excepted, is the Spring Peeper, *Hyla crucifer crucifer*. But is principally by voice that he is known, for though it is a common species, most people have never seen this small creature. He is an inhabitant of woodland swamps, though during the breeding season most ponds have a chorus of peepers on their banks. These small treefrogs are brown or tan in colour with a dark X mark on the back. Very often the marking is not a perfect X but is incomplete or modified so that it appears like an inverted Y. There is a V mark between the eyes and an inverted V joining the dark barred hind legs. The underparts are cream-coloured. This creature is about ¾ of an inch long and has the typical toe disks of a hylid, though it does not climb very much. Actually, its toe disks are very efficient, for it can and does cling to the upright sides of an aquarium for days.

The males are considerably smaller than the females, and during the breeding season are much darker in over-all colour. In fact, sometimes they are so dark that the characteristic markings are all but indistinguishable. The male also has a dark throat, which is lacking in the female. The breeding season is early. The Wrights say April marks the height of the breeding activities, but we hear calls as early as the middle of March in our vicinity. Farther south, the breeding season is much earlier– even as early as November.

The vocal sacs of the male swell into a glistening round flesh-coloured throat bubble as the utters his sweet *Preep, peep, peep*. Generally, unless the day is warm and overcast or rainy, the peepers do not begin calling much before four o'clock in the afternoon. But finding them in daylight, even though the chorus be deafening before your approach, is difficult. For as you draw near, the singers in your immediate vicinity cease their calling. Being small, and choosing stations that are particularly concealed, they are most inconspicuous. We find the same caution at night during the colder weather of early spring. Though herpetologists insist that hearing plays no part in stilling the calls of amphibians, the Spring Peeper has made us skeptical of this dictum.

We regularly go to two ponds for our amphibian watching. One on the edge of the woods, was once a gravel pit, long since abandoned. It is an ideal peeper breeding place, for the pools are surrounded by low shrubs. The footing at night, however, is not very good, for the banks are steep and gravelly. We have gone there on dark moonless nights with dark clothes. Though the peepers were singing when we arrived, the soon stopped as we approached- as quietly as possible, but very quietly because of the gravel. If the peepers didn't hear us, how did they know we were there? We switched off our lights. WE were unable to see each other five feet away. Can we nevertheless assume that the peepers could see us?

Our other observation place is a spring-fed pond in an open situation surrounded by lawn on one side and a mowed field on the other. On bright moonlit nights, further brightness by a street light and our flashlights, we clomp loudly over and see hundreds of peepers in no way disturbed by our presence. The peepers start breeding in this pond two or three weeks later than in the gravel bank.

It is our theory that when the nights are warmer and the peepers are in chorus stage, they are oblivious to noise and even sight of danger in the evening. But at the start of the breeding season, and when the temperatures are down in the 40's, they are much more aware of danger. We feel that hearing plays a part in stilling their calls on cold dark nights. Though experienced herpetologists do not think so, or other explanation occurs to us.

You might think that frog in amplexus would be easy to distinguish from unmated ones. Such is sometimes the case. But it is almost uncanny how much two peepers in amplexus resemble a singe peeper. The female is larger than the male and markedly lighter in colour.

The male clings tightly to her, flattening his back and keeping his legs well drawn up. From a slight distance, in poor light, they appear as one peeper. Any single peeper in the pond is suspect, however, for they usually call from land, though they generally choose the damper spots on the bank. You will be surprised, especially when the first go collecting, to find out how many single peepers turn out to be a pair.

When a mated pair of frogs is capture, the male will very often croak in protest, especially if he is touched. We do not find this true of the male peeper. Though peepers in amplexus are more easily separated than the American Toad, for instance, we have never heard a male utter a peep in protest when handled. Their behaviour when accosted by another male when they are unmated is different. Then they peep loudly in protest, and give a sort of body shrug, which in most cases is sufficient to induce the embracing peeper to cease his unwelcome attentions.

The eggs, about 1,000 in number, are laid singly. They are black on top, white underneath, and are about 1 mm. in size. We well remember the first time we captured a mated pair and confined them in an aquarium to obtain the eggs. On the bottom of a white refrigerator dish was a little dirt from the rock that we had put in with our specimens. The next morning, at first glance, we thought that the peepers had not laid! A closer look showed we were mistaken, but the few grains of dirt and sand made the eggs difficult to distinguish. This brought home, more clearly than previous fruitless searching in the pond had, the extremely small size of a peeper egg. The incubation period of the peeper egg in the field has never been accurately determined. In the aquarium, we find it to be between 4 and 5 days, when kept at a room temperature of between 65°F and 70°F. Since the water in which the peepers lays is much colder–sometimes only a few degrees above freezing—we may expect to find that the field-incubation period is considerably longer.

The tadpole is sandy in colour on its back. The belly is white or cream. In certain positions the characteristic V-shaped indentation of the tail crest by the anus is clearly visible. Hatching evidently exhausts these small mites, for afterward most of them lie inert on the aquarium bottom or pond bottom. In the pond they show up quite clearly against the dark bottom if one knows what to look for. About a week later, the peeper tadpoles have assumed their characteristic shape. When viewed from above, the body is oval is outline. The tail crests are transparent. The belly has become a beautiful iridescent shade of reddish gold with a very metallic sheen. By this time they are eating. When at rest, they no longer lie on their sides on the aquarium bottom, but cling to its sides in an upright position, head up, tails down.

The field-incubation period is 3 months or longer. In our aquarium we find it is about 46 days. As soon as the legs appear and are well developed, the dark markings of the adult peeper appear on the back of the tadpole. The gorgeous iridescent belly is lost. The arms then emerge, and the small creature immediately starts to go toward land. They do not wait for the tail to absorb before assuming a more or less terrestrial existence. Possibly these tailed peepers stay fairly close to the water until their tails are resorbed, though we have not observed them closely enough to be absolutely sure of this. It seems that they do not remain near the pond as long as young toads, but this is purely subjective impression and may not be correct, because there are so many more little toads around a pool that one is bound to see more of them as successive batches transform.

Outside the breeding season, Spring Peepers are scarce and hard to find. They evidently leave the more open situations where they have bred, and retreat to more secluded ones. Finding them becomes more a matter of good luck than anything else. They seem to prefer woodland swamps, and are mor numerous in wet situations where much low brush and shrubbery abounds. Occasionally, one may find them in marshy fields also, but they are not nearly so numerous there.

We know that the peeper is among the first frogs to leave hibernation and start breeding. When does he return to the hibernation site in the fall? We do not have any precise time, and it would, of course, vary with the locality. We do not believe that the peepers have an innate sense of when to hibernate, as does the American Toad. With the peeper, we believe temperature is the sole governing factor, for this year—a rather warm fall–we heard scattered calls in a woodland swamp as late as November 1st. When the Spring Peeper does hibernate, it is on land, under the earth. But even in the North he does not remain there for long–2 or 3 months at most. Then he is out again to tell us that spring will soon arrive.

Our next two swamp dwellers are salamanders. Although they are not closely related, they have several characteristics in common. Both are plethodontids; both are small; and both have but four toes on all four feet, instead of five behind and four in front as do most salamanders. The first of these is the Eastern Four-toed Salamander, Hemidactylium scutatum. It is found throughout the northeasters states and the Middle West as far south as northern Georgia. It is absent from most of Maine and the more northerly portions of New York, New Hampshire, and Vermont. This salamander is from 2 to 3 inches long, the males being smaller. The back is reddish brown in colour; the belly is white distinctly marked with black spots. There are, as we said, but four toes on all the feet, and at the base of the tail there is a constriction. The constriction is an indentation around the tail. It marks the place where the tail will break off it grasped by an enemy.

The Eastern Four-toed Salamander is found in areas of sphagnum bogs, either in the woods or, less frequently, in the open. The breeding season is in the late summer and early fall. The courtship is much like that of the Two-lined Salamander. The pair walks along, the female behind the male, with her chin on his wagging tail. The eggs are not deposited until the following spring. The female selects a cavity under moss or among grasses very close to a pool of water. She lies on her back and lays about 30 eggs. These are deposited singly, but close together. The jelly surrounding each egg is sticky, and it may appear that several eggs have been deposited together, so closely do they adhere to one another. Such is not the case, however, for they are not united by a common jelly envelope. The vitellus is large–up to 3 mm. in size–and is surrounded by 2 jelly envelopes making egg and jelly a little over 5 mm. in total diameter.

The female remains with the eggs for at least part of the incubation period of a month to two months. Dr. Oliver tells us that in the more southerly regions of their range–specifically Virginia–the female broods the eggs for about half of the total incubation period. In New York and Michigan she remains with them until they hatch. This difference in the length of the brooding period is very interesting, and prompts us to wonder whether the eggs laid in Virginia might be smaller than those laid in the North, or whether the salamander may be larger in the South. This question is not such a nonsequitur as it might seem to be. Scientists seem to feel that the exhaustion of the female after laying plays a part in determining the length of the brooding period. *Hemidactylum* is a small salamander, but it lays a very large egg. If we put only the vitelli of the total egg complement together in a line, we find that this small creature has laid eggs that are equal to or exceed her own body length. It is small wonder that she is exhausted after this feat. It may be found that in the South, Hemidactylium is larger, the eggs are smaller, the total complement is less, or a combination of these factors is true. If so, the shorter brooding period in the South may be partially explained.

We say "partially" advisedly, for female exhaustion after laying does not completely explain why *Hemidactylium* broods. It is not at all uncommon to find a communal nest where a number of salamanders have deposited their eggs in close proximity to one another. In this case, not all the layers remain with the eggs. A few only remain on the nursery detail while the others depart. What determines which females shall stay? Is it those that laid their eggs first that remain? Or do the first layers depart as later arrivals deposit their eggs? Does age or size play any part in determining which shall remain with the nest? Someday we may know the answers.

When the eggs hatch, the larvae wriggle down into the water. Unlike all other plethodontids, these larvae are pond larvae similar to those of the Mole Salamanders in body form. The tail fin extends up onto the back, and the gills are large and bushy. The larval period is about 1½ months. The transformed salamander, only about ¾ of an inch in length, leaves the water and takes up its adult life on land in the moist mosses where it was born.

The Dwarf Salamander, *Manculus quadridigitatus*, found along the coasts of the souther states from North Carolina to Texas, is about the same length as the Eastern Four-toed Salamander, but it is more slender and has a very long tail. In colour and shape of head, it looks very much like the Two-lined Salamander. Its back is bronzy tan occasionally marked with darker spots and dashes. On each side there is a dark line that runs the entire length of the body. The belly is yellow. Despite the similarity of colour and markings, the Dwarf Salamander may be distinguished from the Two-lined Salamander by the fact that the former has but four toes on its back feet, whereas the latter has five toes.

The male Dwarf Salamander has cirri extending from its nasolabial grooves, as well as elongated premaxillary teeth that are slanted forward and are therefore exposed during the breeding season. These teeth are found on the triangular-shaped bone (the premaxillary) that forms the tip of the upper snout. The male Dwarf Salamander is not unique in this respect. Many of the plethodontids – the genus *Eurycea* especially– have these monocuspid elongated teeth. Their function has not been fully explained. It is true that during the breeding season the males rub the females with their snouts, and it is felt that the exposed teeth provide extra friction.

For many years herpetologists insisted that amphibian males did not bite or grasp a female in the fashion of many breeding mammals.

This statement is now open to question, for quite recently the male Slimy Salamander, *Plethodon glutinosus glutinosus*, was seen holding the female's body or tail gently in his mouth during courtship. In addition, when a male Slimy Salamander attempted to court other males, quite frequently the courted male bit the courting one severely, though sometimes it was the courting male that attacked the one he was courting when he discovered the latter was not a female.

So far as we know, no one has ever seen any such behaviour as this on the part of the Dwarf Salamander. But since the males of many species of salamander develop specialized premaxillary teeth, perhaps more thorough investigation of their habits may disclose that they use these teeth during courtship.

The Dwarf Salamander breeds and lays its eggs in winter from December though February in the larger pools of swamps or slow-flowing permanent ditches. The eggs are deposited singly or in small groups on the undersides of logs or vegetable debris in water. The total complement may be as high as 50, though smaller numbers are more common. The eggs are about 2 mm. in size, white in colour, and surrounded by 2 envelopes. The larvae are found in March, and after a larval period of 2 or 3 months they complete their transformation and assume a more terrestrial existence.

These are but a few of the swamp and ditch dwellers you may find. Most notable of the omissions are the Chorus Frogs, genus *Pseudacris*. These belong to the family Hylidae. They are small–most do not exceed 1½ inches in length–brightly coloured little animals. They are generally marked with five rows of spots or five stripes on the dorsum. Representative species may be found throughout most of the United States, New England excepted.

8

CAECILIANS

Caecilians are worm-like, mostly secretive, burrowing amphibians confined to the tropical and subtropical regions of the world. Although their name (pronounced "see-sil-e-an") means "blind", most caecilians have small eyes, which in some species are hidden beneath the bones of the skull. The order Gymnophiona is the least diverse and most poorly known of the three orders of amphibians.

Built for Burrowing

Caecilians are the only living amphibians that are completely legless. They are capable of moving snake-like across surfaces when forced to do so, but normally they live underground and move through pre-formed tunnels or create new tunnels by pushing their head through loose mud or moist soil. Caecilians have many adaptations for burrowing. The skull is powerfully constructed with a pointed snout and an underslung lower jaw or recessed mouth, features that allow the head to be used as a ram. The eyes are reduced in size and imprortance, as there is no light in their underground world.

Caecilians have a unique pair of sensory organs called tentacles, one emerging from a groove or cavity on each side of the snout between the eye and the nostril. The tentacles, which are probably organs of taste and/or smell, are admirably suited for sensing the environment of tunnels. Another peculiar feature of caecilians is the manner in which they close their jaws: instead of having a single set of jaw-closing muscles as in all other terrestrial vertebrates (land-dwelling animals with backbones), caecilians have a dual mechanism that consists of two sets of jaw-closing muscles. Even this mechanism may be considered an adaptation for burrowing. Another interesting feature of caecilians is the presence of numerous skin folds, or rings, called annuli, which partially or completely encircle the body, although it is not yet known if the annuli are directly related to locomotion. Many species also have small fish-like scales hidden under the skin folds.

The body muscles of caecilians are arranged in such a manner that the body can act like a rod moving within a tube. The "tube" is the skin and outer layer of body muscles. The "rod" is the head, the vertebral column, and the deep body muscles associated with the vertebral column. With the tube fixed in position in a tunnel, the rod can be pushed slightly forward through the soil to extend the tunnel. When the tip of the rod (the head) has reached its maximum forward progression, the tube is pulled forward in waves and fixed in position again so that the rod can again be pushed through the soil. However, in very loose mud, caecilians use a different kind of locomotion, they simply swim eel-like through the watery medium.

Reproduction

Unlike most other amphibians, all caecilians have internal fertilization; the males have a distinctive protrusible copulatory organ, called a phallodeum which serves to inseminate the females. Like all amphibians, caecilians have eggs and embryos similar to those of fish, and unlike those of reptiles, birds, and mammals. The latter three groups have a pair of special membranes (amnion and chorion) that enclose the embryos. Fish, caecilians, and other amphibians lack these membranes, and the embryos typically develop within a gelatinous egg case. Some caecilian species lay eggs; the eggs hatch into larvae; the larvae later metamorphose into adults. Other species are viviparous: they give birth to living young, and there is no larval stage.

Females of egg-laying species remain with their eggs until they hatch. The function of this parental care is unknown, but presumably the female protects her embryos against predation and perhaps other sources of embryo mortality such as destructive molds and desiccation.

Viviparous species have a number of unusual adaptations. The embryos developing in the female's oviducts soon use up all of the yolk provided in the egg, and the embryos then begin feeding on a substance called "uterine milk", which is secreted by the oviducts. Embryos are provided with numerous tiny teeth, which have distinctive shapes and are shed soon after birth. The function of these embryonic teeth is not understood, but they may have something to do with intrauterine feeding. The gills of embryos developing in the oviducts are enormously developed in some species and are thought to function in gas exchange between the tissues of the female and her embryo. This has not been proved, and other functions are certainly possible.

Food and Enemies

The food and feeding habits of caecilians have not been carefully studied, but the larval caecilians that have been dissected by scientists had eaten a variety of immature insects, earthworms, and other invertebrates. Land-dwelling caecilians seem to feed primarily on earthworms. Beetles and other insects have also been found in their digestive tracts, and occasionally small frogs and lizards.

Snakes, especially coral snakes seem to be the primary predators of caecilians. In the Seychelles islands, chickens, pigs, and the shrew-like tenrecs (introduced from Madagascar) at least occasionally eat caecilians. None of these predators is native to the Seychelles.

Like all amphibians, the skin of caecilians contains numerous poison glands, which presumably function to discourage predators. Many frogs and salamanders with highly toxic skin secretions are brightly coloured, and experimental evidence indicates that the bright colours warn predators of the presence of the toxins and therefore help the amphibian to avoid injury from predator attack. Most caecilians have subdued colours, usually various shades of gray, but some (for example, members of the Rhinatrematidae and Ichthyophiidae families) have bright yellow lateral stripes. Although the function of the bright colours in these caecilians has not been studied, it seems likely that they too will prove to be warning colours.

Primitive Caecilians

Two families are considered to be relatively primitive, which means they are more like the common ancestor of all caecilians than are other caecilian families alive today. These are the Rhinatrematidae of northern South America and the Ichthyophiidae of India and Southeast Asia.

The rhinatrematids (two genera, nine species) of South America have retained the highest number of primitive characteristics: they have a "terminal" mouth (which means the mouth-opening is not recessed on the underside of the snout); their tentacles are in contact with the relatively large eyes; they have numerous skull bones; and they have a tail. These primitive caecilians also have the annuli (skin folds, or rings) subdivided into secondary and tertiary annuli which completely encircle the body, and numerous scales throughout the length of the body. The significance of secondary and tertiary annuli is still uncertain, and scientists know of their existence only by studying the embryonic and larval development of certain species. The specialized dual jaw-closing mechanism of caecilians, is least developed in members of this family.

The life history and ecology of these seldom-seen caecilians is very poorly understood. However, we do know that they deposit eggs in soil cavities and that the eggs hatch into larvae with tiny external gills and gill slits. The larvae live in seepages and streams until they metamorphose into the adult form. The adults live in moist soil, leaf litter, and rotten logs.

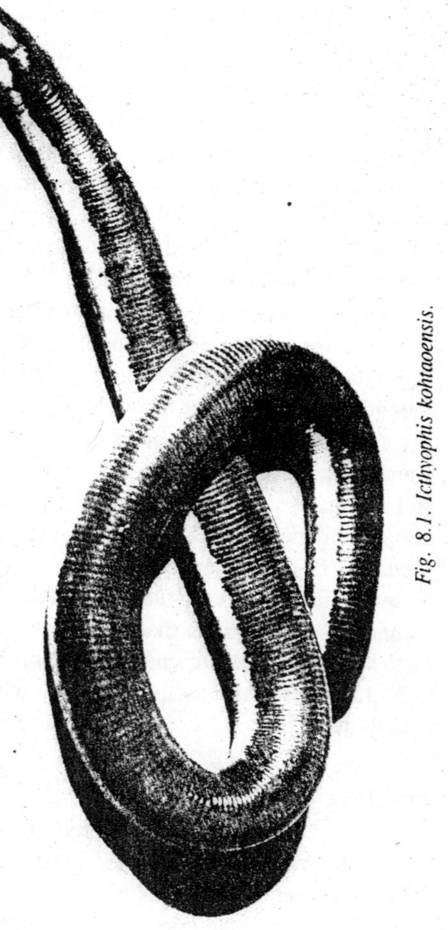

Fig. 8.1. Icthyophis kohtaoensis.

The ichthyophiids (two genera, 37 species) of tropical Asia look very much like rhinatrematids, and until recently they were classified in the same family. They too have terminal mouths, numerous skull bones, a short tail, abundant scales, and primary annuli that are subdivided twice. Relatively advanced traits of the Ichthyophiidae include the position of the tentacle, which is forward of the eye, and a well-developed dual jaw-closing mechanism. The life history of ichthyophiids is similar to that of rhinatrematids. Although few clutches of eggs have been found, all of these were attended by a female.

Transitional Caecilians

A small group of caecilians confined to India is placed in the family Uraeotyphlidae (one genus, four species). Uraeotyphlids are in some ways intermediate between the primitive and advanced caecilians. Primitive traits include the presence of many bones in the skull, a short tail, and numerous scales. Advanced features include a mouth-opening recessed below the snout, tentacles very far forward of the eyes, and the condition of the primary annuli—these skin rings are subdivided only once (instead of twice as in the primitive families), and some of the annuli at the forward end of the body do not completely encircle the body. The dual jaw-closing mechanism is well developed. Although very little is known about the life history of uraeotyphlids, the few observations suggest their way of life is similar to that of ichthyophiids and rhinatrematids, but with a shorter larval period.

Advanced Caecilians

The remaining two families of caecilians—the Scolecomorphidae and the Caeciliidae (including the subfamilies Caeciliinae and Typhlonectinae) may be viewed as relatively advanced groups with specialized form, structure, and life history. Scolecomorphids and caeciliines live on land and are well adapted for burrowing in moist soil; Typhlonectines are aquatic or semi-aquatic. These two families share a number of advanced characteristics, such as a reduced number or skull bones, a recessed mouth, absence of tertiary annuli, and reduced number of scales, and they are tailless. Even so, the evolutionary relationships or these families to each other and to the more primitive families are still not fully understood.

Scolecomorphids (two genera, five species) occur in equatorial East and West Africa. These are among the most bizarre of caecilians. They have large tentacles placed far forward on the underside of the snout in front of the mouth. The vestigal eyes are attached to the base of the tentacles. In resting position, the eyes are under the skull bones, but when the tentacles are protruded the eyes are carried outside the skull along with the tentacles. Scolecomorphids are the only caecilians that lack stapes, the tiny bones that conduct sound vibrations from the environment to the inner ear. Other advanced features of this family include the loss of secondary annuli (only primary annuli that do not completely encircle the body are present); and, as is usually the case with caecilians that have lost their secondary and tertiary annuli, scolecomorphids do not have scales. The body ends in a small, blunt "terminal shield", which lacks annular grooves. The East African species (in the genus *Scolecomorphus*) are viviparous; that is, they give birth to living young. The eggs are fertilized internally and are retained in the oviducts of the female, where embryonic development takes place. The aquatic larval stage has been lost, and the young are born as fully terrestrial juveniles. Nothing is known about the life history of the West African species of the genus *Crotaphatrema*.

The Caeciliinac (23 genera, 89 species) is the most diverse and most geographically widespread of the caecilian subfamilies. Species of this subfamily are found in tropical Central and South America, equatorial Africa, the Seychelles archipelago in the Indian Ocean, and India. Caeciliians are recognized largely by the condition of the skull, which consists of a characteristic number and arrangement of bones. The skull bones are relatively few, and they are strongly joined to transform the skull into a solid ram used in burrowing. The mouth of all caeciliians is recessed, with the exception of *Praslinia cooperi* of the Seychelles Islands, which has a terminal mouth. The position of the tentacle opening varies considerably among caeciliids, in some species, the opening is close to the eye: in others, it is closer to the nostril. Only in the enigmatic *P. cooperi* is the tentacle in the primitive position adjacent to and in contact with the eye.

Caeciliids have a variable number of their primary annuli subdivided into secondary annuli (these secondary annuli always occur at the tail end of the body); or they have only primary annuli. Similarly, aceciliids have a variable number of scales in the skin folds; and species that have no secondary annuli generally also have no scales. None of the caeciliids has a tail; in some, the body ends in a blunt terminal shield.

Caeciliids have a variable life history. A few species, such as *Praslinia cooperi*, deposit eggs on land that hatchinto water-dwelling larvae with a prolonged larval period. Others, such as *Grandisonia larvata* of the Seychelles Islands, have a very brief larval period. Still others, for example, *Afrocaecilia taitana* of East Africa, lay eggs on land and the eggs hatch directly into terrestrial juveniles without an aquatic larval stage. A some, like *Schistometopum thomense*, give birth to living young. Larval caeciliids are found in seepages and streams. The metamorphosed juveniles and adults are terrestrial burrowers. One species, *Hypogeophis rostratus* of the Seychelles, is found in streams, especially at night, as well as on land. The subfamily Typhlonectinae (four genera, 12 species) is found throughout much of tropical and subtropical South America. Relatively primitive typhlone-ctines are semi-aquatic, whereas the more advanced species are aquatic. The skull or typhlonectines is basically like that of the burrowing caeciliids, with relatively few bones that are solidly conjoined and with a recessed mouth. Even the most aquatic typhlonectine species are adept at burrowing in soft mud and gravel in the bottoms and edges of aquatic habitats. Typhlonectids have primary annuli only, which are sometimes wrinkled, giving the appearance of being divided into secondary annuli, and they have no scales. They are tailless, with a terminal shield that lacks rings. The aquatic species have a skin fold or fin that runs along the back from the end of the body toward the head; the body is compressed laterally, which together with the fin increases swimming ability. The tentacular opening of typhlonecrids is very small and variously placed between the eye and nostril, but it is never in contact with the well-developed eye. Unlike other caecilians, typhlonectids apparently do not protrude their tentacles. One recently discovered typhlonectid is peculiar in that it has no lungs (all other caecilians have lungs), and the passages between the nostrils and the mouth cavity are scaled. Typhlonectids are viviparous, producing young that are fully metamorphosed at birth.

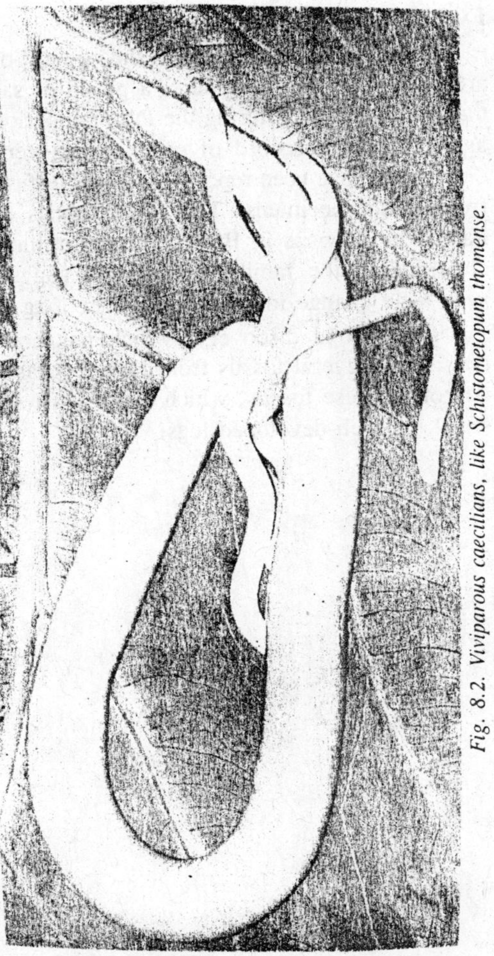

Fig. 8.2. Viviparous caecilians, like Schistometopum thomense.

Evolutionary History

The origin of caecilians is unknown. Some scientists believe they are derived from microsaurs, a group of salamander-like amphibians that became extinct during the Permian period, about 250 million years ago. The fossil record of caecilians is very poor—in fact only two discoveries have been reported and it is difficult to draw any conclusions from these specimens. The first consists of a single vertebra from Paleocene deposits in Brazil; this 60-million-year-old fossil has been classified in the family Caeciliidae, suggesting that there has been very little change in this family over long periods of time.

The second discovery, which has not yet been fully reported, consists of several fossils from Lower Jurassic deposits in northeastern Arizona. These fossils, which are at least 170 million years old, have small but well-developed legs.

9

FROGS AND TOADS

In Europe, where there are relatively few species of the order Anura, the most conspicuous are smooth-skinned, long-legged frogs of the genus *Rana* and short-legged, warty toads of the genus *Bufo*. So it is only natural that every European language should have specific words for these two kinds. But where species are more diverse, this distinction breaks down, and it is neither possible nor necessary to force animals into one or other category. The name "frog" may properly be used for any member of the Anura, whereas "toad" is loosely applied to members of the genus *Bofo* as well as to other frogs of similar body form.

Easily Recognized Animals

A frog cannot be mistaken for any other animal. Early in their evolution, frogs acquired a body structure well suited to jumping: the ankle bones are elongated so that, with the femur and tibiafibula, they form a third major segment that gives the hind legs additional mechanical advantage in jumping. The short, rather inflexible vertebral column with no more than ten free vertebrae followed by a bony rod (the coecyx, representing fused tail vertebrae) is another adaptation to leaping. Given such limitations on its structure, a frog—whether it burrows in desert sand, swims in a high Andean lake, or lives in a rainforest tree—is eminently recognizable.

The diversity in body form displayed by frogs expresses their adaptation to particular ways of life. Almost identical solutions to problems of adaptation have evolved time and time again, in different evolutionary lines. Yet any one group of related species may include a diverse range of adaptive types. For example, the so-called "treefrpg" family, Hylidae, includes mostly species with the tips of the fingers and toes enlarged into adhesive pads (an aid in climbing) and with the eyes positioned to provide binocular vision ahead, as well as visual fields below and above. But some hylid frogs of semi-arid regions are adapted for terrestrial life, with eyes set higher on the head, narrow toe pads, and a tubercle on each hind foot that facilitates burrowing. Other hylids may lack any trace of toe pads and live in marshy fields and ditches. These various forms and structures are matched in several other families of frogs. Indeed, given an unfamiliar frog to identify, even an experienced specialist would first have to examine its internal anatomy to ascertain its family relationship, before going on to determine the genus and then the species.

Distinguishing Species

The number of known species of frogs (nearly 4,400) approaches that of mammals (about 4,670). It is impossible to give an exact number because new species of frogs are continually being recognized and given scientific names. This is largely because of discoveries in poorly known regions, especially tropical rainforests. Also, in recent years means other then morphology (form and structure) have been used to detect differences between species. Many species of frogs prove to be virtually identical I external appearance but can be distinguished by their calls (as indeed they distinguish each other), and with modern electronic techniques for recording and analyzing sounds it becomes possible for biologists to separate these so-called "sibling species". Biochemical techniques, too, make an important contribution by demonstrating genetic (DNA) differences between species whose morphology is virtually the same.

Sight, Smell, and Sound

Frogs are primarily visually oriented animals. The eyes are generally large and protrude from their sockets, providing a broad visual field and compensating for lack of rotatory movement. The eyes can, however, be retracted into their sockets, whereupon they budge against the roof of the mouth and assist in swallowing! The upper eyelid is a cover without independent movement. The lower eyelid can be moved and has a transparent upper portion, the nictitating membrane. The pupil is a horizontal oval in most frogs, but in others it is a vertical oval or may even be triangular or diamond-shaped. The iris commonly resembles the side of the head in colour and may even be part of some aspect of facial pattern. Presumably this helps to conceal the eye, although some frogs have brightly coloured eyes.

There is no evidence that frogs have colour vision. A pineal organ (homologous with the "third eye" of reptiles) is found in some frogs and may be detected as a pale spot atop the head. The organ reacts to light, and experiments suggest that it is involved in sun-compass orientation and other aspects of behaviour.

Frogs possess both the nasal olfactory organ and a separate vomeronasal organ that functions in a similar fashion. These two organs facilitate homing on breeding sites (presumably by recognition of chemical clues) and may serve in identification of prey.

Hearing is important to frogs, for finding mates and in territorial behaviour. The main receptor is a membrane, the tympanum, which is stretched across an oval or round cartilaginous ring behind the eye. A rod of bone, the columella, transmits the vibrations of the tympanum to the inner ear, where sensory cells detect and sort hem, passing the information on to the brain. This system is most effective in detecting higher-frequency airborne vibrations, whereas low frequencies pass into the body from the ground or whatever substrate the frog is resting on. The tympanum may be sexually dimorphic, being much larger in males. Frogs of some species have the tympanum concealed beneath skin, but others lack a tympanum completely.

Activity

Whether frogs are active—that is, engaged in feeding or breeding—depends largely on moisture, temperature, and time of day. Some species are principally nocturnal, others mostly diurnal (day active), while others may be active day and night. Where moisture is adequate, activity may be limited to times when the temperature is above a level critical for the species concerned. For example, in North America the wood frog *Rana sylvatica* emerges from hibernation and breeds in waters barely above freezing, whereas other species of *Rana* in the same immediate area wait to breed until many weeks later because their eggs are less tolerant to low temperatures. In tropical and subtropical regions, where low temperatures are less likely to be important, active periods nevertheless may be controlled by seasonal changes in moisture. This is especially true for species that live away from permanent sources of water and depcnd on rainy seasons for both feeding aand breeding activity.

Breeding

Wifh some specialized exceptions, frogs fertilize their eggs externally. The male approaches the female from behind and grasps her around the body with his arms. The hold may be at the waist

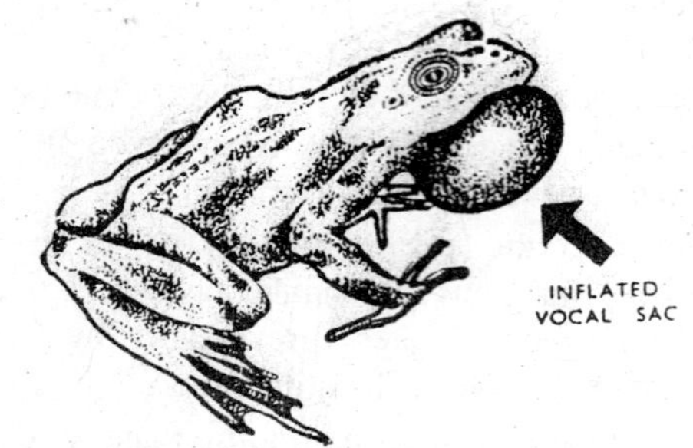

Fig. 9.1. An inflated vocal sac of male frog during breeding season.

(considered the primitive mode) but in most species is just behind the female's arms or (rarely) around the head. Males of many species have horny areas on their hands that help in gripping a slippery mate. The frogs maintain this posture, while the eggs are fertilized as they are extruded. When eggs make contact with water, the jelly membranes surrounding them swell, encasing each egg in a transparent sphere.

Fig. 9.2. Amplexus in frog.

Eggs deposited in water are most commonly grouped in globular masses containing a few to hundreds of eggs. Some species that breed in still, warm water likely to be depleted of oxygen spread the eggs in a single layer on the surface, whereas those breeding in fast-flowing streams may attach the eggs singly to submerged rocks. Eggs deposited

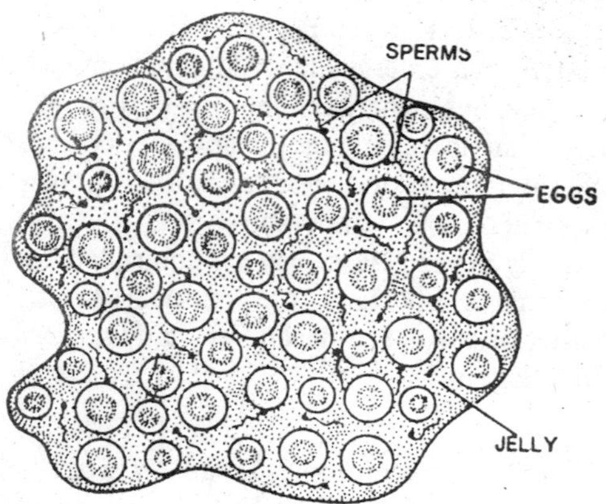

Fig. 9.3. Frog's spawn.

out of water may be in a variety of moist locations such as in tree holes, in plants growing on trees, underneath moist leaf-litter, or in holes in the ground. Thlese are only a few examples of the many known sites and modes of egg deposition.

Larval Life

Frogs resemble other amphibians and differ from reptiles in that most species (75 to 80 percent of frogs) have a larval stage interposed between the egg and the mature body form. This period may be as brief as one week in species breeding in short-lived desert rainpools, or up to two years or more, but generally it averages a few weeks. A typical frog larva, or tadpole, lives in water and has a rather oval body with a strong finned tail, but no clear distinction between head and body. The mouth has a beak and rows of "teeth" made of a chitinous suhstance. Water taken in through the mouth pass‿ over gills concealed within a chamber behind the mouth, before being expelled through the spiracle, a hole usually on the left side of the body. The gills not only serve for respiration, but also filter tiny food

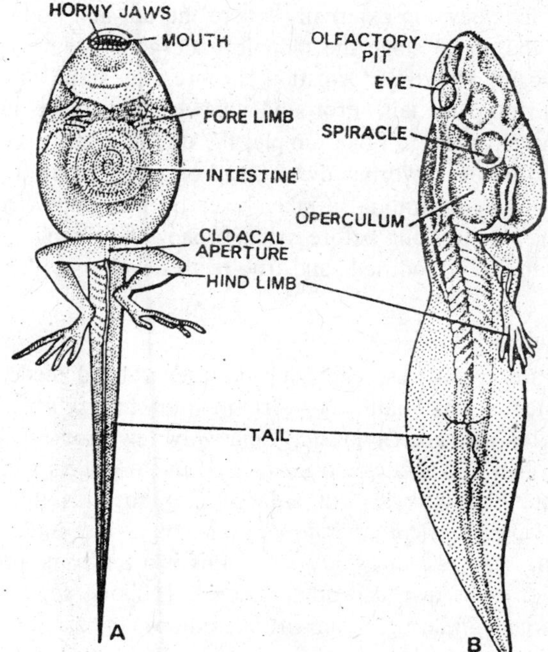

Fig. 9.4. Stage of limb-development from ventral (A) and lateral (B) views.

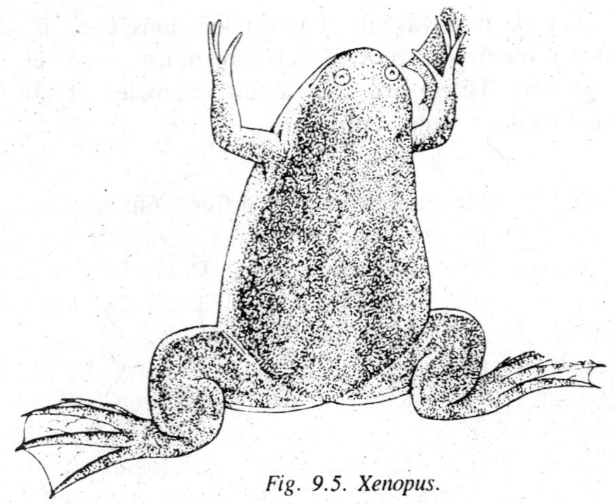

Fig. 9.5. Xenopus.

particles from the water, diverting them toward the stomach and supplementing the algae and detritus obtained by biting and scraping. Also concealed within the gill chamber are the animal's growing front legs. The rear legs develop externally where the tail and body meet. At the end of the alrval stage the tadpole undergoes metamorphosis into the adult form. This transformation is more than simply growing limbs and resorbing the tail. Proround changes take place in both morphology and physiology. For example, the digestive tract shortens, becoming suited to a carnivorous diet instead of a vegetarian diet; the gills disappear, and respiration is taken over by the lungs (tadpoles also may use their lungs long before metamorphosis is completed); the skeleton is extensively modified, and true teeth develop in the upper jaw.

Food and Enemies

All frogs are carnivores, and many are generalized feeders that will take whatever small animals—vertebrate or invenebrate—their capacious mouths can accommodate. Relatively few frogs are large enough to eat other vertebrates, so most of them eat insects and other arthropods, and earthworms. But a large frog such as the North American bullfrog *Rana catesbeiana* can take birds and mice, small turtles and fish, and under crowded conditions is a fearsome predator on smaller frogs of its own and other species. Tadpoles for the most part are vegetarians—filtering organisms from the water, scraping algae from stones, consuming bottom debris. Some species have predaceous tadpoles, however, which capture invertebrates or other tadpoles.

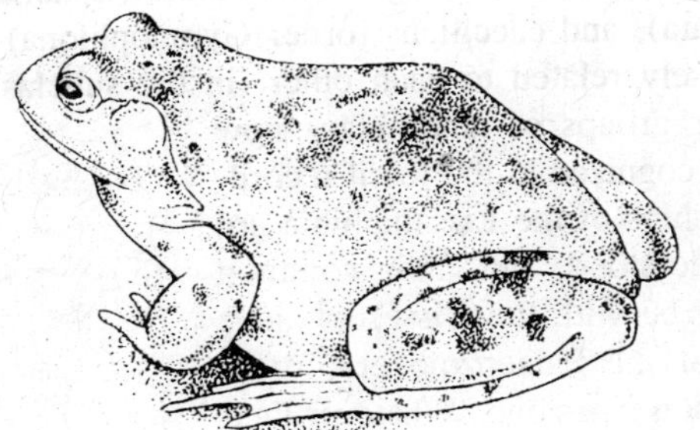

Fig. 9.6. Rana.

Frogs are prey to a host of enemies, ranging from tarantulas to humans. Many snakes live largely or entirely on frogs. Herons skewer frogs in shallow water, bats snatch them from branches over tropical pools, turtles ambush them from under water, big frogs eat smaller frogs, parasitic flies lay eggs on them, and leeches attack them externally and even enter their bodies. The larval stage, too, is a perilous time. Almost any meat-eater finds a tadpole a succulent morsel.

Where Frogs Live

Frogs are native to all the continents except Antarctica but are absent from almost all remote oceanic islands and from Greenland and other Arctic islands. Although widely distributed over the Earth, frogs are far more abundant and diverse in moist areas of the tropics, especially tropical rainforests, than farther north or south; regions closer to the poles are subject to extremes of weather, and dryness and cold are especially hostile to frog life.

The near-absence of native frogs from oceanic islands is rooted in the geological history of the islands, most of which are volcanoes or coral islands built on volcanic footings. There has simply been no way for animals so sensitive to desiccation and heat to reach them (frogs cannot live in salt water). Many of these islands may have suitable habitats; for example, several species of frogs have been introduced to the Hawaiian islands, where they prosper. Some islands—such as the Seychelles group in the Indian Ocean—are unusual in having endemic frogs (species found nowhere else), presumably because these landmasses were formerly part of or close to continents but were moved by geological processes, carrying with them the ancestors of the present frog species.

Ancestors and Relatives

All living amphibians—frogs (order Anura), salamanders and newts (order Caudata), and caecilians (order Gymnophiona)—are thought to be more closely related to each other (in the subclass Lissamphibia) than to other groups of amphibians now exrinct but recognized in the fossil record. The relarionships of the Lissamphibia are not well understood, but their common ancestry may be with the Dissorophoidea amphibians of the Palaeozoic era, more than 245 million years ago. Whatever their exact ancestry, today's amphibians lie on an ancient branch of the amphibian tree and are not on the evolutionary line that led to reptiles and, eventually, mammals.

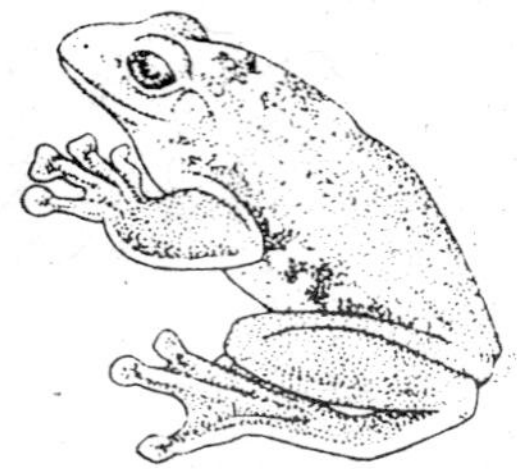

Fig. 9.7. Polypedates.

The oldest frog-like fossil is *Triadobatrachus* found in Madagascar in deposits of the Triassic period, more than 200 million years ago. Early frog fossils are also found in deposits laid down early in the Jurassic period, almost 200 million years ago, in Argentina. By the beginning of the Tertiary period 65 million years ago, there were frogs we can classify in genera that are still represented today. So frogs that differed in no essential respect from those alive today were contemporaties of the dinosaurs, survived the great extinctions at the close of the Mesozoic, and persisted largely unchanged while mammals underwent their great evolutionary expansion.

Archaeobatrachian Frogs

There are four suborders of modern frogs. Some are considered more "primitive" (in evolutionary terms) than others.

The suborder Archaeobatrachia includes four families, Ascaphidae, Bombinatoridae, Leiopelmatidae, and Discoglossidae, thought to be the most primitive living frogs. A clearly primitive character they share is the presence of free ribs (all other frogs have their ribs fused to the vertebrae).

Ascaphid and Ieiopelmatid frogs

There are only four species of Ascaphidae and Leiopelmatidae alive today. The tailed frog *Ascaphus truei* lives in cold, fast-flowing mountain streams of the northwestern United States and southwestern Canada. The so-called "tail" is present only in the male and is an organ adapted for insertion, so that the eggs are fertilized internally before being released from the female's body (in this type of habitat, external fertilization might be unsuccessful) and attached to the undersides of rocks in the water. The sucker-mouthed tadpoles take three years to develop and are well adapted to life in fast-running waters.

The other three species are restricted to New Zealand and are the only native frogs there. One of these, Hamilton's frog *Leiopelma hamiltoni*, occurs on just two small islands and is at risk because of its tiny area of distribution. They are all small frogs, no bigger than 50 millimeters (2 inches) body length. *L. hamiltoni* and Archey's frog *L. archeyi* lay their eggs on moist ground under rocks, logs, or vegetation; the hatchlings are in a relatively late state or larval development and climb upon the back of the attending adult male. Hochstetter's frog *L. hochstetteri* lays eggs in shallow hollows with seepage; the tadpoles are more adapted for swimming and remain in the water near to where they hatched until metamorphosis, although not feeding.

Subfossil remains, which date back to the period before humans colonized New Zealand, reveal that there were additional species of *Leiopelma*, one twice the size of the living species. Fossils from early and late in the Jurassic period (up to almost 200 million years ago) found in southern Argentina, are considered to belong to the family Leiopelmatidae.

Bombinatorids and discoglossids

The bombinatorid and discoglossid frogs are only slightly more numerous but are less geographically restricted than the ascaphid and leiopelmatid frogs. All but two of the 16 species live in the Northern Hemisphere north of the tropics. Species of the genus *Discoglossus* are semi-aquatic frogs found in Europe, North Africa, and the Middle East. Different species of fire-bellied toads (genus *Bombina*) are small to moderate-sized frogs—up to about 70 millimeters (almost 3 inches) body length in a large Asian species—that live in the Far East and Europe, where they have a mostly aquatic existence in shallow water habitats. When disturbed they arch their back and throw up their arms and legs, showing the bright red or yellow colours beneath. Because the frog's skin secretions

Fig. 9.8. Bufo.

are distasteful, this is considered a form of warning behaviour. The midwife toads (genus *Alytes*) are largely terrestrial frogs of Europe and North Africa, known for the male's care of the eggs; these adhere to his back and thighs and are carried about in and out of water until hatching time, when he enters the water, allowing the tadpoles to swim away. The two species of *Barbourula* are exceptional among discoglossids, not only in their geography—one in the southern Philippines, the other south of the Equator in Borneo—but also in being wholly aquatic. Their adaptation to life in streams includes having broadly webbed fingers and toes. The fossil record shows discoglossids to have existed in Europe as far back as the Jurassic, about 150 million years ago, and the family was present in North America in the late Cretaceous period, more than 65 million years ago.

Pipoid Frogs

Members of the suborder Pipoidea, with two families, the Pipidae and the Rhinophrynidae, are among the most peculiar of frogs. The pipids (30 species, 5 genera) have flattened bodies, are tongueless (all other frogs have tongues), and except for frogs in the genus *Pipa* have claw-like structures on three of the toes. All species are highly aquatic, wilh large, fully webbed feet. They don't have vocal cords but produce clicking sounds under water by using bony rods in the larynx. The size range is from about 40 to 180 millimeters ($1^1/_2$ to 7 inches) body length. Frogs in the genus *Pipa* live in South America east of the Andes and in eastern Panama. The Surinam toad *Pipa pipa*, which is almost 180 millimeters (7inches), is bizarre in both appearance and breeding habits. The eggs becomes embedded in pockets in the skin of the mother's back, where they develop to emerge as tiny frogs. In another *Pipa* species, tadpoles emerge from the chambers and complete their development on their own.

In Africa south of the Sahara Desert there are three genera. The African clawed frogs, genus *Xenopus*, were for a time medically important as animals used in testing for human pregnancy: urine from a pregnant woman injected into a female frog causes it to extrude eggs. Development of chemical tests has spared platannas (as they are called in South Africa) from this indignity, although they continue to be important animals for laboratory research because of the ease with which they may be kept and bred in captivity.

Fossils of pipid frogs from the early Cretaceous period, perhaps 145 million years ago, have been found in Israel, and from the late Cretaceous, more than 65 million years ago, in Africa and South

America. Clearly, their distribution dates from a time when the southern continents were part of Gondwana, before Africa and South America became separate landmasses.

Some scienrists classify the Mexican burrowing toad *Rhinophrynus dorsalis* with the pipid frogs; others place it in its own suborder (Rhinophrynoidea). It is the only living representative of the family Rhinophrynidae. This rather globular, small-headed, burrowing frog has a prominent spade-like digging tubercle on each hind foot, grows to about 75 millimeters (3 inches), and ranges from southern Texas to Costa Rica. It emerges after heavy rains to call and breed in temporary ponds. Fossils show that in the late Paleocene to the early Oligocene epochs (about 37 million years ago) related species lived as far north as southern Canada.

Pelobatoid Frogs

The suborder Pelobatoidea comprises three families, Megophryidae, Pelobatidae, and Pelodytidae, although some scientists treat them as one family. These frogs are considered rather primitive, but in some ways are transitional to more advanced frogs, and have in common the supposedly primitive habit of the mating male clasping the famle around the waist or groin, rather than higher on the back, just behind the arms. The eyes of all species have vertical pupils, and all have free-living, aquatic tadpoles.

Megophryids, pelobatids, and pelodytids

The family Megophryidae has fewer than 90 species recognized in nine genera found in Asia, from Pakistan and western China to the Philippinnes and through Indonesia to the Greater Sunda Islands in the Malay Archipelago. Their hahitats vary from forested tropical lowlands and uplands to Himalayan peaks. *Megophrys* species inhabit the forest floor; they have funnel-mouth tadpoles which develop in still waters. *Megophrys montana* is notable for its cryptic colouration and form; not only does it resemble dead leaves in colour and pattern, but one subspecies has pointed projections from the eyelids and nose that disrupt the frog-like outline. Also inhabiting the forest floor are *Leptobrachium* species; these frogs breed in small streams, and the tadpoles live amid rocks and gravel. Mountains as high as 5,200 meters (17,000 feet) are the home of *Scutiger* species, which lead a largely terrestrial existence but breed in cold mountain streams. An odd feature of the *Scutiger* male is that his mating grasp of the female is enhanced by two spiny patches on the chest in addition to the more usual spiny areas on the first two fingers.

"Spadefoot toads" is the common name for the three genera in the family Pelobatidae: *Pelobates*, found in Europe, some areas of Mediterranean northwestern Africa, and western Asia; and *Scaphiopus* and *Spea*, found in North America from southern Canada to southern Mexico. They are toad-like burrowing frogs of moderate size, up to 100 millimeters (4 inches) body length, and their common name comes from the prominent digging tubercle on each hind foot. The four species of *Pelobates* are partial to areas of sandy soil, where they burrow and also breed, in pools (often temporary ones) and ditches. *Scaphiopus* and *Spea* too spend most of the year underground. When the soil temperature is warm enough, the sound of rain falling will prompt them to emerge and they migrate to temporary pools of water to breed. Embryonic and larval development is rapid—less than two weeks may elapse between egg-laying and metamorphosis. But even this may not be rapid enough if additional rain doesn't fall to offset evaporation in the fierce desert heat. Some spadefoot tadpoles develop a different morphology—enlarged jaw muscles and a large beak—from others of their species, and they cannibalize their pondmates, thus making the most of the available protein. The rains that call out the spadefoots also bring forth hordes of termites and other insects, and the frogs can get enough food in a few nights to survive many more months underground.

Pelobatid fossils occur as early as the late Cretaceous of North America and Asia (more than 65 million years ago), and the modern genera *Scaphiopus* and *Pelobates* appear in the Tertiary of North America and Europe, respectively, 35 to 47 million years ago.

The family Pelodytidae has two living species in the single genus, *Pelodytes*; one in Europe; and one further east, in the Caucasus region (between the Black Sea and the Caspian Sea). These terrestrial frogs look like small (50 millimeters, or 2 inches) more-typical frogs, being long-legged and lacking the digging spade. The earliest fossils are from the middle Eocene epoch in Europe (47 million years ago). From North America, where the family is extinct, there is a middle Miocene fossil some 15 million years old.

The Last Suborder: Neobatrachia

The suborder Neobatrachia, thought to represent the more advanced frogs, includes more than 4,200 species, or about 96 percent of all living species of frogs. There are 17 families.

Leptodactylid Frogs of the Americas

The largest family in this suborder is the Leptodaetylidae, a diverse assortment of almost 1,000 species classified in 48 genera, inhabiting

South America, the Caribbean islands, Central America, and Mexico: only six species cross into the southern United States. The world's southernmost frog is a leptodactylid, *Pleurodema bufonina*, which reaches the Strait of Magellan.

Within the vast array of leptodactylids are virtually all modes of life open to frogs. In the genus *Eleutherodactylus*, which has over 600 species, many spend their active hours in trees or shrubs, but others forage on the leaf-litter of the forest floor, and all but one *Eleutherodactylus* have direct-developing eggs (the exception is viviparous). Species of several other genera have rather cryptic lives on or within forest leaf-litter; others that live in arid regions remain underground, encased in a protective "cocoon", until rainfall permits a brief emergence for feeding and breeding: and many species have more familiar habits in and around streams and ponds. Wholly aquatic species inhabit lakes in the Andes.

Morphology and lifestyle are related, so the variations in their body form are a reflection of their ecological diversity. For example, tree-dwelling *Eleutherodactylus* are rather small, slender frogs with features found in other treefrogs, such as enlarged disks on fingers and toes, and large forward-directed eyes. Burrowing species of other genera are short-legged, with a prominent spade on each hind foot for digging.

Leptodactylid frogs of several genera lay their eggs in foam "nests"; according to the species involved, such a nest may float on open water, be in a cavity adjacent to water, or be in a burrow or other sheltered site on land. The nest is constructed while the frogs mate, the male using his feet to whip the mixtue of eggs and seminal fluid (and water, if present) into a froth. The nest protects the fertilized eggs and then the tadpoles from enemies, from overheating (if exposed to the sun), and from desiccation. A nest in a cavity can shelter the tadpoles until rains flood the site and the tadpoles escape to continue their development, as occurs in some species of *Leptodactylus*. In species of *Adenomera*, which create their nests in moist cavities on land, the tadpoles remain in the nest without feeding until they metamorphose. Hatchlings from nests on open water may merely swim off, but some surprising exceptions are known. For example, the female *Leptodactylus ocellatus* remains with the nest, and when the tadpoles hatch they stay together in a school and the female protects them aginst predators.

The presence of leptodactylids in South America throughout the Cenozoic (beginning 65 million years ago) is documented in the fossil record. A most remarkable fossil from the dominican Republic on the

Caribbean Island of Hispaniola is an *Eleutherodactylus* preserved whole in amber, at 37 million years old.

Myobatrachid Frogs of Australasia

The myobatrachid frogs of Australia and New Guinea, numbering about 120 species in 23 genera, are the eastern hemisphere counterparts of the leptodavtylids of North and South America. Indeed, some people classify them as members of the Leptodactylidae. Australia has a large proportion of arid land, where rainfall is sparse and seasonal, so it is not astonishing that many of the myobatrachids are burrowers and are seldom seen or heard except when they emerge to mate and deposit eggs in temporary pools. The more conventional burrowers such as members of the genera *Heleioporus*, *Notaden*, and *Neobatrachus* are squat, short-legged frogs with a digging spade on each hind foot. The turtle frog *Myobatrachus gouldii* and the sandhill frog *Arenophryne rotunda*, both of southwestern Australia, are exceptional in that they deposit their eggs deep in moist sand, as deep as 1 meter ($3^1/_4$ feet), where they undergo direct development without a free-living tadpole stage. Both are also unusual in that they burrow head-first (rather than rump-first as most burrowing species do) into the sand or soil.

In contrast to the Leprodactylidae, the Myobatrachidae family does not include any treefrogs (this niche in Australia and New Guinea is reserved for members of the Hylidae and Minohylidae), but still there is much diversity in addition to the burrowers, especially in the tropical and temperate rainforests of eastern Australia. Examples include small frogs of the genus *Taudactylus*, at home in torrential creeks of the northeastern rainforests; and tiny *Ranidella*, about 25 millimeters (1 inch) in length, which are abundant in most of Australia's moist habitats. The latter pose a problem for herpetologists because of their morphological similarity—some species can be distinguished only by differences in their calls.

Two species of *Rheobatrachus*, the gastric brooding frogs, are highly aquatic stream-dwellers that scarcely became known to science before they disappeared for no obvious reason, possibly having become extinct. Another species with a less strange but nevertheless odd form of parental care is the pouched frog *Assa darlingtoni*, a ground-dweller of forests on the Queensland-New South Wales border. The male of this secretive species has a brood pouch on each side of the body. When the eggs (which are laid on land) hatch, the male straddles the mass and the tadpoles make their way into the pouches. Further development and metamorphosis take place there, and the froglets emerge some seven to ten weeks later.

Frogs in several Australian genera make a foam nest for the eggs, but unlike the Americam leprodactylids—where the male's feet do the job—in Australia the mating female beats the foam with her hands, assisted by flanges on some of the fingers. Some create their foam nests on open water, others in water-filled burrows, and several create them on land. At least four recent genera of myobatrachid frogs (*Lechriodus*, *Limnodynastes*, *Crinia*, and *Kyarranus*) were present in Australia in the mid-Tertiary, 15 to 25 million years ago.

Toads and Harlequin Frogs

Toads (family Bufonidae) are native to temperate and tropical zones, deserts and rainforest, mountains and proiries, everywhere except the Australian region. Madagascar, and oceanic islands. Of about 400 species in the family, more than half belong to one of the 33 genera, *Bufo*. Toads of this genus range in size from 25 millimeters (1 inch) body length for some African species to 25 centimeters (10 inches) for giant toads *Bufo blombergi* and *B. marinus* of the wet forests of South America—large enough to fill a dinner plate.

No matter what their size, all *Bufo* species conform to a standard appearance: heavy-set, short-legged, with numerous wart-like glands on the body and legs, and a prominent, rounded or elongate parotoid gland behind the eye. Also, they are toothless, although this is evident only on close examination. Habits tend to be rather similar among the species, too. They are ground-dwellers, typically hiding in holes during the day and emerging at night to hop about looking for invertebrates which they snap up with their long sticky tongue.

Breeding is remarkable only for the immense numbers of eggs that many species produce. The eggs typically are laid in paired strings (one from each ovary) and may number 20,000 or more from a toad only 70 or 80 millimeters (about 3 inches) long. All frogs have a variety of skin glands, and in many species they produce defensive secretions, bad-tasting or even deadly poisonous. Among the most effective are toad poisons, concentrated especially in the parotoid glands; a dog that mouths a giant Colorado River toad *Bufo alvarius*, for example, may be fatally poisoned. This does not mean that you'll be poisoned if you handle a toad (or get warts, for that matter!), but it is a good idea not to rub your eyes after handling any frog, until you've washed your hands.

Atelopus is a Central and South American genus (more than 60 species) of small, often brightly coloured, slow-moving toads, whose sucker-mouthed tadpoles live in rapidly flowing water. These diurnal toads produce potent toxins, so the conspicuous colour patterns may serve to warn potential predators. Some of the remaining genera—all of which have fewer than ten species—are much like *Bufo* in structure and habits, but others diverge. For example, *Nectophrynoides* of Africa is noted for giving birth to metamorphosed froglets; *Pedostibes* of Southeast Asia is a tree-dweller; and *Ansonia*, also of Southeast Asia, lives on the ground but breeds in riffles or cascades of water, and its tadpoles are adapted to torrents.

Bufonid fossils occur in the upper Paleocene of South America (more than 57 million years ago) and in the later Tertiary of North America, Europe, Asia, and Africa.

Hylid Treefrogs

The treefrog family Hylidae, has more than 770 species in 40 genera. More than 500 of these species live in the Americas, especially the tropics. There are more than 140 species in the Australia-New Guinea area; and a mere 12 closely related species of the genus *Hyla* are distributed across temperate regions of Eurasia, scattered from Spain to Japan, one of which also occurs in northwestern Africa.

Most species in the family are arboreal or at least climbing forms, which show the characteristic adaptations to this way of life: fingers and toes with expanded tips having adhesive properties, and eyes placed somewhat laterally and forward-directed, enhancing vision downward and binocular perspective. (Terrestrial and aquatic frogs generally have the eyes more atop the head and aimed less forward, thus giving them a larger horizontal visual field and permitting aquatic species to rest with only their eyes above water.) Treefrogs of the genus *Phyllomedusa* in tropical America have the first finger opposable to the other three, like a thumb, permitting them to grasp twigs and stems. Like other large families, the hylids have undergone adaptive radiation into several major ecological niches. There are no thoroughly aquatic species, but a number of terrestrial species are typically associated with ponds and marshes, and in wet weather they may range more widely. These are long-legged frogs, powerful jumpers with webbed toes; the tips of their fingers and toes are pointed or only slightly expanded, in contrast to those of their climbing relatives. Some of these occur in the same genus as that of tree-dwelling species the large Australasian genus *Litoria*, for example, which also includes frogs adapted to habitats other than solely arboreal or solely aquatic. There are burrowing "treefrogs" too. Members of the Australian genus *Cyclorana* spend much of their lives underground, and one species is noted for the large amount of water it can store to tide it over periods of drought. These, and the North American *Pternohyla*, form a "cocoon" while underground for protection against desiccation.

The breeding habits of hylids are, with some notable exceptions, fairly conservative, and most species have aquatic tadpoles. Eggs of torrent-dwelling species are fixed firmly to rocks. Those in still water may be attached to aquatic vegetation, or spread in a thin film on the surface, thus ensuring adequate oxygen in warm waters suffering from oxygen depletion. A common tendency in regions of high humidity is to place the eggs on vegetation emerging from the water or even on leaves high in trees overhanging the water, so that when the tadpoles hatch they drop into the water. Some forest species breed in water-filled tree holes, and others in water-holding epiphytic plants such as bromeliads.

Tadpoles of many hylid species are midwater pond-dwellers, with laterally placed eyes (the better to see below as well as above) and a broad-finned tail tapering to a filamentous tip. These tadpoles may be seen hanging at an angle in the water with the tail tip vibrating while they filter microscopic food particles. Stream- and torrent dwelling tadpoles, such as those of *Nyctimystes* in New Guinea, are quite different: depressed body, low tail fins, dorsal eyes, and mouthparts formed into an oval sucker.

Several tropical American genera (all classified in the subfamily Hemiphractinae) have the peculiar habit of brooding the eggs on the female's back: in the genus *Hemiphractus*, the eggs are exposed on her back: in the genus *Gastrotheca*, they are completely enclosed within a pouch; and there's a whole range of body-behaviour adaptations in between. One species of this subfamily deserves special mention apart from its breeding habits: *Amphignathodan guentheri* is the only living species of frog with teeth in its lower jaw. Other frogs may be toothless or possess teeth on the upper jaw, or on the upper jaw and the roof of the mouth, and some have large fang-like bony projections at the anterior ends of the lower jaws, but this species is unique in having true teeth.

Fossil hylid frogs are known from the Paleocene of South America (more than 57 million years ago) and the middle to later Tertiary of North America, Europe, and Australia.

Glass Frogs

The glass frogs, family Centrolenidae, are a group of more than 120 species, most of them tree-dwellers, inhabiting moist forests from southern Mexico to Bolivia, plus southeastern Brazil-northeastern

Argentina. The "glass frogs" name derives from a scarcity of pigment in the skin of the abdomen, which makes the internal organs visible. Most of the species are small (maximum size about 30 millimeters, or $1^1/_4$ inches body length) and green. Two larger species attain 75 millimeters (3 inches). Three genera are recognized: *Centrolene*, *Cochranella*, and *Hylinobatrachium*.

Males of most centrolenids call from leaves overhanging streams and then remain near the eggs laid on leaves at these calling sites. Egg masses placed in such situations are safe from many predators but are parasitized by flies which lay their eggs on the mass so that the maggots consume the frog eggs. Frog larvae that survive the parasites fall to the stream, where they live in gravel or debris. These tadpoles are elongate with muscular tails and very low fins (broad fins are useful only to tadpoles that swim in open water). Some larger centrolenids live and breed in rocky waterfalls, where the egg masses are stuck to rock surfaces. No fossil centrolenid is known.

Dendrobatids

The family Dendrobatidae has some of the most colourful and interesting frogs. Most are rather small, the smallest less than 15 millimeters ($^1/_2$ inch) body length, although one species reaches 62 millimeters (almost $2^1/_2$ inches). There are more than 170 species in eight genera. They inhabit moist tropical regions in Central and South America, from Nicaragua to southeastern Brazil and Bolivia. Unlike the majority of frogs, almost all dendrobatids are diurnal (active during the day). They lay small numbers of eggs in moist sheltered places, and a parent (usually the male) guards the eggs. When the tadpoles hatch they wriggle onto the parent's back, are carried to water, and released to complete their development. Some species have a peculiar variant of this behaviour: the female releases the tadpoles into a water-holding bromeliad plant, and she returns occasionally to deposit an unfertilized egg as food for the tadpoles.

The dendrobatids fall into two main groups. One includes a large number of mostly dull-coloured specks of the genera *Colostethus*, *Mannophryne*, and *Nephelobates* which live alongside streams or on the forest floor and, with one known exception, are non-toxic. The second group consists of the poison frogs (genera *Dendrobates*, *Phyllobates*, *Epipedobates*, and *Minyobates*) which are very colourful and whose skin glands excrete alkaloid poisons that act on the nervous system. In addition there is the genus *Aromobates*, with a single species *A. nocturnus*, unique on several counts: it is the largest dendrobatid, it gives off a foul, presumably protective odor but is not poisonous, and it is a nocturnal stream-dweller.

The toxicity of the poisonous dendrobatids varies greatly from species to species. Presumably the poisons are a defense against predators, and the bright colour patterns act as a warning. One species deserves special mention: *Phyllobates terribilis* is so poisonous that it is unsafe even to handle. The toxins in one frog's skin would be sufficient to kill more than 20,000 laboratory mice, and less than 200 micrograms introduced into a human's bloodstream could be fatal. The toxicity of this species and others less-poisonous has long been known to certain groups of Indians of western Colombia, who use the frogs to poison their blow-gun darts (not arrows, as the popular literature often states). The toxin of *P. terribilis* is so abundant and potent that merely rubbing the point of a dart across a living frog's back is sufficient to make it deadly in hunting.

No fossils of dendrobatid frogs are known.

Ranids or "True" Frogs

The "true" frogs, family Ranidae, have the widest distribution of any frog family: North America (even in Alaska), Central America, and northern South America; Europe and across Asia sooth of the Arctic Circle, through the East Indies to New Guinea, the extreme north of Australia, and the Fiji islands; and most of Africa, and Madagascar.

The northernmost species of frog is the common frog of Europe, *Rana temporaria*, nearly matched by the moor frog *R. arvalis*. These two, and the wood frog *R. sylvatica* of North America, range north of the Arctic Circle. The largest frog, the goliath frog *Conraua goliath* of West Africa, is a ranid, but the family runs almost the gamut of body sizes. Ranid frogs are most diverse in Africa, where there are 16 endemic genera, whereas Asia has about 12 genera. Europe and the Americas have only *Rana* species, with none held in common.

The genus *Rana* includes about one-third of the 650-odd species credited to the family. These are the classic "frogs" as compared to "toads": typically living in and on the margins of water, they are relatively smooth-skinned, powerful jumpers with long legs and extensive webbing on the feet. In most cases the eggs are laid in the water, followed by a tadpole stage lasting several weeks. The ranids present many examples of adaptive radiation. Frogs of the Asian genus *Amolops* have enlarged finger and toe disks (like those of treefrogs), which facilitate clinging to rocks beside the swift-flowing streams where their tadpoles live. The tadpoles, in turn, avoid being swept away by fixing themselves in place with a ventral sucker, rather than a sucker-shaped mouth as in other tadpoles of fast waters.

The burrowing-frog niche is exploited by the genus *Tomopterna*, called sand frogs in South Africa. Like frogs of other families with similar habits, these are squat, short-legged, somewhat wrinkled animals with a prominent digging tubercle or spade on each hind foot. Also in South Africa is the Hogsback frog *Anhydrophryne rattrayi*, named for the area where it was first found. This small frog lives in the forest leaf-litter, where the male digs a nest chamber in moist soil with his nose. The eggs develop directly to froglets, so there is no need for frogs of this species to frequent streams or ponds. At the aquatic extreme are small frogs of the genus *Occidozyga*, which live in swamps and pools in Asian rainforests. Their large, fully webbed hind feet are better adapted to swimming than to leaping.

Because oceanic islands are typically barren of native frogs, the presence of two ranid species of the genus *Platymantis*—one a tree-dweller, the other terrestrial—in the Fiji islands is noteworthy. Frogs of this genus all live on islands, ranging from the southern Philippines to New Guinea, and eastward through the Solomon Islands before making the great jump to Fiji. Four closely related but morphologically diverse genera also inhabit the Solomons and share with *Platymantis* the direct mode of embryonic development. The history of how these frogs attained their present distribution can never be known for sure, but it undoubtedly involves passive distribution on islands moving very slowly over millions of years of tectonic activity. Numerous fossils from North America and Europe, none older than Oligocene (37 million years) are referred to the genus *Rana*.

Rhacophorid Treefrogs

The rhacophorids, most of them treefrogs, are relatives of the alrgely aquatic and terrestrail ranids, and inhabit temperate and tropical parts of Africa and Asia, including Madagascar and Japan. This is a family of modest extent, with about 300 species in 12 genera, ranging in size from 15 to 120 millimeters ($^1/_2$ to $4^3/_4$ inches) body length. The flying frog *Rhacophorus nigromaculatus* of Southeast Asia is a member of this family.

The African genus *Chiromantis* has some interesting adaptations. In addition to the digital disks common to all arboreal frogs, *Chiromantis* has the inner two fingers opposable to the outer two, providing a firm grip on twigs. The frogs have unusual resistance to desiccation and can spend dry periods fully exposed. They lay their eggs in a tree above water. As the eggs are produced, the mating frogs use their feet to heat the eggs and accompanying liquid into a froth which hardens, protecting the developing eggs. Several pairs of frogs may work together building a communal nest. The larvae remain for a time in the nest before dropping into the water to complete development. Foam nests feature in the life history of most rhacophorid frogs.

Small tree-dwelling animals are unlikely candidates for fossilization, and there are no paleontological records for the Rhacophoridaea—as indeed there are none for many other families.

Reed and Lily Frogs

The hyperoliids (Hyperoliidae) are another family of modest size—about 300 species, in 19 genera—related to the ranids. They are mostly small species, about 15 to 80 millimeters ($^1/_2$ to 3 inches) body length, living in Africa and Madagascar, with one endemic species on the Seychelles islands in the Indian Ocean. A modest adaptive radiation has produced tree-dwelling frogs and terrestrial frogs as well as a majority that climb but tend to remain for the best part of the time in low vegetation near water.

The genus *Hyperolius* includes half the species in the family. These small frogs live mostly in marshy or swampy areas, resting on reeds and sedges, on which many of the species deposit their egg masses. They are often boldly coloured and patterned, with considerable individual variation in markings. This, and a general similarity of body form and structure, make it difficult to distinguish between the species, but a knowledge of the mating calls helps biologists to differentiate between them.

The genus *Leptopelis* includes both relatively large tree-dwelling frogs and burrowing frogs that rarely climb. Species of the genus *Afrixalus* lay their eggs on a leaf and then fold the edges of the leaf together, cementing them over the egg mass with secretions from the oviduct. Curiously, this may be done either in or out of water, in the latter case, the hatchling lad poles must fall into the water to survive.

Hyperoliids are unknown as fossils.

Squeakers

The small family Arthroleptidae (about 70 species in seven genera), distributed in Africa south of the Sahara, is placed within the Ranidae by some authors. The voice of *Arthroleptis* species is the reason why people have named them "squeakers". These are small frogs that live on and within leaf-litter of the forest floor. The eggs, deposited in cavities or burrows in moist earth, undergo direct development; in some species the froglets are completely metamorphosed when they hatch, whereas in others the tail remains to be absorbed.

One arthroleptid is unique among all frogs. This is the so-called "hairy frog" *Trichobatrachus robustus* of Cameroon and Equatorial Guinea. In the breeding season, males develop vascularized hair-like structures on the flanks and thighs. They are reported to sit under water on egg masses in streams, and apparently the "hairs" serve to augment respiration through the skin, increasing the time the frogs can remain submerged.

No fossil arthroleptids are known.

Shovel-nosed Frogs

The shovel-nosed frogs, family Hemisotidae, of Africa south of the Sahara, are 11 moderate-sized species (up to 80 millimeters, or 3 inches body length) in the single genus *Hemisus*. They are odd-looking frogs, round-bodied with short legs and a small pointed head with the tip of the snout hardened. Shovel-nosed frogs are burrowers, generally living in open country near pools, and are seldom seen above ground. The eggs are laid in an underground cavity, and the female remains with them until they hatch. Using her snout, she digs a burrow leading to water nearby, and the larvae then swim out and assume a more normal tadpole existence. Their bizarre appearance notwithstanding these frogs are thought to be related to the ranids and are treated as a subfamily of Ranidae by some authors.

Microhylid Frogs

The family Microhylidae occurs in the Americas from the southern United States to Argentina, equatorial and southern Africa, and eastern India and Sri Lanka, through Southeast Asia to New Guinea and northern Australia. More than 320 speeies are recognized in 66 genera, the largest number of genera of any family of frogs. Microhylids comprise almost half the species of frogs in New Guinea and a sizable proportion in Madagascar, but are less significant elsewhere. Most are small frogs—several species are less than 15 millimeters ($^1/_2$ inch) body length—but others reach 80 to 90 millimeters (3 to $3^1/_2$ inches). Morphology varies greatly, from rotund burrowers to typical treefrogs. A majority of the species live in moist tropical regions, but the evolutionary radiation of the group has placed species in arid habitats as well, and in a variety of terrestrial and arboreal niches. Some species are streamside frogs, but there do not seem to be any primarily aquatic microhylids.

Ground-dwelling microhylids live in both arid and humid tropical habitats. Among the most peculiar in arid regions are the rain frogs of the African genus *Breviceps*. (In arid regions they are likely to be seen only when it rains.) These frogs have small heads, short limbs, and round bodies, a shape accentuated by their habit of puffing up with air when disturbed. The arms are so short that the male can nor clasp the female around the body when mating (the usual method of maintaining contact between mating frogs), so instead the male and female become stuck together by secretions from skin glands on the male's ventral surface, giving an effect not unlike that of two golf balls glued together. The eggs are laid in an underground chamber prepared or enlarged by the female. The tadpoles do not feed but live on yolk provided in the egg, and the female remains with the nest until the tadpoles metamorphose and leave.

The moist leaf-litter of tropical forests is prime habitat for many microhylid species. Some burrow in the deep litter or soil and rarely emerge on the surface. Others come to the surface at night to wait for or actively seek food. The litter also serves as a daytime retreat for small climbing species that ascend into low vegetation at night to feed and advertise for mates. Microhylids with wide bodies but narrow, pointed heads commonly feed on termites and ants. Other species with more normal frog proportions have the catholic tastes of other frogs; one New Guinean species even eats other frogs. Many microhylids, notably in Madagascar and New Guinea, are tree-dwellers.

All the 120 or so species of microhylids found in New Guinea and Australia have direct development, skipping the tadpole stage. The tree-dwellers therefore do not need to leave the trees to seek pools or streams in which to breed, but may find appropriate arboreal sites—such as an epiphytic plant called the ant plant, which has a chamber that holds moisture.

Microhylid tadpoles differ from those of other families in features of their anatomy. With rare exceptions, they lack the horny beak and denticles ("teeth") of other frog larvae. In some species the mouth is formed into a funnel shape and is used, from below, in ingesting food from the water surface. (This adaptation occurs also in tadpoles of other families.)

The only pre-Pleistocene fossils of this family are from the early Miocene of Florida, about 24 million years ago. They are classified in the living genus *Gastrophryne* which inhabits this region near the present-day limits of the distribution of microhylids.

Four South American Families

Some small families of frogs (lacking fossil records) are recognized not so much because of their distinctiveness but because their species cannot be fitted unambiguously into any of the larger groups. Four South American families—Allophrynidae, Brachycephalidae, Rhinodermatidae, and Pseudidae—are examples. *Allophryne ruthveni*, the only member of its family, is a small arboreal frog of northern South America variously considered most closely related to bufonids, leptodactylids, or hylids.

The three species of brachycephalid frogs live in the Atlantic forests of southeastern Brazil. They are very small frogs—*Psyllophryne didactyla* grows to less than 10 millimeters (less than $^1/_2$ inch) body length, so is not only one of the two smallest frogs, but is also one of the smallest four-legged animals. The two species of the genus *Brachycephalus* are only slightly larger. Tiny frogs tend to have fewer digits than usual, and brachycephalids, for example, have only three functional toes on each foot. At least one (and probably all three) species of brachycephalids have direct embryonic development, hatching as tiny frogs. In the family *Rhinoderma* are noted for their habit of oral brooding. They are small ground-dwelling frogs, about 30 millimeters ($1^1/_4$ inches) long, found in the cool temperate forests of southern Chile and adjacent Argentina.

The paradox frog *Pseudis paradoxa* gained its name because the tadpoles can reach remarkably large size, up to 250 millimeters (10 inches) in length, yet after metamorphosis the largest the frogs get is about 70 millimeters ($2^3/_4$ inches). The five species in the family Pseudidae—two *Pseudis* and three *Lysapsus*—are almost totally aquatic, although an ability to survive dry periods buried in mud has been reported for the paradox frog. The family ranges through tropical lowlands over much of northern and eastern South America from Colombia to Argentina.

Ghost Frogs and Seychelles Frogs

There are another couple of small families whose evolutionary affinities have been disputed. The so-called "ghost frogs", family Heleophrynidae, comprise the genus *Heleophryne* with five species confined to the Cape and Transvaal regions of extreme southern Africa. The common name may have been coined because one of the species is found in a place called Skeleton Gorge; certainly the frogs are not vaporous or otherwise ghostly.

Heleophrynid frogs are up to about 60 millimeters ($2^1/_4$ inches) long and are rather flattened, with prominently enlarged tips to the fingers and toes. They are therefore well adapted to fit into crevices and cling to rock surfaces along the cool, shaded mountain streams that are their habitat and where their tadpoles live. Like other tadpoles adapted for life in swift-flowing water, the tadpoles of ghost frogs have their mouthparts modified into a large sucking disk which allows them to cling to slippery rocks while feeding.

The suggestion that the heleophrynid frogs should possibly be classified within the Australian family Myobatrachidae implies a relationship going back many millions of years to when Africa and Australia were part of the Gondwanan supercontinent.

Evidence from chromosomes and behaviour suggests that the family Sooglossidae, of the Seychelles, may also be related to the Myobatrachidae of Australia. There are three species—two in the genus *Sooglossus* and one in *Nesomantis*. They are small terrestrial frogs, up to 40 millimeters ($1^1/_2$ inches) body length, and they diviate from typical frog behaviour in their method of breeding. Eggs are laid on the ground rather than in water and follow two modes of development: direct to small frogs in one species of *Sooglossus*; in the other species, tadpoles are carried on the back until they metamorphose. In *S. sechellensis*, the tadpoles are carried not by the male, as is usual, but by the female.

Conservation

A few species of frogs are listed as endangered by one agency or another, but it is not individual species so much as endangered habitats that need to be conserved. Destruction of rainforests in tropical regions has undoubtedly eliminated many species of frogs before they even became known to scientists. Wetlands in temperate areas and isolated sources of water in arid regions also merit special attention. Island faunas, too, are especially vulnerable to habitat destruction. Even where no specific cause can be identified, there are many instances of species apparently having disappeared or virtually so—for example, gastric brooding frogs in Australia and frogs of the family Ranidae native to southern California.

10

SALAMANDERS AND NEWTS

Salamanders have been the subject of countless myths and legends since remote times. The origin of the salamander name is an Arab-Persian word meaning "lives in fire". In fact, until a few centuries ago it was believed that the black and yellow fire salamander *Salamandra salamandra* of Europe could pass unscathed unscathed through flames, a belief still held in some areas. Such modern names as the English "fire salamander" and the German "*Feuersalamander*" came from this ancient legend. In recent times salamenders have become a favourite of terrarium and aquarium keepers around the world because of the ease with which they can be raised in captivity.

What in a Name?

Salamanders belong to the order Caudata (from the Latin word *caudatus*, meaning "provided with a tail") and are also referred to as caudates; as the scientific name implies they retain a tail even after matamorphosis from the larval stage. In this they differ from other amphibians—caecilians have not ail or only a rudimentary one, and frogs and toads have a tail only in the alrval stage.

"Salamander" is a broad term applicable to any member of the Caudata, whereas "newt" is more restrictive. Members of 10 genera *Cynops*, *Euproctus*, *Neuregus*, *Notophthalmus*, *Pachytritonm*, *Paramesotriton*, *Pleurodeles*, *Taricha*, *Triturus*, and *Tylototriton*, which are mostly aquatic, at least during the breeding season, are commonly called newts; the others, whether amphibious or exclusively aquatic or exclusively terrestrial, retain the salamander name. "Newt" has a curious origin: it derives from the Anglo-Saxon word *efete* or *evete*, used to indicate newly metamorphosed newts, becoming *ewte* in Middle

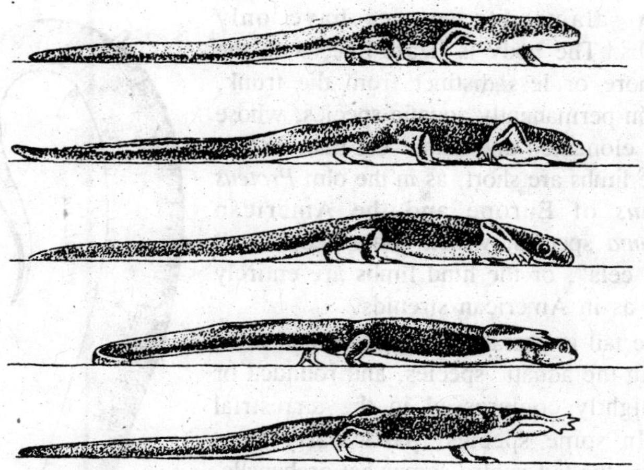

Fig. 10.1. A newt, Triturus.

English, and then *newte*: an ewte = a newte. Salamanders and newts are found almost exclusively in the Northern Hemisphere—in Europe, central and northern Asia, northwestern Africa, North America, and Mexico. In the Southern Hemisphere there are only a few members of the families Hynobiidae and Salamandridae in Southeast Asia, as well as the Chinese giant salamander, and species of lungless salamanders are found as far south as central Bolivia and southern Brazil. Caudates are found from sea level to about 4,500 meters (14,700 feet), wherever there is sufficient humidity during at least part of the year. Some specialized species dwell entirely in subterranean waters, others are tree and/or rock dwellers.

There are very few fossil remains testifying to the origin of this order. The earliest known true salamanders are three specimens dating back to the late Jurassic, about 150 million years ago. In 1726 the fossil remains of a giant salamander were curiously attributed by J.I. Scheuchzer to that of a human "witness to the Universal Flood", named *Homo diluvii testis*, and it was not until a century later that the remains were correctly identified and named *Andrias scheuchzeri*.

Like Lizards without Scales

Superficially salamanders resemble lizards but are immediately distinguished by their complete lack of scales. Even Linnaeus, the father of modern taxonomy, erroneously assigned some species of salamanders to the lizard genus *Lacerta*. Salamanders also resemble lizards in usually laving four limbs; but again, like some lizards, certain salamander species have only forelimbs. The body is lizard-like, with the head more or less distinct from the trunk, except in permanently aquatic species, whose head is elongated and rather eel-like; in this case the limbs are short, as in the olm *Proteus anguinus* of Europe and the American *Amphiuma* species, which some people call "congo eels", or the hind limbs are entirely absent, as in American sirenids.

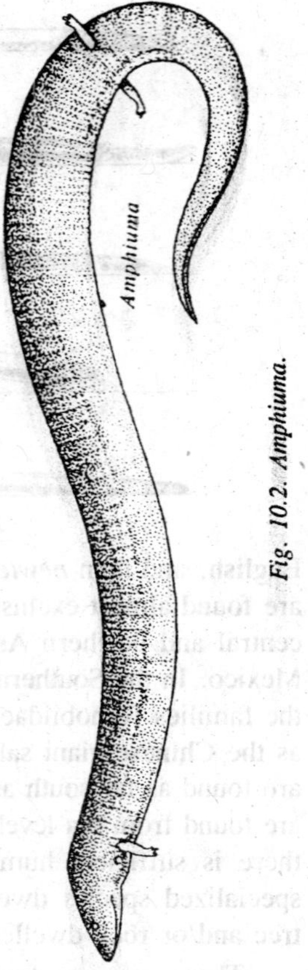

Fig. 10.2. Amphiuma.

The tail is laterally compressed and often rested in the aquatic species, and rounded or only slightly compressed in the terrestrial ones. In some species, particularly tree-dwellers, the tail is also somewhat prehensile. When threatened by a predator, some lungless salamanders do what many lizards can do: voluntarily sever their tail (known as autotomy) as a decoy; the tail then regenerates. In most species the adults are 10 to 20 centimeters (4 to 8 inches) total length.

Skin

Salamanders and newts continue to grow even after reaching sexual maturity, which is one of the reasons why the superficial horny layer of the skin is periodically shed. Depending on the species, the old skin comes off (as frequently as once a week in some cases) either in fragments or in one piece. This slough, or exuvia, is then usually devoured by the salamander.

Unlike that of reptiles, the skin of salamanders contains three different types of glands distributed rather uniformly over the entire body surface: mucous, granular, and mixed. The secretion of the mucous glands, and the mucous part of the mixed glands, is homogenous and frothy, neutral or alkaline, and sticky. On land its primary function is to protect the skin from drying out, permitting respiratory exchanges which could not occur through a dry surface. In water, it helps to maintain the body's internal osmotic pressure (the salt and water balance in the body fluids) and simultaneously acts as a lubricant during swimming. The granular glands, and the granular part of the mixed glands, produce a granular secretion containing various types of poisons and often giving off a specific odor; they are located especially on the upper part of the head behind the eyes (paratoid glands), on the tail, and on the sides of the back. The mixed glands, as the name implies, produce both a mucus and a granular secretion and are located over almost the entire body. Some members of the families Plethodontidae, Ambystomatidae, and Salamandridae possess a fourth type known as the hedonic glands, which resemble the granular and mixed glands and are located in various zones. Their secretion, containing pheromones, plays an important role during courtship and mating; particularly noticeable is the chin gland of most plethodontid males.

In some species the skin is smooth and in others it is bumpy, or even wart-like at the poison gland outlets. The skin of amphibious species is smooth when the animals live in the water but can become roughish when they live on land. The skin is rich not only in glands but also in pigment cells. In many species the body colour is brownish, yellowish, or grayish, sometimes with a barely contrasting pattern. Several have gaudy colouration on their back or underside throughout the year in both sexes (as in the fire salamander and various lungless salamanders), or else only in the males, particularly during the season of courtship. The courtship period is also the time when many male newts develop conspicuous skin folds over their body, tail, fingers, and toes. Species that live permanently underground lack pigment and thus are white or pink. The colour of the larvae usually differs from that of the adults.

With the exception of the two species of the Asian genus *Onychouactylus*, whose toes have a horny sheath which turns into a little blackish claw at least during the mating season, salamanders and newts do not have nails.

Skeleton

Some parts of the skull begin as cartilage and then turn into bone during growth. Others, not derived from cartilaginous elements, are known as membranous bones. The shape, number, and arrangement of these two types of bones are important keys for classification.

The spine is generally divided into five regions: cervical, dorsal, sacral, sacro-caudal, and caudal. The cervical region consists of a single vertebra joined to the skull at four points (instead of two points in all other amphibians). The dorsal region usually has 13 to 20 vertebrae, although the minimum number is 11 (in the mole salamander *Ambystoma talpoideum*) and the maximum 63 (in *Siren* species). The

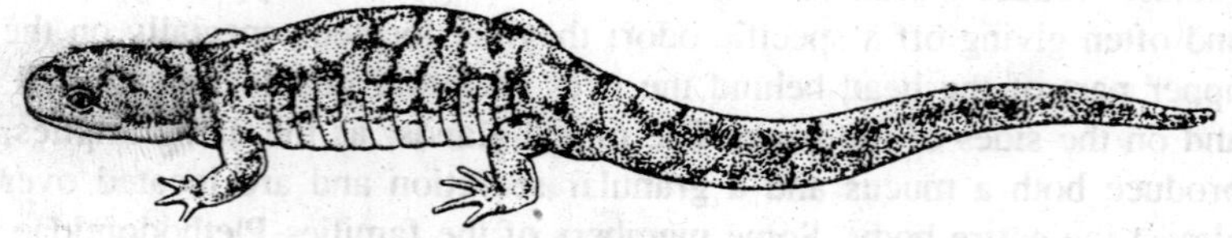

Fig. 10.3. Ambystoma.

sacral region consists of one vertebra; and the sacro-caudal from two to four. The caudal (tail) region varies from 20 vertebrae to more than 100 in some *Oedipina* species of Central and South America.

Those with an eel-shaped body, such as *Siren*, *Proteus*, and *Amphiuma*, have ribs corresponding with only the dorsal vertebrae closest to the head, but in other species the ribs occur along all or almost all the dorsal tract and sometimes also in the sacro-caudal region. In two types of newts, the sharp-ribbed newt *Pleurodeles waltl* and the alligator newt *Echinotriton andersoni*, the tips of the ribs can actually perforate the skin, increasing the likelihood that the newt's poison will enter the body of any predator.

The limbs have the basic structure common to all vertebrates; the front legs are linked to a thoracic girdle which does not articulate with the vertebral column, and the back legs to a pelvic girdle articulating with the single sacral vertebra. Most species have four fingers on each forelimb and five toes on each hind limb.

Digestive System

Salamanders have no salivary glands. Those that spend all their life in water have what is called a "primary" type of tongue, a fleshy fold on the floor of the mouth with very little mobility because it has no intrinsic muscles; it is not used to capture prey. All other salamanders have a fairly mobile and well-developed tongue, which is used—when they hunt on land—to procure food. In some lungless salamanders it is mushroom-shaped and can be darted onto the prey, just as chameleons do. The teeth, usually small, are implanted on the margins of both the upper and lower jaws, as well as on the roof of the mouth; the sirenids have them only on the latter. The teeth of some lungless salamander males have a sexual function (discussed later). Salamander larvae also have true teeth, in contrast to those of frogs and toads which have only horny structures.

The esophagus is fairly short and leads into the stomach with no regional differentiation and then an almost-straight intestine with little distinction between the small and large components. The terminal part of the rectum enlarges to from a cloaca, which contains the outlets of the urinary and reproductive tracts; its opening is located underneath the base of the tail.

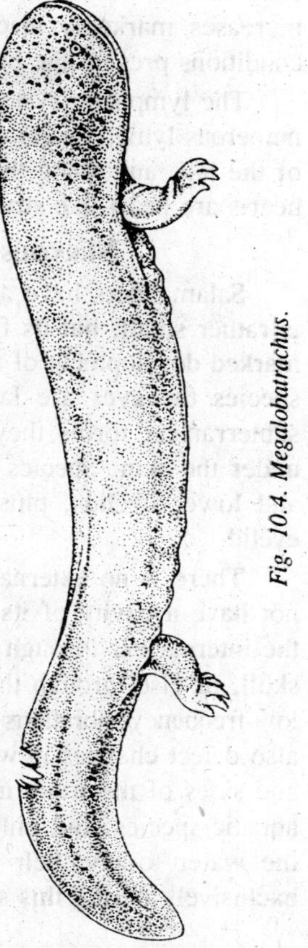

Fig. 10.4. *Megalobatrachus*.

Respiratory and Blood Circulation

The lungs of salamanders (in species that have them) are sac-like. Usually they are identical in size, although in some species the right lung is slightly shorter, and in amphiumids the left lung is rudimentary. The lungs of the spectacled salamander *Salamandrina terdigitata* and a few other genera, such as *Euproctus*, are tiny. All plethodontids and some hynobiids are lungless; respiratory exchanges occur only through the skin and through the mucous membranes of the mouth and throat. Respiration through the skin also plays an important role, at times essential, even in species with lungs, for instance when they hibernate underwater. The olm and other species that never lose their gills (known as perennibranchiates), as well as larvae, and neotenics (those species which occasionally retain larval features in the adult form—see page 505) can breathe either through their skin or through external gills which stick out in bright red tufts on the sides of the head. Only amphiumas have internal gills. The colour of the gills is due to the rich supply of blood, and respiration through the skin is enhanced by the marked vascularization of the skin.

The heart consists of a single ventricle in which the arterial and venous blood mix only partially. The left and right auricles are distinct in the lunged forms, and not completely separate in the perennibranchiate and lungless forms. The latter obviously do not have lung veins, and they have a smaller left auricle. Red blood cells are large (the eel-like amphiumas of the southeastern United States have the largest of all vertebrate animals), usually elliptical, and usually have a nucleus. According to the species, their number ranges from about 30,000 to 100,000 per cubic millimeter of blood; but as recently discovered in

Triturus newts, the number of circulating red blood cells probably increases markedly also in other salamanders when environmental conditions prevent the blood from being sufficiently oxygenated.

The lymphatic system is fairly well developed. Salamanders have numerous lymph hearts under the scapula (shoulder-blades), at the root of the tail, and along the body flanks just under the skin. The lymph hearts are heart-like structures that improve the lymphatic circulation.

Nervous System and Sense Organs

Salamanders have a relatively simple nervous system. The brain is rather small, but its front lobes are well developed, paralleling the marked development of the olfactory organs (detecting smell). In most species the eyes are large, with round pupils, but in permanently subterranean forms they are reduced in size and sometimes hidden under the skin. Species with well-developed eyes usually have upper and lower eyelids, plus a nictitating membrane known as the third eyelid.

There is no external ear, only a vestigial middle ear which does not have a cavity of its own. Vibrations are probably transmitted to the internal ear through the jaws or the bones connecting them to the skull, or else through the forelimbs and pectoral girdle. In the water, low-frequency vibrations are also perceived by lateral line organs (which also detect changes in water pressure and currents) found on the head and sides of the trunk in all larvae and in adults of many prevalently aquatic species. But only female spectacled salamanders, which enter the water to lay their eggs, develop lateral organs that are active exclusively during this short period.

Reproduction

Some salamander and newt species are exclusively aquatic, and some exclusively terrestrial. Ohters are clearly amphibious, and the females, or both sexes, return periodically to water to reproduce. This usually occurs in the spring and can involve large migrations.

In the most primitive families, Cryptobranchidae and Hynobiidae, males fertilize the eggs outside the female's body. Fertilization is internal in other salamanders, although for sirenids observations still have to be confirmed. In species with internal fertilization the male deposits a gelatinous spermatophore (a capsule containing sperm) during mating, which can have a shape peculiar to the species or genus, with a whitish mass of sperm adhered to its apex. The spermatophore, or only the sperm cap, is sucked in by the cloacal lips of the female.

Ovulation and fertilization may either follow insemination or be delayed for several months, in some cases up to two and half years, as in the fire salamander. When fertilization is delayed, sperm are stored in special diverticula of the female genital tract and subsequently fertilize the eggs at the time of ovulation.

Rutting takes place on land in exclusively terrestrial species and in those such as the fire salamander and spectacled salamander, whose females go to the water only to lay their larvae and eggs, respectively. Rituals linked to courtship, such as nuptial dances, are fairly complex and differ from species to species. It can involve, as in *Euproctus* newts, a sort of embrace which is obviously not accompanied by copulation, as salamanders have neither a penis nor any other form of intromittant organ.

The eggs do not have a protective shell but are normally surrounded by a gelatinous layer whose development and consistency differ from species to species. On land eggs can be laid in several different places

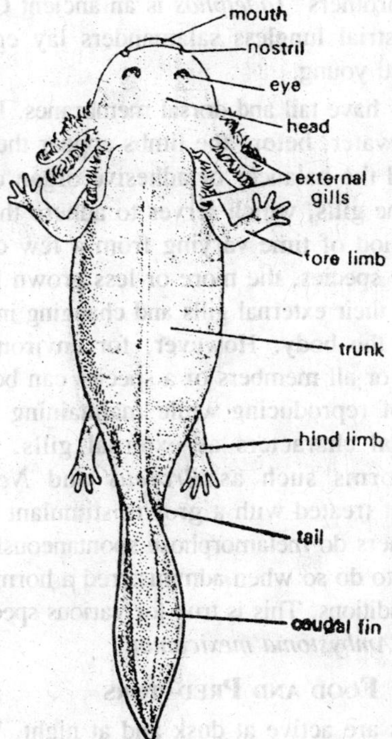

Fig. 10.5. Axolotl larva.

as long as there is sufficient humidity and protection. In water they are attached to rocks or submerged logs or roots. Some tree-dwelling lungless salamanders lay their eggs in bromeliads, in the water cupped in the base of the leaves. Some species, such as the olm, have parental care, a task usually carried out by the female but also by the male in the giant salamanders, the mud salamander *Pseudotriton monlanus* and the clouded salamander *Hynobius nebulosus*. The number of eggs depends on the species and the size of the individual, varying from five to six in some terrestrial salamanders to a maximum of about 5,000 in some *Ambystoma* mole salamanders which lay their eggs in water.

Several salamanders are ovoviviparous—instead of laying eggs the female gives birth to well-developed larvae or to young that almost perfectly resemble the adults. The fire salamander usually gives birth to well-developed larvae but can in some regions give birth to already metamorphosed young. In some ovoviviparous species adelphophagy takes place within the maternal genital tract; this consists of the more-developed embryos absorbing the eggs and smaller embryos, thus feeding on their siblings or "brothers" (*adelphos* is an ancient Greek word for brother). Many terrestrial lungless salamanders lay eggs that thatch already metamorphosed young.

The larvae usually have tail and dorsal membranes. In some species that reproduce in the water, before the limbs appear the larvae have a stick-like organ called the balancer or adhesive organ on the sides of the head in front of the gills, which serves to adhere the larvae to the substrate. After a period of time varying from a few days to several years according to the species, the more-or-less grown larvae undergo metamorphosis, losing their external gills and changing in other external and internal parts of the body. However, for environmental and/or genetic reasons, some or all members of a species can become sexually mature and capable of reproducing while maintaining such larval or juvenile morphological characters as external gills. The so-called perennibranchiate forms such as *Proteus* and *Necturus* never metamorphose, even if treated with a growth stimulant such as iodine or thyroid extract. Others do metamorphose spontaneously after a short time or when induced to do so when administered a hormonal treatment under experimental conditions. This is true for various species of *Triturus* newts and the axolotl *Ambystoma mexicanum*.

Food and Predators

Most salamanders are active at dusk and at night. When it is too cold or too dry many species take refuge under rolling vegetation or deeply buried rocks, in rock crevices or deep in the ground, becoming active only when the external environment is again suitable. They are all carnivorous and usually feed on small invertebrates such as insects, spiders, crustaceans, mollusks, and worms. The larger species also prey on small vertebrates. Cannibalism is not rare and may occur in both larval and adults.

Unlike frogs and toads, salamanders do not usually make any sound as they have no larynx, or only a rudimental one, and no vocal cords. However, if disturbed or excited some species are capable of producing a weak squeak (the fire salamander, *Aneides*, *Dicamptodon*, and some species of newt), a faint yelp (sirenids) or a tiny shout (*Pleurodeles* newts). Salamanders are preyed upon by other amphibians and by water tortoises, snakes, lizards, fish, birds, mammals, and large invertebrates. When in danger they defend themselves by secreting toxic or sticky substances, or by resorting to autotomy of the tail (as in some lungless salamanders) or by assuming a typical reflex posture which differs according to the species. One posture involves arching the body with the tail perpendicular to the body (or sometimes rolled up) to show off the gaudy ventral colouration that seems to warn the predator of a toxin in the skin. This particular posture is known as the unken reflex.

To humans the economic importance of salamanders is insignificant and is virtually limited to species of particular beauty and/or rarity, which are sought after as pets. Some species are used in biological research. The giant salamanders from Japan and China and the axolotl (the neotenic form of the mole salamanders) are appreciated as food, and the latter also as a supposed aphrodisiac. Others are used as live bait by fishermen, which has caused a great reduction in many populations of the lungless seal salamander *Desmognathus monticola* throughout the southeastern United States, and in some places have pushed them to the edge of extinction. The hynobid *Batrachuperus pinchonii* of eastern China is venerated as a divinity under the name "White Dragon", and the pool on Mount Omei where the species lives is the destination of pilgrimages, so much so that a sanctuary has been built nearby. This does not, however, prevent it from being dried and ground to a powder as a remedy for stomach disorders, a fate suffered also by its close relative, the Japanese clawed salamander *Onychodactylus japonicus*, which in traditional Japanese medicine has been used to rid the patient's body of worms.

Humans have had a direct impact on some populations through the capture of certain species for sale or study, and an even greater impact indirectly through the capt8ure of certain species for sale or study, and an even greater impact indirectly through the disruption or destruction of their habitat. Predation, especially on larvae, by fish stocked in pools for sport fishing has a very deleterious effect, but an even worse threat may come from acid rain.

Most Primitive Families

The suborder Cryptobranchoidea includes the most primitive living salamanders, the only ones that have external fertilization. Their eggs are always deposited in two groups, each contained in a gelatinous sac. In the adults, two bones of the lower jaw (the angular and prearticular) are clearly separated from each other. In other respects the two families of this suborder, Cryptobranchidae and Hynobiidae, differ greatly in appearance.

Hellbenders and Giant Salamanders

The cryptobranchids (family Cryptobranchidae) ways live in running water. They are corpulent and have large skin folds along the flanks, which increase the body surface and thus enhance the absorption of oxygen from the water. Their metamorphosis is incomplete, so the adults still have gill slits or grooves and no eyelids. All species have four fingers on each forelimb and five toes on each hind limb.

The hellbender *Cryptobranchus alleganiensis*, which can reach a total length of 75 centimeters (30 inches), inhabits central to northeastern United States. The genus *Andrias* is today represented by the Chinese giant salamander *A. davidianus*, growing to 180 centimeters (almost 6 feet), and the Japanese giant salamander *A. japonicus*, whose length never exceeds 150 centimeters (almost 5 feet). They feed on various invertebrates such as crustaceans but occasionally also eat small aquatic vertebrates. Each female lays up to 450 eggs, in paired rosary-like strings. The male releases his sperm on the eggs and seems to guard them until they hatch 10 to 12 weeks later.

Hynobiids

In the family Hynobiidae none of the 36 species grows larger than 25 centimeters (10 inches) total length. They live in Asia, with the exception of the Siberian salamander *Salamandrella keyserlingi*, which has spread westward as far as European Russia. The family consists of seven genera: *Batrachuperus*, *Hynobius*, *Liua*, *Onchodactylus*, *Pachyhynobius*, *Ranodon*, and *Salamandrells*. They all reproduce in water (for example, streams, ponds, and tarns) but outside the breeding season are terrestrial. The female lays two spindle-like capsules, each containing 35 to 70 eggs, according to the species. Observers have often seen examples of parental care.

Some *Hynobius* and *Salamandrells* and all the *Batrachuperus* species have four-toed feet. Other types have five toes on each hind limb.

Eight Advanced Families

The suborder Salamandroidea, which includes the most advanced salamanders, differs from the primitive families in having the angular bone fused with the prearticular bone in the lower jaw, at least in the living species. With the probable exception of sirens, the sperm are taken into the female's body to fertilize the eggs.

There are eight families in this suborder, with some 50 genera and a total of about 380 species. Members of this suborder occur everywhere salamanders exist: all regions of the world except Antarctica, Australasia, Oceania, and Africa south of the Sahara.

Sirens

The family Sirenidae has so many peculiar characters that some authors classify it in a separate suborder, Sirenoidea, or even a different order, Trachystomata. Being neotenic (see page 505), sirens have gills throughout their lifetime, and they lack eyelids; they have small eyes, no hind limbs and an eel-shaped body.

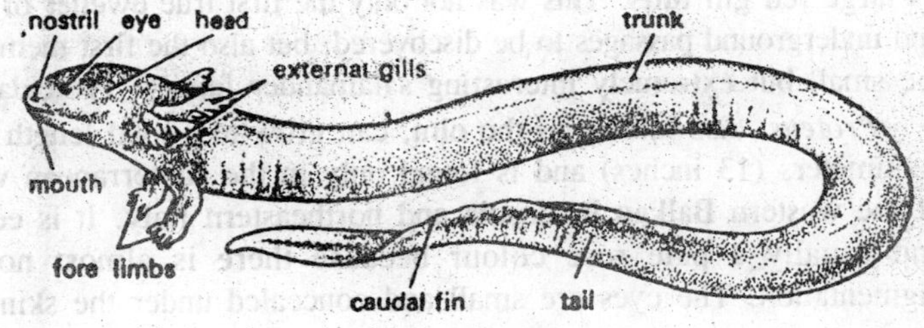

Fig. 10.6. Siren.

Sirens live in southeastern United States and adjacent regions of Mexico, spending their life in ponds and swamps with rich aquatic vegetation and a muddy bottom; like eels, they can cover short distances on land at night during rainy periods. The greater siren *Siren lacertina*, which grows to 95 centimeters (37 inches) total length, and the lesser siren *S. intermedia*, reaching 68 centimeters (27 inches) have three pairs of gill slits and four fingers on each hand. The two dwarf sirens, genus *Pseudobranchus*, which grow to 25 centimeters (10 inches), have only one pair of gill slits and three-fingered hands.

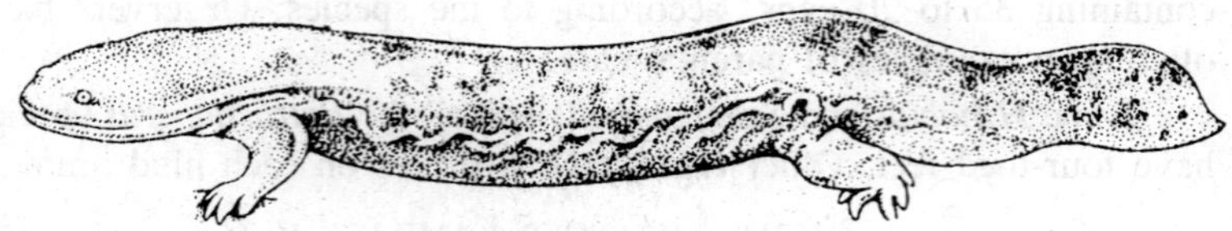

Fig. 10.7. Cryptobranchus.

Lillie is known about their reproductive biology, but judging by their uro-genital apparatus, one may suppose that they have external fertilization; their eggs have been found isolated or in small clumps attached to submerged plants. Like the lungfishes, if their habitat dries they can survive for weeks or months embedded in the mud, enveloped by a kind of mucus cocoon with only the tip of the snout jutting out.

Waterdogs, the Mudpuppy, and the Olm

Proteus, the sea god who had the power of assuming whatever shape he pleased, inspired the Viennese naturalist Laurenti in 1768 to use the name *Proteus anguinus* for a curious amphibian discovered 24 years earlier near Ljubljana (now in Yugoslavia). Its discoverer, Baron J.W. Valvasor, had considered it to be the juvenile form of an animal destined to change into a dragon. The nobleman was actually not far from the truth, as the little animal, fortuitously swept above ground by the flood of a subterranean river, really looked like a larva with its large red gill tufts. This was not only the first true dweller of caves and underground passages to be discovered, but also the first member of the small but extremely interesting salamander family, Proteidae.

Proteus, also known as the olm, can grow to a total length of 33 centimeters (13 inches) and is found only in the subterranean waters of the western Balkan Peninsula and northeastern Italy. It is eel-like and usually a pale rose colour because there is almost no skin pigmentation. The eyes are small and concealed under the skin. The

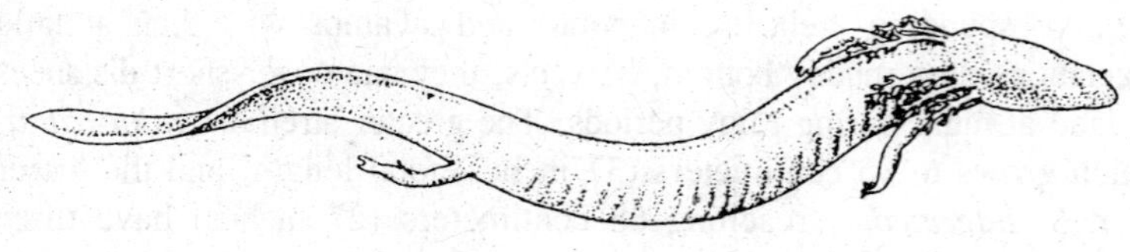

Fig. 10.8. Proteus.

forelimbs and hind limbs bear three and two digits, respectively. Larvae and juveniles, darker and with larger exposed eyes, are noteworthy for feeding on bacteria, protozoans, and organic matter contained in the slime. The family includes five other species, all in genus *Necturus*, which inhabit central and eastern United States. Because of the popular misconception that they can bark, four are called waterdogs, and these do not exceed a total length of 28 centimeters (11 inches); the fifth is the mudpuppy *N. maculosus*, sometimes a little more than 40 centimeters ($15^3/_4$ inches) long. All members of the Proteidae family are neotenic and consequently never lose their gills; they have two gill slits. They also lack eyelids and upper jaws. The eggs, which are tended by the parents, are attached to submerged stones or logs.

Three Families of Mole Salamanders

The mole salamanders spend their life almost entirely underground, emerging from their subterranean world only to reach the ponds or streams in which they reproduce. The breeding season is usually spring, but at least one species, the marbled salamander *Ambystoma opacum*, produces during fall. Fertilization, often preceded by a nuptial dance, is immediately followed by the laying of up to 200 eggs, generally in water but alternatively buried where rising water will flood the nest.

The family Ambystomatidae comprises about 30 species in the genus *Ambystoma*; they are found from southern Alaska and Canada throughout the United Stales and most of Mexico. Pacific mole salamanders are members of the family Dicamptodontidae, with three species in the genus *Dicamptodon* and of the family Rhyacotritonidae, with four species in the genus *Rhyacotriton*, all confined to northwestern United States. These three families differ in that two (Dicamptodontidae and Rhyacotritonidae) have a lacrymal bone, while Ambystomatidae members do not. They also differ in their habitat preferences during the breeding season: the former prefer mountain brooks, while the Amhystomatidae prefer still water such as lakes. Some species, such as the axolotl, are neotenic.

Two species, *Ambystoma laterale* and *Ambystoma jeffersonianum*, hybridize in eastern North America. It was thought that this hybridization had produced two all-female species that reproduced by parthenogenesis. However, later evidence indicates that hybrid populations are maintained by breeding with one of the diploid bisexual species, and the hybrid offspring are the result of gynogenelic (that is, they develop without the male's chromosomes entering the genetic complement of the new individuals) or hybridogenetic reproduction.

The resulting hybrid populations are thus quite genetically diverse and not self-maintaining, so do not meet the criteria for distinct species. All mole salamanders have a rather squat body, less than 35 centimeters (13 inches) in total length. Some have bright colour patterns, which contrast with the dark colour of the ground.

Amphiumas

If it were not for the presence of four tiny limbs, the amphiumas of the southeastern United States could easily be mistaken for eels, especially as they are neotenic and thus live mostly in water and, like eels, burrow in the mud; they can also move across wet ground. Adults have one pair of gill slits and inner gills, characters that increase their resemblance to eels. They have no eyelids or tongue. They lay their eggs under different kinds of shelters or on wet mud, in long strings each containing up to 150 eggs or more. The female remains coiled around them until they hatch, which takes about five months. On hatching the larvae must make their way to bodies of water, usually when it rains. The family (Amphiumidae) includes three species: the one-toed amphiuma *Amphiuma pholeter*, at 30 centimeters (12 inches) total length; the two-toed amphiuma *A. means*, at 116 centimeters (46 inches); and the three-toed amphiuma *A. tridactylum*, at 106 centimeters (42 inches).

Salamandrids

The family Salamandridae includes some 60 species in 15 genera. Members of 11 genera more or less linked to water are commonly called "newts". The other four genera, predominantly or exclusively terrestrial, have the popular name "salamanders" in English, and are found in Europe, northwestern Africa, and southwestern Asia. The gold-striped salamander *Chioglossa lisitanica* is endemic to Spain and Portugal, while the spectacled salamander is endemic to peninsular Italy. The genus *Merteniella*, with two species in Turkey, some of the Greek islands, and the Caucasus, has males that bear a spur on the tail base. In the genus *Salamandra* there are two mountain-dwelling species, which give birth to perfectly metamorphosed young—the black, or black and yellow Alpine salamander *S. atra* from northern Albania to the western Alps, and Lanza's salamander *S. lanzai*, always black, endemic of the southwestern Alps—as well as the largest member of the family, the fire slamander *S. salamandra*. This magnificent black and yellow animal, sometimes longer than 30 centimeters (12 inches), has different subspecies throughout most of Europe extending to Iran and northwestern Africa.

Among the newts only three species of *Notophthalmus* and three of *Taricha* inhabit North America, the black spotted newt *Notophthalmus meridionalis*, the striped newt *N. perstriatus*, and the eastern newt *N. viridescens*, inhabit eastern North America, while the alrger (up to 22 centimeter or $8^3/_4$ inches) rough-skinned newt *Taricha granulosa*, red-bellied newt *T. rivularis* and Californian newt *T. torosa* may be found only in the marginal western United States. Two *Pleurodeles* species, live in Europe or Asia. The distribution of the genus *Euproctus*, sometimes referred to as brook salamanders, is interesting to biogeographers: while *E. asper* occurs in the Pyrenees mountains between France and Spain, *E. montanus* is endemic to Corsica and *E. platycephalus* to Sardinia, both islands of continental origin which detached from the south of France about 20 million years ago.

Only a few species are dull in colour. Usually these salamanders and newts have lively and contrasting colours, some of them in gaudy patterns. Especially famous for the bright pattern of the breeding males are the 12 *Triturus* newts, and above all the marbled newt *T. marmoratus*, the alpine newt *T. alpestris*, and the extraordinary banded newt *T. vittatus*. The male banded newt develops a vertically striped and incredibly high dorsal crest. With the exception of a few species that give birth to live young, the Salamandridae lay their eggs in water. In *Triturus* and some *Cynops* newts the female secures each egg in the cavity of a leaf, which she folds between her hind limbs, ensuring that the egg is protected and oxygenated. Neoteny occurs also in the newts.

Salamandridae differ from the other salamander families in several skeletal characters too complicated to be discussed here, but they always have four well-developed limbs, with four fingers and generally five toes; only the spectacled salamander and some individuals of the newt *Echinotriton andersoni* have four-toed feet. They range in size from 7 to 30 centimeters. ($2^3/_4$ to 12 inches) in total length.

Lungless Salamanders

Plethodontidae, the largest family of the order Caudata, includes more than 60 percent of the known living species—about 270 in some 30 genera—which are found in the Americas from Nova Scotia and southeastern British Columbia to central Bolivia and eastern Brazil. Only a few species are European, inhabiting Sardinia, the southeast of France, and Italy from the Maritime Alps to the central Apennines.

After metamorphosis they may be distinguished from all the other salamanders and newts by having a special structure, the nasolabial groove, which extends from the nostril vertically to the upper lip and enhances chemoreception (the sensory reception or chemical stimuli). They always have four limbs, four fingers, and five (rarely four) toes. Their colouration differs according to the species; some of the cave-dwelling forms resemble the olm in having a long flesh-coloured body and scarlet gill tufts. The smallest are *Thorius* species, some of which reach only 2.7 centimeters (1 inch); the largest is *Pseudoeurycea bellii* at 32.5 centimeters ($12^1/_2$ inches).

As indicated by their common name, these salamanders are always lungless. They breathe through the mucous membrane in the mouth and throat and through the skin, both well supplied with many blood vessels. To keep their skin wet and thus able to absorb oxygen, these animals are linked even more to wet habitats than the lunged salamanders. They shelter in caves, crevices in rocks, spaces between roots and stones, or under logs, and will venture out only when it is humid enough and the temperature is mild. Like most other salamanders they do not tolerate heat well.

Some species, such as cave salamanders, genera *Hydromantes* and *Speleomantes* (a European genus formerly known as *Geotriton* or *Hydromantes*) are mostly rock-dwelling; others, such as *Bolitoglossa*, are tree-dwellers. Some climbing salamanders (genus *Aneides*) and dusky salamander (genus *Desmognathus*) like to climb on trees or rocks. The shovel-nosed salamander *Leurognathus marmoratus*, and the neotenic and cave-dwelling *Typhlomolge* and *Haideotriton* species are totally aquatic. The many-lined salamander *Stereochilus marginatus* and some *Gyrinophilus* species are mostly aquatic.

Each species of lungless salamander has a different type of nuptial dance. Males of species that have a chin gland also have enlarged teeth, with which they scarify the skin of their partner in order to "vaccinate" her with the aphrodisiac secretion of the chin gland.

The lungless salamanders are found almost exclusively in America, where they probably originated, although they also occur in Europe. According to some scientists, the ancestors of the seven European species (all belonging to *Spelcomantes*, a genus closely related to the Californian *Hydromantes*) reached western Europe by crossing the Bering land bridge and then became extinct all over Asia and most of Europe; other scientists, including the writers of this chapter, suppose that the ancestors of today's European species had colonized western Europe before its detachment from North America about 50 million years ago.

11

Reproduction

In frog sexual dimorplus is not well distinct. However during breeding season some characters appear. The female is bigger than male. The identifying features of male which distinguish it from the female are a *darkened thumb pad* which changes thickness and colour intensity as the breeding season approaches; a distinct low, guttural *croaking sound* with the accompanying swelling by air of the lateral vocal sacs located between the tympanum and the forearm; a more slender and *streamlined body* than of the female; and the *absence of coelomic cilia* except in the peritoneal funnels on the ventral face of the kidneys. Males of a many species carry additional features such as brilliant colours on the ventral aspects of the legs (*R. sylvatica*), black chin (*B. fowleri*), or the size of colour of the tympanic membrane.

Male Reproductive Organs

Testes

There is a pair of testes present just above the dorsally placed kidneys and remain suspended by a double fold of peritoneum known as the *mesorchium*. This mesentery surrounds each testis and is continuous with the peritoneal epithelium which covers the ventral face of each kidney and lines the entire body cavity. The testes are whitish and ovoid bodies lying ventral to and near the anterior end of each kidney. The *vasa efferentia*, ducts from the testes, pass between the folds of the mesorchium and into the mesial margin of the adjacent kidney. The number of vasa efferentia leaving the testis is about a dozen. During the breeding season, or after slight compression of the testis of the hibernating frog these ducts become the more apparent due to the presence in them of whitish masses of spermatozoa in

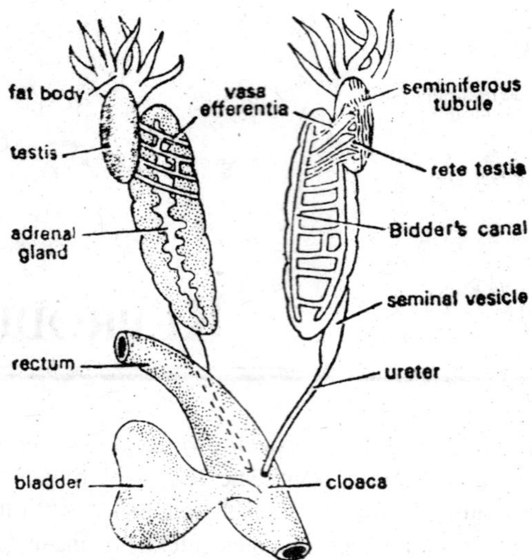

Fig. 11.1. Frog—Male urinogenital organs.

suspension. The ducts are very small in diameter, tough walled, and interbranching. They are lined with closely packed cuboidal cells. All vasa efferentia of one testis unit together inside the kidney of the related side and form a cerenal duct the *Bidders Canal* that transports sperms through the kidney. The Bidder canal is a longitudinal tube situated along the median edge of the kidney communicated with the writer by transverse collecting tubes.

The presence of spermatozoa in the kidney also can be achieved artificially by injecting the male frog with the anterior pituitary sex-stimulating hormone. The spermatozoa are produced in inside *seminiferous tubules*. These are closely packed, oval-shaped sacs, which are separated from each other by thin partitions (septula) of supporting (connective) tissue known as *interstitial tissue*. This tissue presumably has some endocrine function. The thickness of this tissue is much reduced immediately after breeding or pituitary stimulation. The interstitial tissue is continuous with the covering of the testes known as the tunica *albuginea*.

Spermatogenesis

When the photoperiodism is appropriate during breeding season the given cells undergo spermatogenesis but shortly after the breeding season the *spermatogonium*, which has ceased all mitotic activity, enters upon a period of rest but not inactivity. During this period the nucleus

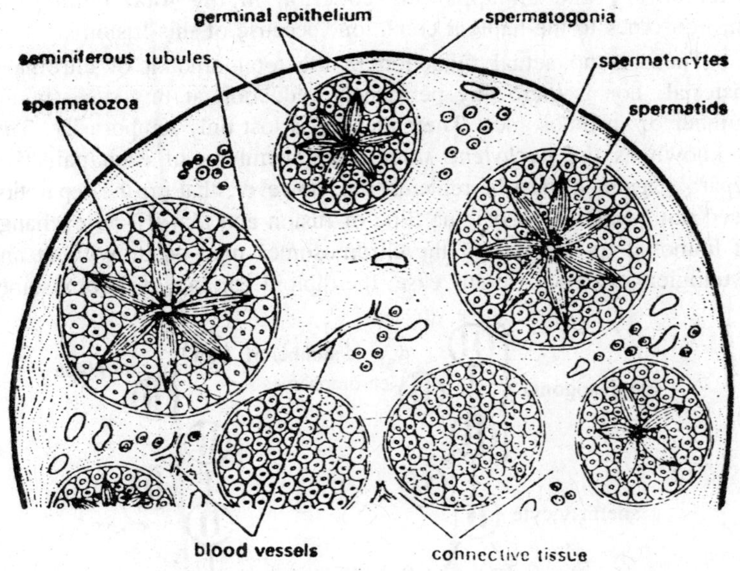

Fig. 11.2. Frog—T.S. of a part of testis.

passes through a sequence of complex changes which represent an extended prophase. This is in anticipation of the two maturation divisions that finally produce the haploid spermatid which metamorphoses into a spermatozoon. The Spermatogenia show several cytolosical charges.

The nucleus of the spermatogonium contains chromatin which appears as relatively coarse lumps distributed widely over an achromatic reticulum. Both the cytoplasm and the nucleus grow and the chromatic granules become finely divided and arranged into contiguous rows, bound by an achromatic thread, together known as chromosomes. This is the *leptotene stage* of spermatogenesis. Shortly the chromosomes become arranged in pairs which converge towards that side of the nucleus where the centrosome is found. The opposite ends of the paired chromosomes merge into the general reticulum. This is the *synaptene stage* or *boughal stage*. The chromatin granules become telescoped together on the filaments so that the aggregated granules, known as *chromosomes*, appear much shorter and thicker.

Pairs of chromosomes become intertwined and the loose terminal ends become coiled and tangled together. This is the *contraction* or *synizesis stage*. Then the members of the various pairs become laterally fused. While there is no actual reduction in total chromatin, there is a temporary and an apparent reduction in the total number of chromosomes to the haploid condition, because of this fusion.

There is no actual reduction in the total amount of chromatin material, nor is there any permanent reduction at this stage in the number of chromosomes. Their identity is lost only temporarily. This is known as the *pachytene stage*. The members of each pair then separate again. It must be remembered, however, that (a) the separation need not be along the original line of fusion and that (b) an exchange of homologous sections of the chromosomes may occur without any cytological evidence. In any case, the diploid number of chromosomes

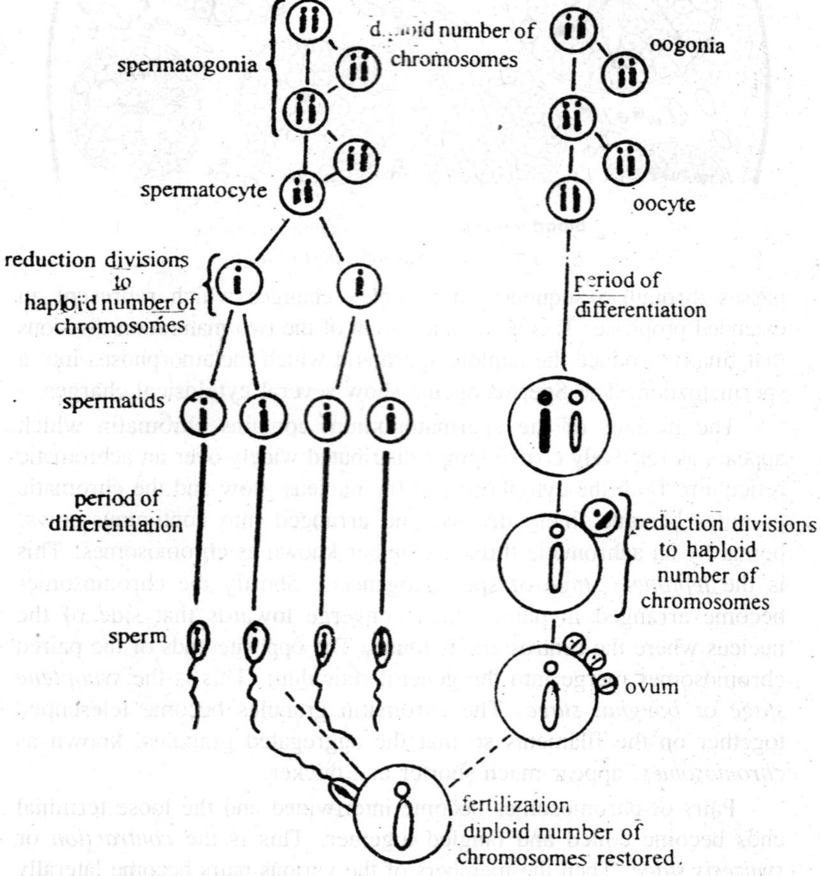

Fig. 11.3. Diagramatic representation of gametogenesis.

reappears and this is then known as the *diplotene stage*. During theses changes in the chromatin material of the nucleus, the volume of the nucleus and the cytoplasm are considerably increased, the nuclear membrane break down, and the chromosomes assume bizarre shapes and various size. They may be paired, curved, or straight; "V" and "C" and reversed "L" shapes, and grouped as tetrads. This is known as the *diakinesis stage*. The chromosomes are then lined up on a spindle in anticipation of the first of the two maturation divisions.

Frog is a seasonal breeder and the spermatogenesis in is completed within the testes. The walls of the seminiferous tubules produce *spermatogonia* which go through mitotic divisions and then the series of nuclear charges without mitosis. This results in the appearance, toward the lumen of each tubule, of clusters of mature *spermatozoa*. The spermatogonia are found close to the basement membrane of the seminiferous tubule. These then await their turn to undergo the maturation changes necessary for the production of spermatozoa which will be ready for the breeding season. The elongated and filamentous tail of the clustered mature *spermatozoa* project into the lumen of each seminiferous tubule. At the peak of spermatogenetic activity, all the stages of maturation from the spermatogonium to the spermatozoon are present.

The spermatogonia are always located around the periphery of the seminiferous tubule and are small, closely packed cells, each with a granular, oval nucleus. In between the spermatogonia may be found

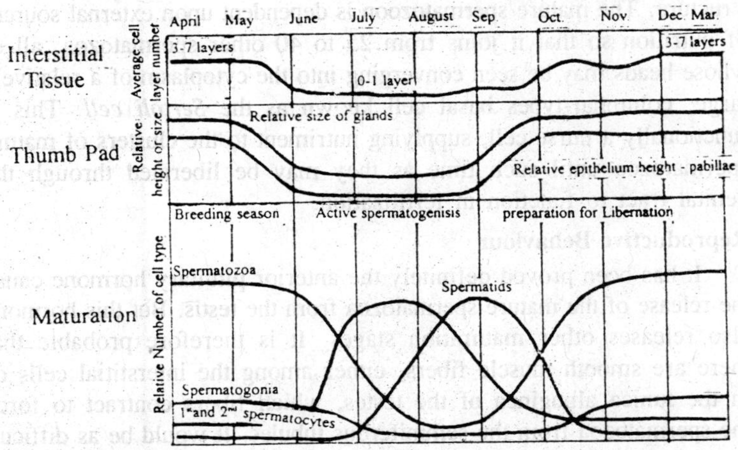

Fig. 11.4. Normal cyclic changes in the primary and secondary sexual characters of the frog, Rana pipiens.

occasional very large cell, the *primary spermatocytes*. These tend to be irregularly spherical, possessing large and vesicular nuclei. The cells are so large that they may be seen under low power magnification of the microscope. Apparently they divide to form secondary spermatocytes almost immediately, for they are so few and far between. The *secondary spermatocytes* are about half the size of the primaries, and lie towards the lumen of the tubule. They generally have a darkly staining nucleus, and the cytoplasm may be tapered toward one side. The *spermatid*, following another division, is even smaller and possesses a condensed nucleus of irregular shape. Clusters of spermatids appear as clusters of granules, the dark nucleus being almost as small as the cross section of a sperm head.

The metamorphic stages from spermatid to spermatozoon are difficult to identify with ordinary magnification, and are often confused with the spermatids themselves. During this change the inner of two spermatid centrioles passes into the nucleus while the outer one give rise to the tail-like flagellum. The mature *spermatozoon* averages about 0.03. mm. in length. It has an elongated, solid-staining head (nucleus) with an anterior acrosome, pointing outwardly towards the periphery of the seminiferous tubule. The short middle piece generally is not visible but the tail appears as a gray filamentous extension into the lumen, about four or more times the length of the sperm head. In any cross section of this testis, bundles of sperm heads or tail may be cut at right angles or tangentially, giving misleading suggestions of structure. The mature spermatozoon is dependent upon external sources of nutrition so that it joins from 25 to 40 other spermatozoa, all of whose heads may be seen converging into the cytoplasm of a relatively large, columnar-types basal cell known as the *Sertoli cell*. This is functionally a nurse cell, supplying nutriment to the clusters of mature spermatozoa until such time as they may be liberated through the genital tract to function in fertilization.

Reproductive Behaviour

It has been proved definitely the anterior pituitary hormone cause the release of the mature spermatozoa from the testis. But this hormone also releases other maturation stages. It is therefore probable that there are smooth muscle fibers, either among the interstitial cells or in the tunica albuginea of the testes, which fibers contract to force the spermatozoa from the seminiferous tubules. It would be as difficult to physiologically demonstrate the presence of these fibers in the testis as it is simple to demonstrate them in the contracting cyst wall of the ovary. Responding to sex stimulation, the spermatozoa became free from their Sertoli cells and are formed from the lumen of the seminiferous tubule into the related *collecting tubule*. These collecting tubules are small and are linked with closely packed cuboidal cells. They join the *vasa efferentia* which leave the testis to pass between the folds of the mesorchium and thence into the *Malpighian corpuscles* of the kidney. From this point the spermatozoa pass by way of the excretory ducts, the uriniferous tubules, and into the *mesonepheric duct* (ureter) which may be found attached to the lateral margin of the kidney. Within the excretory system the spermatozoa are immotile, due to the slightly acid environment. They are carried passively down the ureter to the slight dilation near the cloaca, known as the *seminal vesicle*. Within the vesicle the spermatozoa are stored briefly in clusters until amplexus and oviposition occur.

At oviposition the male ejaculates the spermatozoa into the neutral or slightly alkaline water where they are activated and then are able to fertilize the eggs as they emerge from the cloaca of the female. During the normal breeding season amplexus is achieved as the females reach the pounds where the males are emitting their sex cells. During amplexus there are definite muscular ejaculatory movements on the part of the male frog, coinciding with oviposition on the part of the female. Amplexus may be maintained by the male for many days, even with dead females. As soon as the eggs are laid and the male has shed his sperm, he goes through a brief weaving motion of the body and then releases his grip to swim away. The frogs completely neglect the newly laid eggs.

Accessory Reproductive Organs

In the male frog the ureter is not directly connected with the *bladder*, as it is in higher vertebrates. It is possible that the bladder in the Anura may be an accessory respiratory and hydrating organ, particularly in the toads, where water may be stored during migrations onto land. The male frog also has a duct, homologous to the oviduct of the female, known as the "rudimentary oviduct" or *Mullerian duct* This duct normally has no lumen, and is very much reduced in size so that it may be difficult to locate. There is experimental evidence that this duct may be truly a vestigial oviduct since it respondents to ovarian or female sex hormones by enlarging and acquiring a lumen.

At the anterior end of the testes of some Anura (e.g., toads) there may be found an undeveloped ovary known as *Bidder's organ*. This structure is said to respond to the removal of the adjacent testis or to the injection of female sex hormones by enlarging to become structurally like an ovary. Occasionally isolated ova have been found within the seminiferous tubules of an otherwise normal testis, suggesting the similar origin and the fundamental similarity of the testis and the ovary. Finally, attached to the anterior end of the testis of the hibernating frog may be seen finger like *fat bodies* (corpora adiposa) which represent stored nutrition for the long period of hibernation, and for the pre-breeding season when food is scarce. Under the microscope these fat bodies appear as clusters of vacuolated cells, and are not to be confused with the mesorchium. It is believed that they, as well as the gonads, arise from the genital ridges of the early embryo. The fat bodies tend to be reduced immediately after the breeding season, only to be built up again as the time for hibernation approaches.

Female Reproductive Organs

Secondary Sexual Characters

The mature female frog is generally larger than the male of the same age and species, the *Rana pipiens* female measuring from 60 to 110 mm. in length from snout to anus. The sexually mature female has a body length of at least 70 mm. It can be identified by the absence, at any season, of the dark thumb pad; the inability to produce lateral cheek pouches resulting from the croaking reaction; a flabby and distended abdomen; and the presence of peritoneal cilia. These cilia are developed in the female in response to the prior development and secretion of ovarian hormones.

Ovaries

There is a pair of large flat, irregular multi-lobed ovaries attached to the dorsal body wall by a double-layered extension of the peritoneum known as the *mesovarium*. This peritoneum continues around the entire ovary as the *theca externa*. Each lobe of the ovary is hollow and its cavity is continuous with the other 7 to 12 lobes. The ovaries of the female are found in the same relative position as the testes of the male but the peritoneum extends from the dorso-mesial wall rather than from the kidneys, as in the male. The size of the ovary varies with the seasons more than does the size of the testis. From late summer until the spring breeding season the paired ovaries will fill the body cavity and will often distend the body wall. They contain from 2,000 (*Rana pipiens*) to as many as 20,000 ova (*Rana catesbiana*), each measuring about 1.75 mm. in diameter (*Rana pipiens*).

The mature ova are highly pigmented on the surface of the animal pole, so that the ovary has a speckled appearance of black pigment

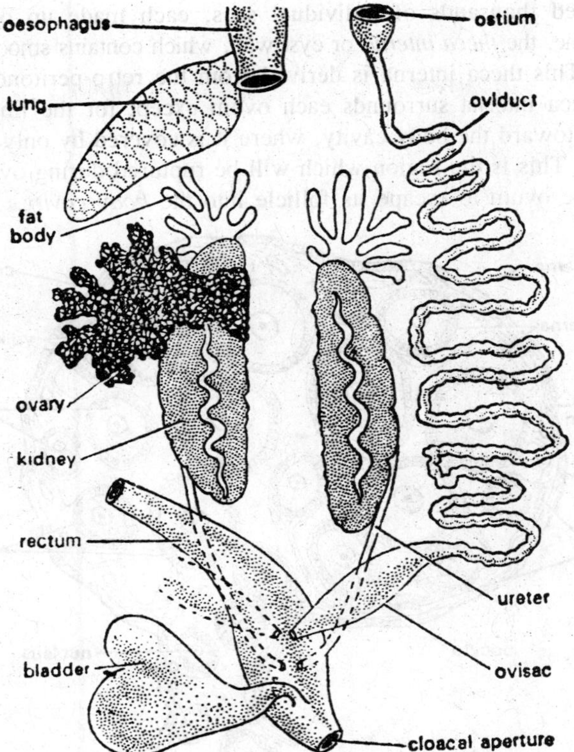

Fig. 11.5. Frog—Female urinogenital organs.

and white yolk, representing the animal and the vegetal hemispheres of the ova. All reproductive organs undergo cyclic changes. There is no appreciable change in the size of the ovary during hibernation, not is there any observable cytological change in the ova. However, if a female is forced to retain her ova beyond the normal breeding period by isolating her from males or by keeping her in a warm environment and without food, the ova will begin to deteriorate (cytolize) within the ovary.

Immediately after the spring breeding season, when the female discharges thousands of *mature ova*, the remaining ovary with its *oogonia* is so small that it is sometimes difficult to locate. There is no pigment in the tissue of the ovary (in the stroma or in the immature ova), and each growing *oocyte* appears as a small white sphere of protoplasm contained within its individual *follicle sac*. The history of the ovary shows that within its outer peritoneal covering, the *theca externa*, are suspended thousands of individual sacs, each made up of another membrane, the *theca interna* or cyst wall, which contains smooth muscle fibers. This theca interna is derived from the retro-peritoneal tissue. This theca interna surrounds each ovum except for the limited area bulging toward the body cavity, where it is covered by only the theca externa. This is the region which will be ruptured during ovulation to allow the ovum to escape its follicle into the *body cavity*.

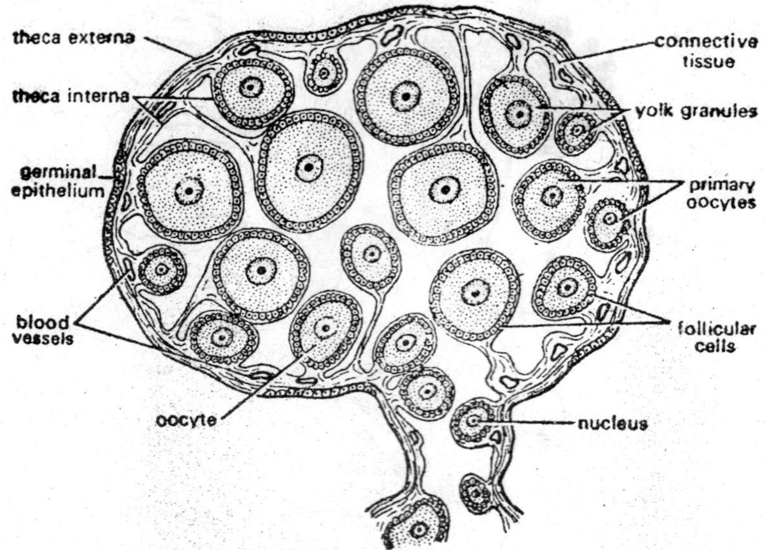

Fig. 11.6. Frog—T.S. of a part of ovary.

The theca interna, plus the limited covering of the theca externa, and the follicle cells together comprise the *ovarian follicle*. These two membranes make up the rather limited *ovarian stroma* of the frog ovary, and they contain both blood vessels and nerves. Within each follicle are found *follicle cells*, with their oval and granular nuclei derived originally from oögonia. These follicle cells surround the developing oöcyte and are found in close association with it throughout those processes of maturation which occur within the follicle. Enclosed within the follicle cells, and closely applied to each mature ovum, is the non-cellular and transparent *vitelline membrane*, probably derived from both the ovum and the follicle cells. This membrane is developed and applied to the ovum during the maturation process so that it is not seen around the earlier or younger oögonia. Since the bulk of the ovum is yolk, the membrane is appropriately called the *vitelline membranes*. It is sometimes designated as the primary egg membranes. After the ovum is fertilized this membrane becomes separated from the ovum and the space between is then known as the *perivitelline space*, filled with a fluid. The fluid may be derived from the ovum which would show compensatory shrinkage.

As the oocyte matures and enlarges, the follicle cells and membranes are so stretched and flattened that they are not easily distinguished. The ovum will mature in any of a variety of positions within its follicle, the exact position probably depending upon the maximum blood supply. As one examines an ovary the ovum the will be seen in all possible positions, some with the *animal hemisphere* and others with the *vegetal hemisphere* toward the theca externa and body cavity. It is believe that the most vascular side of the follicle wall will tend to produce the animal hemisphere of the ovum and hence give it its fundamental symmetry and polarity. The frog's egg is of *mesolecithal type* having a moderate amount of yolk. The eggs are *telolecithal* having folk at its lower *vegetal pole*. There is a thin outer layer of *cytoplasm*, more concentrated toward the animal hemisphere and in the vicinity of the *germinal vesicle* or immature nucleus. Surrounding the entire ovum is a non-living surface coat, also containing pigment. This *pigment* is presumably a metabolic by-product. This coat is necessary for retaining the shape of the ovum and in aiding in the morphogenetic processes of cleavage and gastrulation.

Body Cavity and the Oviducts

There is a pain of oviducts suspended from the dorsal body wall by a double fold *peritoneum*. Its anterior end is found between the heart and the lateral peritoneum, at the apex of the liver lobe. At this anterior end is a slit-like infundibulum of *ostium tuba* with ciliated and highly elastic walls. The body cavity of the female is almost entirely lined with cilia, each cilium having its effective beat or stroke in the general direction of one of the ostia. These cilia are produced in response to an ovarian hormone and therefore are regarded as secondary sex characters. They are found on the peritoneum covering the entire body cavity, on the liver, and on the pericardial membrane. There are no cilia on the lungs, the intestines, or the surface of the kidneys except in the ciliated peristomial (peritoneal) funnels which lead into the blood sinuses of the kidneys.

The abundant supply of cilia of the female means of that ova ovulated from any surface of the ovary will be carried by constant ciliary currents anteriorly towards and into one another of the ostia.

As soon as the ovum leaves the ovary it is nude except for the non-living, transparent, and closely applied *vitelline membrane*. Thus far it has been impossible to fertilize these body cavity *ovum* and have them develop. When they are placed in a sperm suspension some will show surface markings which resemble very closely the normal cleavage spindles and the cleavage furrows but none have developed as embryos as yet. These body cavity ova are often quite disported, due to the fact that the ovulation process involves a rupture of the follicle and forcing out of the ovum from a very muscular follicle.

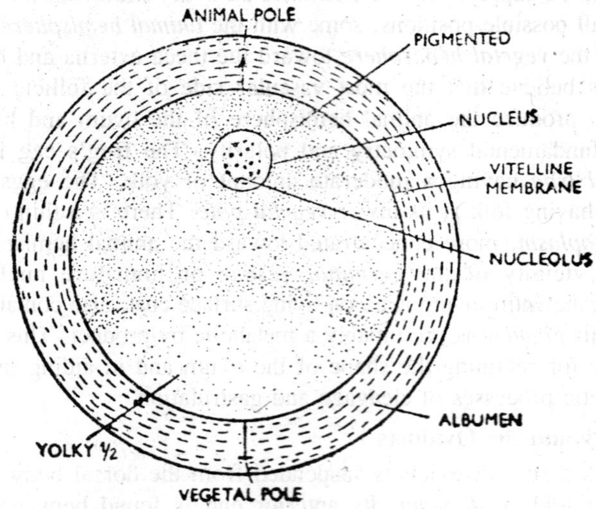

Fig. 11.7. Unfertilized egg of frog.

The ovum is literally squeezed from the follicle, through a small aperture. The process looks like an Amoeba crawling through an inadequate hole. *Ovulation* which is the emergence of mature one from ovary to the body cavity takes several minutes at a laboratory temperatures, and is not accompanied by halmorrhage. By the time the ovum reaches the ostium (within 2 hours), as the result of ciliary propulsion, it is again spherical. Ciliary currents alone force the ovum into the ostium and oviduct. The ostial opening is very elastic and does not respond to the respiratory or heart activity. The ova are simply forced into the ostium, from all angles, stretching its mouth open to accept the ovum. As soon as the ovum enters the oviduct and begins to acquire an albuminous (mucinjelly) covering, it becomes fertilizable. One can remove such an ovum from the oviduct by pipette or by cutting the oviduct 1 inch or more from the ostium, and can fertilize such an egg in a normal sperm suspension.

The physical (or chemical) change which occur between the time the ovum is in the body cavity and the time it is removed from the oviduct, which make it fertilizable, are not yet understood. As soon as the ovum is propelled through the oviduct by ciliary currents, it receives coatings of *albumen* (jelly). The initial coat is thin but of heavy consistency, and is applied closely to the ovum. The ovum is spiralled down the oviduct by its ciliated lining so that the application of the jelly covering is quite uniform. There are, in all, there distinct layers of jelly the outermost one being much the greater in thickness but the less viscous.

The intermediate layer is of a thin and more fluid consistency. There is hyperactivity of the glandular elements of the oviduct just before the normal breeding season, or after anterior pituitary hormone stimulation, so that the duct is enlarged several times over that of the oviduct of the hibernating female. The presence of the jelly layers on the oviducal or the uterine egg is not readily apparent because it requires water before it reaches its maximum thickness. Ova sectioned within the oviduct show the jelly as a transparent coating just outside the vitelline membrane. As soon as the ovum reaches the water, however, imbibition swells the jelly until its thickness becomes greater than the diameter of the ovum. The function of the jelly is to protect the ovum against injury, against ingestion by larger organisms, and from fungus and other infections. Equally important, however is the evidence that this jelly helps the ovum to retain its metabolically derived heat so that the jelly can be said to act as an insulator against heat loss. *Bernard* and *Batuschek* (1891) showed that the greater the wave length of light the less heat passed through the jelly around the frog's egg, in comparison with an equivalent amount of water and under similar conditions.

Originally, and erroneously, the jelly was thought to act as a lens which would concentrate the heat rays of the sum onto the ovum, but since the jelly is largely water, which is a non-conductor of heat rays, this theory is untenable. One can demonstrate that the temperature of the ovum is higher than the temperature of the immediate environment, even in a totally darkened environment. So the jelly has certain physical function in addition to those as yet undetermined function which aid in rendering the eggs fertilizable. The eggs takes about 2 to 4 hours, at ordinary temperatures, to reach the highly elastic *uterus*, at the posterior end the oviduct and adjacent the *cloaca*. Each uterus has a separate opening into the *cloaca*, and the ovulated ova are retained within this sac until, during amplexus (sexual embrace by the male), they are expelled into the water and are fertilized by the male. Generally the ova are not retained within the uterus of more than a day or so. There may be quite a few hours between the time of appearance of the first and the last ova in the uteri.

Oogenesis

The process farming ova is called *oogenesis*. During the process the ova develop from oögonia which divide repeatedly. These pre-maturation germ cells divide by mitosis many times and then come to rest, during which process there is growth of some of them without nuclear division. These become ova while those that fail to grow to become follicle cells. However, there are pre-prophase changes of the nucleus of the prospective ovum comparable to the pre-prophase changes in spermatogenesis. The majority of oögonia, therefore, never mature into ova, but become follicle cells. The process of maturation involves contributions from the nucleus and the cytoplasm. *First,* chromatin nucleoli aid in the synthesis of yolk, and *second,* the breakdown of the germinal vesicle allows an intermingling of the nuclear and the cytoplasmic components. Only a small portion of the germinal vesicle is involved in the maturation spindle so that it may be at this time that the nucleus exert its initial influence on the cytoplasm.

All cytoplasmic differentiations must be initiated at a time when the hereditary influenced of the nucleus are so intermingled with it. Growth of the developing oocytes is achieved largely by the accumulation of *yolk*. As soon as growth beings the cell no longer divides by mitosis and is known as an *oocyte* rather than an oögonium. The growth process is aided by the *centrosome*, which is found to one side of the *nucleus*, and around which gather the granules or *yolk platelets*. The *chromatin* filaments become achromatic and the *nucleoli* increase in number, by fragmentation, and become more chromatic. Many of the nucleoli, which are concentrations of nucleo-protein, pass through the nuclear membrane into the surrounding cytoplasm during this period.

It is not clear whether this occurs through further fragmentation of the nucleoli into particles of microscopic or sub-microscopic size, and then their ejection through the nuclear membrane. It may occur by the loss of identity (and chromatic properties) by possible chemical change and subsequent diffusion of the liquid form through the membrane to be resynthesized on the cytoplasmic side of the membrane. During the growth of the oocyte, further nucleoli appear within the nucleus, only to fragment and later to pass out into the cytoplasm. The presence of chromatic nucleoil in the cytoplasm is closely associated with the accumulation of yolk. The granules within the cytoplasm (extruded fragments of nucleoli) function as centres of yolk accumulation and have therefore been named "yolk nuclei." This is an unfortunate name, for the structure is a nucleus only in the sense that it is centre of aggregation. It is not a true cell nucleus.

The centrosome and other granular centres lose their identity and the yolk granules then become scattered throughout the cytoplasm. The source of all yolk for the growing ova is originally the digested food of the female. This nutrition is carried to the ovary by way of the blood system and conveyed to the nurse of follicle cells and thence to the oocyte. The yolk is at first aggregated around yolk nuclei, then concentrated to one side of the nucleus. Finally, it assumes as ring shape around the nucleus between an inner and an outer zone of cytoplasm. Subsequently the nucleus is pushed to one side by the ever-increasing mass of yolk so that eventually there is an axial gradient of concentration of oval *yolk platelets* from one side of the egg to the other.

The smaller platelets are found in the vicinity of the nucleus, in the animals hemisphere. The larger platelets are located toward the vegetal hemisphere. There is an increase averaging from 200 to 700 per cent in the total lipoid substance, neutral fat, total fatty acids, total cholesterol, easter cholesterol, free cholesterol, and phospholipin content of the ovaries of *Rana pipiens* occurring during the production and growth of ova. The primary oocyte may show a slight flattening of the surface directly above the region of the nucleus. These growth changes and the unequal distribution of pigment, yolk, and cytoplasm are first indications of *polarity* or a gradient system within the ovum.

When the polarity is well-established the cytoplasm, the superficial melanin or black pigment, and the nucleus are all at the animal hemisphere. The light coloured yolk is more concentrated toward the vegetal pole. The eggs is then regarded as a *telolecithal* egg. During this phase of egg maturation there is a drain on the metabolism of the frog which requires an excess of food intake because the materials for the growth of ovum must be synthesized from nutritional elements received from the vascular system of the female. For *Rana pipiens* this period of most active feeding comes during the summer when the natural foods, insects worms, etc., are the most abundant. During the growth of the oocyte in general there are important changes occurring within the nucleus (germinal vesicle) of the ovum. Thirteen pairs of *chromosomes* may be seen in synizesis (contraction), converging toward the centrosome at the "yolk nucleus" stage. A little later the nuclear membrane develops sac-like bulges, the *nucleoli* are scattered, and there is a colloidal chromosome core which almost fills the entire nucleus. The chromosome themselves are small and almost invisible. When the *primary oocyte* is about half its ultimate size, there appear definite sacs on the nuclear surface.

The fragmented nucleoli are located at the periphery of the lobulated nuclear membrane, and the chromosome frames have become relatively large. The chromosomes, by this time, have reached their maximum length and possess large lateral loops. Finally, in the fully grown nucleus of the primary oocyte the nuclear sacs are very prominent, and the nucleoli appear in clusters in the center of the ovum surrounding the *chromosome frame*. This frame is a gel structure which give rise to the *first maturation spindle*, containing 13 pairs of slightly contracted chromosomes. Before the time of hibernation the ova that are to be ovulated for the next spring are in the fully grown primary oocyte stage, having their full complement of yolk, cytoplasm, and pigment. Externally more than one-half of the ovum appears densely black, due to surface pigment granules, while the rest is creamy white.

The nucleus is prepared for the maturation divisions. Such an ovum measures about 1.75 mm. In diameter. The surface layer of the amphibian ovum is formed before fertilization and it is definitely not hyaline, as it is in some Invertebrate eggs. It contains many small yolk grains and irregular accumulations of spherical, black pigment granules. With each cleavage, subsequent to fertilization, this superficial coat is divided between the blastomeres, being an integral part of the living cell. There is no clear-cut demarcation between this surface coat and the inner cytoplasm and yolk. It is believed that these growth changes of the egg are under the influence of the basophilic cells of the anterior pituitary gland, which cells are greater in number at this time than at any other. During the growth period the vitelline membrane appears on the surface of the oocyte as a thin, transparent, non-living, and closely adherent membrane. It is formed presumably by a secretion from the egg itself, aided by the surrounding follicle cells. It appears to be similar in all respects to the membrane of the same name found around the eggs of all vertebrates.

Ovulation and Maturation

Ovulation, or the liberation of the ovum from the ovary is brought about by a sex-stimulating hormone from the anterior pituitary gland. Just before and during the normal spring breeding period there is a temporary increase in the relative number of acidophilic cells in the anterior pituitary. It is believed that this is not coincidental but a causal factor in sex behaviour in the frog. However, until such time as extracts of specific cell types can be made this will be difficult to prove conclusively. Attempts on the part of the male to achieve amplexus are resisted by the female not sexually stimulated. However, such a female can be made to accept the male by injecting the female with whole anterior pituitary a glands from other frogs. It is very probable that environmental factors such as light, temperature, and food may act through the endocrine system to prepare the frogs for breeding when they reach the swampy marshes in the early spring, after protracted hibernation.

It must be pointed out, however, that there are frogs in essentially the same environments which breed in July (*Rana clamitans*) and August (*Rana catesbiana*), so that the causal factors appears to be either complex or possibly different for different species. The pituitaries of the hibernating frogs to contain the sex stimulating factor, but apparently to a lesser degree than the glands of frogs approaching the breeding season. The injection of six glands from adult female frogs will cause an adults female of *Rana pipiens* to ovulate as early as the last week in August, some eight months before the normal breeding period. One or two such glands will accomplish the same results if used early in April.

Another explanation for this may be offered, namely that the ovary itself may become more sensitive to such stimulation as the breeding seasons approaches. The process of ovulation involves the rupturing and the emergence of ova from their individual follicles. The surface of the ovum separated from the body cavity by only the non-vascular *theca externa*, is first ruptured and then the ovum slowly emerges through the small opening. Since the ovum is known contain a peptic-like enzyme, it is believed that the pituitary hormone may activate this enzyme to digest away the tight and non-vascular covering. Then by stimulation of the smooth muscle fibers of the cyst wall (*theca interna*) the process of emergence is completed.

The relation of the pituitary to smooth muscle activity has long been established clinically. It is true that the ovarian stroma show undulating contractions at all seasons, irrespective of sex activity. An ovum will emerge at any time from a surgically ruptured follicle. If an ovulating female is etherized and the body cavity is opened, the ovary may be removed and placed in amphibian Ringer's solution and the ovulation process may be observed directly. This will go on for several hours after all connections with the nerve and blood supply are cut off.

From the initial rupture of the theca externa until the ovum drops free into the body cavity there is a lapse of from four to 10 minutes at ordinary laboratory temperatures. The *first maturation division* occurs at the time of ovulation. The heterotypic chromosomes are placed on the spindle of the amphiaster whose axis is at right angles to the ovum surface. Movement of the chromosomes is identical with that found in ordinary mitosis. The outermost group of telophase chromosome are pinched off, with a small amount of cytoplasm and no yolk, to comprise the *first polar body*. The innermost telophasic group of chromosomes remain within a clear area of the ovum as the nuclear mass of the secondary oocyte. These changes occur as the ovum leaves the ovary and before it reaches the oviduct. Possibly the same forces which bring about follicular rupture also influence this maturation process. The *second maturation division* begins without any intermediate rest period for the chromosomes, at about the time the ovum enters the oviduct. There may be a variation in time up to 2 hours for ova to reach the ostium, depending upon the region the body cavity into which they are liberated.

Thus the stage of maturation of different ova within the oviduct may vary considerably. There is a longitudinal division of the chromosomes of the ovum which are lined up in metaphase angles to the ovum which are lined up in metaphase on the *second maturation spindle*, the axis of which is at right angles to the ovum surface. Since the spindle is primarily protoplasmic, and is made up in part of fibers the space occupied by the spindle will be free of yolk. Since it is peripherally placed, and represents a slight inner movement after the elimination of the first polar body, the surface layer of the ovum is slightly de-pigmented just above the spindle region. This situation is exaggerated in aged ova a relatively large de-pigmented area of the cortex appearing towards the center of the animal hemisphere. Maturation is not completed until or unless the ovum is activated by sperm or stimulated by parthenogenetic means. However, every ovum reaching the uterus is in methaphase of the second maturation division, awaiting the stimulus of activation to complete the elimination of the *second polar body*.

12

COPULATION

It is in spring that the average person is most conscious of the amphibian. The Spring Peeper, *Hyla crucifer*, is considered by many to be more reliable harbinger of warm weather than is the robin. One old saying is that there will be but there more freeze-ups after the peepers are heard calling. This is akin to the superstition that if a ground hog sees his shadow on the second of February, there will be six more weeks of winter. Though the peepers are weather prophets far superior to our finest meteorologists, unfortunately they are not to expert that they can forecast just how many more frosts will occur before it is safe to set out the tomato plants. A great many amphibians breed in the spring, but by no means all of them do. Some breed in the summer, some in the fall, some in the winter. However, in the North we are most apt to notice the breeding activity in the spring.

Spring must be interpreted very loosely, for sometimes the peepers and Wood Frog are calling before March 21st even as far north as New England. The nights are often cold and frostly, and you may wonder why these little creatures have come out of hibernation so early. We do not begin to recognize all the conditions that cause the amphibians to emerge from hibernation and to commence courtship and mating, any more than we understand all the reasons behind bird migration. But though there is much to be learned on this subject, it is interesting helpful to discuss a few of the factors that have an important bearing on the matter. The prime factor governing the amphibian's emergence form hibernation is the weather. If the ground is frozen solid, the animal hibernating beneath the frost line would not be able to break through the frozen crust, even should he so desire, which he does not. For those that hibernate in the detritus of pond bottoms, ice on top prevents their escape. But the amphibian is a cold-blooded creature, whose internal temperature is nearly the same as his external surroundings, and very cold surroundings make him sluggish and incapable of the violent movement necessary to burrow out of hibernation. Therefore, until the temperature of the earth or pond water rises to a certain level, the amphibian remains in the dormant condition known as hibernation, where heartbeat and respiration are slowed to the minimum.

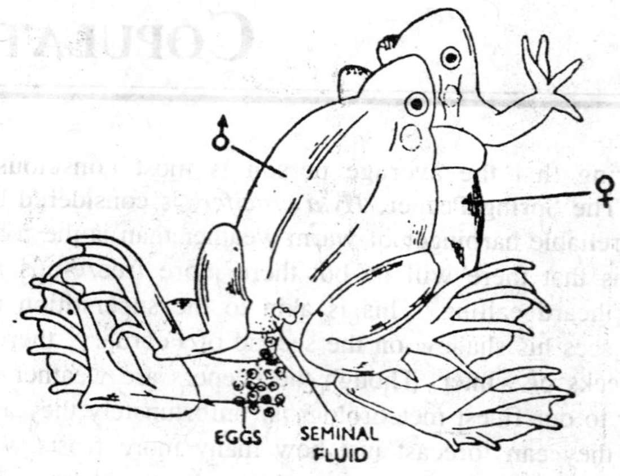

Fig. 12.1. False copulation in frog.

All amphibians are extremely sensitive to temperature changes, and notice the slightest gradations that are imperceptible to us. Furthermore, each amphibian has a temperature at which he must live for optimum comfort and activity. The varies from species to species, and helps to explain why the Spring Peeper emerges from hibernation and begins to breed earlier than the toad. The temperature of the water is important not only because it must be warm enough to arouse those sleeping beneath it to activity but also because it must be warm enough so that any eggs laid in it will not be frozen and killed. Humidity also plays a tremendous part in determining courtship and mating time. You will always find that the peak of breeding activities in all species occurs on evenings of high humidity—the choruses of peepers, toads, treefrogs, and frogs are at their loudest on rainy overcast days and especially nights.

Naturally, the humidity of the air has more influence on those amphibians who are more terrestrial than on those who are partially or completely aquatic. But it affects them all to some extent. Perhaps there is considerable truth in the fisherman's adage that fish bite better on a rainy or dull day. Certainly it would be true if one fished for amphibians. The weather does not, of course, completely govern courtship and mating. It is also regulated by the animal's ductless glands—especially the anterior lobe of the pituitary gland. The secretions of this gland stimulate the sexual glands, which in turn rouse the animal to mate. We usually think of ponds and lakes as being the place where amphibians congregate to mate and lay their eggs.

It is true that the great majority of all amphibians do deposit their eggs in water. The salamanders may choose a quiet pond, as do the Newt and many Mole Salamanders. They may prefer the swift-running water of a brook, as do many Lungless Salamanders. Of they may select a moist place in the woods or in the mud. Doubtless, this dry-land laying is a surprise to many. And no wonder, for it is far more difficult to find a few eggs under a stone or log in the woods than it is to see the thousands upon thousands clustered in various parts of the pond. It may be even more of a surprise to learn that some salamanders lay their eggs in trees.

Frogs and toads lay in much the same places as do the salamanders. There are pond layers, river layers, and even tree layers, though none of our native frogs lays in trees. With the frogs and toads, courtship is always conducted very close to the place where mating and egg laying will occur. How do they court one another? Let us take a specific example, the Spring Peeper. The peepers hibernate in the woods, beneath the earth. When the ground becomes thawed, they emerge from their winter burrows and heads for ponds and swamps. On dull afternoons and in the earnings, the male sits on the bank of the pond and utters his sweet birdlike call. He takes air into his lungs and closes his mouth and nostrils. He sends the air into the vocal sacs located in the throat region. They balloon out into a glistening bubble that is almost as large as the body of the peeper. The air is sent back and forth between lungs and vocal sacs, causing the vocal chords to vibrate. An enormous sound for so small a creature emerges: *Preep, peep, peep*... he sings sweetly. Other males join in, and soon these is a deafening chorus of peepers all calling to the females. A female approaches one of the singing males. She is silent, for only the males sing in the breeding season. But the male senses her approach. He turns towards her, climbs on her back, and embraces her, placing his arms just behind hers in her "armpits." The couple are now ready to enter the pond and lay the eggs.

Courtship in frogs and toads consists mainly of the male's calls. The female approaches the calling make. He may be so intent on his singing that she may even have to touch him to make him aware of her. Female toads frequently mudge males, though in other species of frogs and toads, movement in the vicinity of a calling male is sufficient to make the male cease his calling and turn to embrace the moving object. Since the peak of the breeding activities are carried out at night, sight plays no part in the recognition of a mate. Sound and hearing and feeling are the things that count. The females are guided to the males by their calls. The males hear or feel a nearby female. "But," you may ask quite reasonably, "how do the makes know they are embracing a female and not another make? And how can they be sure that they have a mate of their own species, and not a female of another species, if they cannot see her?"

First of all, the males distinguish the females by their silence. Should a male embrace another male—and this happens quite frequently—the male who is thus treated squeaks or croaks in protest, and struggles to free himself or such unwanted attentions. The female, however, is completely silent during the breeding season. Some female frogs can call at other times of the year, but when the breeding season comes they are mute. Second, different species breed at different times. They are not all in or around the ponds at once. However, there are several separate species there at the same time. The males of like species tend to congregate in groups apart from others. Since each species has its own distinct song, the females naturally go towards the sound of their own kind calling. Should the male embrace a female of a different species, however, he can tell by feel that she is not of his own kind. She must be the correct size—larger than he, and swollen with unlaid eggs, but not so large that she is of a different species. The way she moves when he embraces her may also tell him whether or not she is of his own kind. It sometimes happens, however, that mistakes are made in the excitement of the movement.

Toads, whose nuptial embrace is strong, have been known to grasp a fish and not let to go until the fish was crushed to death. This is rare, but it is probably due to the fact that there weren't enough females to go around and that the urge to reproduce was so strong that the male seized the first object that nudged him. About the only romantic part of the courtship of frog and toads from a human standpoint are their songs. These may be as sweet as are the calls of birds, and many people mistake the trill of the American Toad and some of the treefrogs for birds. They may be plaintive, as is the lamblike bleating of the Narrow-mouth Toad. They may be insectlike, as are the calls of the Cricket Frogs. They may be squirrel-like, or lke the loise of riveters; they may sound like a banjo, like a snore, or like a foghorn. It all depends upon the species, and there is but one male frog in all the world that, so far as we know, lacks a voice with which to call to his mate. This curious exception, which we shall discuss more fully later, is the Tailed Frog, *Ascapus truei*, found in parts of the northwestern United States. When we look at courtship in the salamanders, we perceive an altogether different picture.

In the first place, because the salamanders lack vocal chords they are silent. Not for them are the pleasures of serenading a lady love on a warm humid night. Neither have they any ears; they "hear" through their front legs or, in certain aquatic species, the lower jaw, which are especially adapted to receive vibrations from the ground. So even if the males could sing, neither the singer nor the one sung to could hear the sounds. They are like the frogs and toads in that sight plays no part in the recognition of a mate. How then do the salamanders find their mates, and recognize them? So far as we know, scent is the main factor in recognition of sex and species. Females and males smell differently, and each species evidently has its own distinctive fragrance. We humans with our dull sense of smell cannot detect most of the odours that attract the salamanders to one another. The salamanders have their own built-in perfume factories.

The hedonic glands, as these "factories" are aptly named, are located mainly at the base of the tail and on the underside of the head on the males. Though the females lack any readily differentiated glands, their skin secretions evidently possess the odour that enables the males readily to identify them as to species and sex. But once a salamander finds a mate, he does not embrace her as do the frog and toads. The salamander has more finesse, and carries on a real courtship, for a very good reason. The courting varies with the species, of course. But in general he rubs his chin against her head so that she may smell his perfume. He caresses her with his tail, and he may even invite her to dance if that is the accepted thing with his particular species. The dance may consist of his transporting her on his back, or of many other "engaged" couples joining together and making various figure eights around each other.

In certain species the female may straddle the male's tail, placing her head on the base of the tail, and the two will then waddle off together in this position. It is the male's object to make himself irresistible to the female, and to excite her so that mating may take place. In most salamanders the eggs are fertilized internally, but internal fertilization is accomplished in a manner far different from that employed by mammals. When the male salamander feels that the female is sufficiently excited, he deposits, either on land or in the water, according to the habits of his species, a small jelly-covered package. This is known as the spermatophore, and each one contains the male's sperm. Next, the female comes and picks up one of these spermatophores with the lips of her cloaca, and it is placed inside her body in a special receptacle, known as the spermatheca, where the sperm remains to fertilize the eggs before they are laid.

The salamander has such as elaborate courtship because if he did not put the female in a receptive frame of mind, she might not pick up the spermatophone. There is nothing to force her to do so. In order that the eggs may be fertilized, the males most court the females. Many species of salamanders lay their eggs a day or two after they have picked up the spermatophore. But there are others that delay.

The Mudpuppy, *Necturus maculosus maculosus*, generally mates in the fall, but does not lay her eggs until the following May or June. The same is true of the Red-backed Salamander, *Plethodon cinereus cinereus*, a terrestrial species. As we pointed out, most salamanders fertilize the eggs internally. However, there are a few exceptions. In this country there are two families of salamanders that fertilize the eggs externally. The first of these families is the Hellbenders, or Cryptobranchidae. They are completely aquatic salamanders. The male digs out a large next in the water under a rock. The females enter, and as they lay their eggs the male sheds his sperm over them. The other family of salamanders, the Sirens, or Sirenidae, so far as is known, fertilizes its eggs externally also. All anatomical evidence points to this conclusion. However, no once has yet observed the Siren breeding. It may seem amazing that no one has yet seen and accurately described these vital statistics of an animal found, not in darkest Africa; but in the United States. On the other hand, anyone who has ever tried to observe animals in the wild will realize how long it takes to obtain data on even the most inconsequential phase of their life cycles.

Mating in frogs and toads is an different matter altogether. Most frogs and toads fertilize their eggs externally. The male frog mounts the female, puts his arms around her, and grasps her firmly either just in back of her arms or in her groin. "Firmly" is a mild word for this nuptial embrace. If you catch a mated pair of toads. You will find it almost impossible to separate them, so tight is the male's grasp and so strong is the clasping reflex. However, it does not seem to brother the female in the least. Possibly, she enjoys this bear hug. In any case, if the pair mate on land they soon enter the water. The female does not, however, always lay her eggs immediately. She may lay them anywhere from 3 to 26 days after pairing, the couple remaining in amplexus all during this time. What is it that governs when the eggs shall be deposited? Is it some physiological factor? Do some species require prolonged clasping in order to lay? Or does the length of time between coupling and laying depend mainly on the weather? Perhaps all of these have some bearing on the matter.

As the female lays her eggs, the male sheds his sperm on top of them. After all the eggs are laid, the female is considerably slimmer, for she may have had as many as 20,000 eggs inside her body. As soon as she no longer feels so pleasingly plum, the male releases her. Does he return to the bank and recommence his calling in the hope of attracting another female? According to one authority, frogs and toads fertilize the egg complement of but one female each year, but this may be wrong, at least for some species. It is an extremely difficult thing to prove either way. What of the female who finds no mate? Does she lay her unfertilized eggs just the same? Evidently not. As with many unmated birds, the eggs of the female are gradually resorbed into the body. A few may emerge from the oviduct, but they are never deposited in the characteristic mass of the species as they are when the male clasps the female.

There are several outstanding and interesting exceptions to the way most frogs and toads mate. We shall discuss three of them. The first example is the primitive aquatic Surinam Toad known as *Pipa pipa*. This large flat-bodied creature lives in the ponds of South America. The mating call is a rapid clicking. This metallic sound is not produced by air vibrating the vocal chords, as in all other frogs, but by the cartilaginous disks of two bones "cracking" as the bones are moved. In much the same manner, some people on "crack" the joints of their fingers, jaws, or knees. Clasping before laying is prolonged, and lasts at least 24 hours and sometimes longer. During this period the skin on the female's back becomes swollen and puffy.

For many years it was believed that *pipa* had a protrusive oviduct that was averted when laying began. However, it has recently been proved that this is not true. When the couple are ready to lay, their pair roll over and, while they are in an upside-down position, there to five eggs emerge and are fertilized. The pair then right themselves. The entire roll and righting take about 11 to 14 seconds, and the pair is upside-down for about one second of this entire time. When the female is ready to lay more egg, the pair once more roll over, until, after many such turns, the entire complement of from 40 to 114 eggs are laid. The eggs are pressed into the spongy skin on the female's back by the make's ventral surfaces; the first eggs are placed near the annus, later ones successively farther forward toward the female's shoulders. One day after deposition, the eggs are half brief in the skin of the female, and by the tenth day the top membranes of the eggs are level with the skin of the back.

Our own Tailed Frog, *Ascapus truei* is voiceless. Since he does not seem to be able to hear, and since he spends most of his time in swift-running mountain streams, a voice would be of little use to him anyway. He possesses something far more valuable, namely, a tail. Actually, it is not a real tail, though it certainly looks like one. It is an organ for copulation. The male crawls along the bottom of the stream searching for a mate. When he finds a female, he embraces her and then inserts his "tail" into an extension of her cloaca, and thus her eggs are fertilized internally. The Tailed Frog alone possesses an intromittent organ because of his habitat.

External fertilization of eggs would be precarious, to say the least, in the fast-flowing water where the Tailed Frog makes his home. No sooner would the sperm be shed than the water would carry it downstream before it could possibly fertilize all the eggs of the female. So in order that the species may continue to reproduce. Nature has provided the Tailed Frog with a "tail" so that he may fertilize his eggs internally. There are two African toads with the jawbreaking names of *Nectophrynoides vivipara* and *Nectophrynoides tarnieri*. There toads also fertilize their eggs internally. But the mystifying thing in their case is how they go about it. They have no copulatory organ as has the Tailed Frog. Scientists have long been puzzled as to how the sperm is introduced into the female's body. Do they, like the female salamanders, pick up a spermatophore which the male deposits. There is no indication that they do. But scientists know that the eggs are fertilized internally because both species give birth to fully transformed young. The entire larval period is spent in the mother's body, and when the young are born they are miniatures of their parents.

Nectophrynoides vivipara and *Nectophrynoides tornieri* and the only two known species of frogs and toads that are ovoviviparous, though there are several salamanders that are normally ovoviviparous and several that are occasionally so when existing conditions, such as extreme cold or high altitude, make egg laying impractical. Let us look at some of the external differences between the sexes. With the sole exception of the Tailed Frog, these differences are secondary sex characteristics in frogs and toads. They are to a frog what a beard is to a man. With the Tailed Frog, the male is easily distinguished from the female by his longer "tail." Her "tail" is an extension of her cloaca, and is much shorter and more blunt than is his. With the rest of the frogs and toads, we find great variety is secondary sex characteristics.

About the only secondary sex characteristic that most male frogs and toads possess in common are their vocal sacs—and even here there are one or two exceptions that we shall examine in a movement. These vocal sacs are of *different* sizes and shapes, according to the species. They may be in throat region and when in use swell out to a glistening bubble, as is true in the peepers and many toads. They may swell out in kidney-shaped masses on the side of the head, as in the Leopard Frog. Or there may be no localized swelling, but a general enlargement of the throat, as in the Bullfrog. Vocal sacs might seem absolutely necessary to the male frog, as without them he would be unable to call to the females in the breeding season. But that is evidently not so, for some of our frogs, such as the California Toad, *Bufo boreas halophilus*, and a few others, lack vocal sacs altogether.

The Tailed Frog lacks them also, but he is believed to be silent. However, the California Toad is known to have a voice—a trill, similar in character to that of our northeastern American Toad, *Bufo americanus americanus*, but lower pitched. Neither does the sound carry so well, but whether this is because of the lack of vocal sacs, as one eminent scientist believes, or because lower-pitched sounds do not seem to carry so well as those that are shrill, has not been proved. The vocal sacs, even though they are usually present, are not very helpful to us when we pick up a frog and want to know whether or not he is a male. Unless we see his vocal sacs swelled up, we would not even know he had them. The females of some species lack vocal sacs, and in others they are greatly reduced in size. And, as we have said, the females are silent during the breeding season, though at other times of the year they may make calls similar to the males of their kind, but not so loud or so resonant.

Size is a factor in sex recognition of frogs and toads. The female is a general rule, larger than the male. It is just as well that she is, because she must carry him about on her back during mating and egg laying. In the water, though the additional weight does not matter so much, the female nevertheless does the swimming for both. Many males keep their hind legs drawn up on the female's back, while others allow their legs to float out behind. In a few species, if the couple is alarmed, the male may try to aid the female by using his feet in swimming motion, but their efficiency is impaired by his position on top of her. Many, however, like the American Toad, make no attempt at all to help the female in her race to escape.

It seems curious to us that the male refuses to release his hold on the female and that each does not go its separate way when threatened with danger. But such is not their habit. Some species separate more easily than others, but none does it voluntarily, and often it is almost impossible to force the mated pair apart without injury, so strong is the embrace of the male. Whereas the female is usually larger in overall size, the male may have much larger and heavier forelegs. Perhaps, he has need for powerful arms the better to grasp the female. The male toad, *Bufo boreas*, found in the Northwest and in California, has enlarged forelimbs.

His are insignificant, however, compared with a South American frog, *Leptodactylus acellatus*, whose arms are fully three times the size of those of the female? In some species of frogs, the eardrum or tympanum, of the male is considerably larger than the eye-whereas in the female it is the same size. A male Bullfrog can easily be distinguished from a female by this characteristic. Why the male should need a large ear than the female has never been explained. You would think that if either of them should have a larger one, it would be the female that would have most use for it. And why, of all species, should the Bullfrog, with his tremendous foghorn cry that can be heard from a distance of half mile or more, need a larger ear. The answer probably lies in the Bullfrog's evolution. At one time male perhaps did need larger ears, and though they are no larger useful they have not as yet been discarded. In certain species of frog and toads, the males have much larger thumbs. These may be swollen at the base or at tip. This enlargement of the thumb gives the male better gripping power. Does it also help to cushion the female's skin from being overly bruised by the male amphibian's strong grasp? We have never heard this suggested as a reason—it would probably be considered rank sentimentality by most scientist—but we can see no reason why it should not be so.

There is also a difference in the shape of the toe webbing of the two sexes in certain species. A male Wood Frog and be distinguished from a female in the breeding season by his convex toe webs. The female's are concave. Of course, you have to catch the frog and spread his hind toes apart in order to see this. It can't be observed from a distance. Some reptiles and many birds present striking colour dimorphism between the sexes. This is not true of frogs and toads. Colour differences, when they exist, are usually confined to the underparts, and especially the throat region. Some males have darker or more brightly coloured throats than the females of the same species. Of all the secondary sex characteristics exhibited by the frog and toads, none is more startling than those developed by the male Hairy Frog. *Astylosternus robustus*, of Africa. All male frogs and toads need a great deal more oxygen during the breeding season than at any other time. This is due in part to their vociferous calls, and in part to the fact that their metabolic rate is heightened.

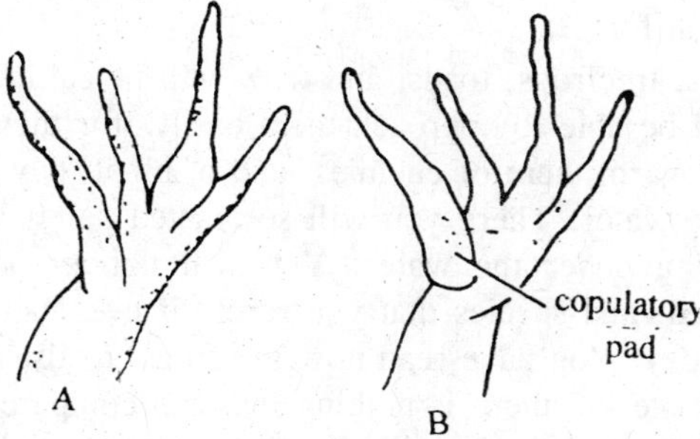

Fig. 12.2. A—Foot of frog showing web, B—Hand of male frog showing nupial or amplexusory pad.

Most male frogs land toads, therefore, have larger lungs to satisfy the necessary requirements of their bodies for oxygen. But the Hairy Frog is an aquatic frog with very small lungs. Since he is a large fellow, he needs a great deal more oxygen in the breeding season than his diminutive lungs can supply. Therefore, he broadens his skin area at that time, so that he will have a greater surface through which to absorb the extra oxygen needed, by growing long "hair" on his hind legs and lower sides. Of course, they are not really hairs, any more than gills and feathers, which they resemble at times. Scientists call these "hairs" vascular villosities. Be that as it may, they look like hair and they work like gills, enabling the Hairy Frog to obtain the additional oxygen necessary during the breeding season.

All in all, the secondary sex characteristics of frogs and toads are not very useful to us in helping us to tell whether our frog is male or female. It is unfortunate, but in this case there is no short cut. In order to distinguish between the sexes, you must know the secondary sex characteristics of each species. Fortunately, the same state of affairs does not exist in the salamanders, in the breeding season. There are the larger overall size of the females, the enlarged hind legs and or tail of the male, differences in the colours of the two sexes, and differences in the teeth. These correspond to those secondary sex characteristics in frogs and toads. But in almost all salamanders, there are differences around the vent, especially during the breeding season, though at other times these may be negligible. The male's vent is large, protuberant, and has tiny nipple-like projections. The female's vent is smooth or in folds. Before we go on to the eggs. Let me urge you to chose a warm humid overcast night in May for an outing. Wear a pair of galoshes, take a strong flashlight, and head for the nearest pond or swamp. You should have an experience that you will not soon forget.

Peepers, treefrogs, toads, and frogs will be calling from all sides. You should be able to approach them easily, for they depend mainly on sight to warm them of enemies, and at night they cannot see you. Look in the water. There you will see mated pairs, salamanders and frogs, calling under the water. You will notice species you never knew existed, and species that you rarely if ever catch a glimpses of during the day. You have read how the frogs, toads, and salamanders court and mate but there is nothing that can compared with the thrill of actually witnessing these things with your own eyes and ears. Even if you aren't a keen field naturalist, you will be amazed at all you will discover, and will be very glad that you went. Perhaps you may even be so impressed and enthusiastic that you will want to repeat the experience.

13

EGGS LAYING

The beginning of most life is hidden from us. We are unable to see the egg being fertilized; its development from one-celled organism to an embryo of many cells is screened from our view until birth or hatching. This is true of the shelled eggs of reptiles and birds. They are fertilized within the parent's body and are then surrounded with a shell before laying. The mammalian egg is shell-less with two exceptions; but it is microscopic in size and inaccessible for observation, since it remains inside the mother's body during its development. If we really want to find out what happens between laying and hatching we must turn to the amphibian egg.

The spermatozoa of the amphibian are microscopic in size, so that we are unable to see (with the naked eye) the egg being penetrated by the sperm. But all stages after fertilization are visible to use; furthermore, most of the changes that occur can be seen without a microscope. To begin with, the eggs are round, though they do not remain so. They may be black or brown on top. The top part is known as the animal pole; it is the section that is alive and will grow. The vegetal pole at the bottom may be white or yellow. It consists of the food on which the egg "feeds" as it develops, for you must remember that all living things; be they animals or that most miraculous of all living entities, a fertilized egg cell, must have nourishment if they are to grow.

The living protoplasm of the animal pole is lighter than the non-living deutoplasm of the vegetal pole. Therefore, it always is uppermost. You can easily test this by turning an amphibian egg so that the animal pole faces down. You will notice that very little time elapses before it has rotated so that the protoplasm is again facing up. A few amphibian eggs are yellow or white, and are called unpigmented eggs. But though the separation of animal and vegetal poles is not readily visible in these, it exists nonetheless. Because the eggs require a certain amount of insulation from shocks land enemies, as they descend through the female's oviduct they are coated with albuminous jelly.

Depending upon the way the oviduct is constructed in different species, there may be one or more jelly envelopes, or envelopes surrounding the eggs may be lacking, and the mass of jelly may constitute the sole protection. You can get a clear picture of what an egg with its jelly envelope looks like if you make a small dot and draw a circle around it. The jelly may be clear, or it may be milky white or greenish. Once the eggs are deposited in the pond, the jelly swells. The capacity of the jelly to absorb water is a decided benefit to both the layer and the egg. It means that the eggs and jelly, when inside the female amphibian, require little space. But one in contact with the water, the jelly swells to twice or three times its original bulk, giving far greater protection to the egg than would be the case if it did not enlarge. Of course, there are instances on record of frogs that delayed too long in laying their eggs.

The water content of the body is absorbed by the jelly, which expands rapidly and beyond all bounds. The female then bursts open and dies. Fortunately such circumstances as these occurs seldom, and may be considered freaks of nature. The eggs vary in size according to the species that lays them. But the size of the adult is no reliable indication of the size of the egg that will be laid. The Tailed Frog is only about 2 inches long. But it lays one of the largest eggs of all our native-frogs. Each vitellus is approximately 5 mm., or roughly 1/4 inch in diameter. Our largest frog, the bullfrog, which may be as long as 8 inches, lays small eggs of about 1.5 mm. The smallest egg in this country is laid by the Southern Chorus Frog. *Pseudacris nigrita nigrita* and is approximately .7 mm in diameter. We shall allow you to figure out the equivalent measurement in fractions of an inch. Frankly, it's beyond our mathematical ability, and we are quite content to leave egg measurements in millimeters.

The average salamander lays a far larger egg than do the frogs and should you ever wish to watch the development of the amphibian egg, you would be well advised to choose those of a salamander rather than those of a bullfrog or a toad. The number of eggs laid also varies greatly, and generally speaking, salamanders lay fewer eggs than do frogs and toads. The Bullfrog and the American Toad lay as many as 20,000. The Rocky Mountain Toad, *Bufo woodhousei woodhousei*, produces upto 25,000 eggs, the largest complement laid by a native salientian. The Robber Frogs of the genus *Syrrhophus* may lay as few as 5 eggs but they are the exception.

Most frogs and toads lay eggs in the hundreds or thousands, while salamanders lay in the tens or a hundred. Because of the salamander's more secretive ways, larger egg size, and internal fertilization, they do not need to lay so many eggs to ensure continuance to the species as do the frogs. The way the eggs are laid also differs with the species. The Red-spotted Newt lays each egg separately, fastening each on the leaf of a water plant and folding the leaf around it thus hiding it very effectively from enemies. The Spring Peeper also lays its eggs individually, but this species merely deposits each egg on the pond bottom. The Bullfrog and the Green Frog both lay theirs in spreading mass on the surface of the water.

The Wood Frog and the Spotted Salamander lay theirs in a round or oval mass attached to submerged vegetation. The American Toad and most other toads lay theirs in two long strings of jelly. The Hellbenden that large aquatic salamander, lays its eggs in rosary like strings, as does the Tailed Frog. The Spadefoot Toads lay their eggs in bands or cylinders. It is, of course, the jelly that gives the egg mass its shape, for the eggs themselves are round, with the exception of those of the Colorado River Toad who occasionally, though not always, lays wedgeshaped eggs. The form of the jelly is governed in part by the way the oviduct is shaped, and in part by the natural egglaying movement, which differ in the various species.

Fig. 13.1. Amplexus and oviposition.

Eggs of amphibians are identified one from another by their colour, size, number of jelly envelopes, if any; shape of the eggs mass, total complement of eggs, and the places in which they are laid. All of this sounds complicated as do most things scientific at first glance. However, it is not so formidable as it seems, and if you are one who enjoys finding out just exactly what species of amphibian has laid those eggs in the southeast corner of your pond, do not be overawed. Consult one of the two handbooks listed in the Bibliography, and forge ahead. You may not find that you can identify every egg that you discover with complete scientific accuracy, but you will be able to identify most, and make a well-informed guess about the others. The eggs are laid in many places.

A large majority amphibians deposit their eggs in ponds, either spread on the surface of the water or, more often, submerged. Those species, such as the Bulling that lay on the surface often lay a large number of eggs, and they also lay them later in the year. One of the reasons for this is that the surface-film eggs would freeze and decompose were they laid in cold weather. Furthermore, this type of eggs has a very high mortality rate from another natural cause-hard pelting rain.

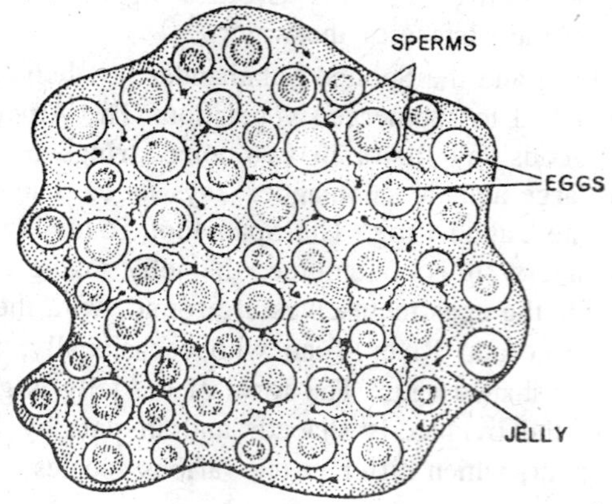

Fig. 13.2. Frog's spawn.

The force of the rain may be hard enough to break the surface film of the water, and cause the eggs to sink to the bottom of the pond. And there they are not likely to develop because there is less oxygen on the pond bottom than at its surface, and the surface-film eggs enquiries more oxygen than it may be able to obtain on the bottom. More salamanders lay in the swiftly running waters of brooks and rivers than do frogs. Undoubtedly, this is primarily because most of them fertilize their eggs internally, whereas the frogs, with the exception of the mountain-stream-dwelling Tailed Frogs, are unable to do so. Those salamanders that lay in rivers either scoop out a nest, with its entrance downstream, for protection from the onslaught of the cascading waters, or they attach their eggs, either singly or in a mass, to the underside of a rock.

In many parts of this country, permanent bodies of water are non-existent. Where this situation occurs, you might think that there would be no water-laying amphibians. But amphibians such as the Spadefoots, the Narrow-mouths, and some of the toads must lay in shallow water, and in dry regions they lay in mud puddles that are formed from a rain that may not recur for a year or more. Naturally, the eggs hatch rapidly, and the tadpoles are transformed in a short time into adults. The habit of laying in the shallow water of mud puddles and swamps is not, however, confined to arid regions. None of our north-eastern toads will deposit their eggs in deep water even though it is available. If you wish to find toad's eggs, the best place to look is a swamp or a mud puddle. We have even found toad's eggs laid in a large mud puddle in the middle of a road.

Lacking swamps in your vicinity, go to a small pond. Search around its shallowest edges, for it is there that you will find the toad's strings of jelly. Very few people who do not make a study of amphibians are aware that some of them deposit their eggs on moist or dry land. This is not particularly surprising, for the chances of finding these land-laid eggs just by accident are slim, hidden as most of them are beneath rocks or logs. As a matter of fact, diligent hunting is generally required to locate them even if you know the most likely spots to search. There are quite a few native salamanders who lay on land. The Marbled salamander, *Ambystoma opacum*, lays in a moist spot near the water, as does the Dusky Salamander, *Desmognathus fuscus*. The Red-backed Salamander, *Plethodon cinereus*, is woodland salamander, and it lays its eggs in cavities or rotted logs. For some reason the Amphiuma, *Amphiuma means*, though it is an aquatic salamander, also lays its eggs in muddy spots on land. There are many reasons for normally terrestrial amphibians to lay on land, but why a salamander that spends its life in the water should do so is a mystery.

As to frogs, there are fewer of them, and most of them are not so common as the land-laying salamanders. The land-laying frogs in this country are confined to several species within one family, the Robber Frogs, or Leptodactylidae, of which we will say more in a subsequent chapter. We have but one native species of amphibian that deposits its eggs in trees. This is the Oak Salamander, *Aneides lugubris lugubris*, of California. It occasionally lays on the ground, but just as frequently lays in trees. None of our native frogs lays in trees, but there are many foreign frogs that do. All the treefrogs of the family Hylidae that live on the island of Jamaica lay in trees. Perhaps this habit arose because there are very few natural ponds and lakes in Jamaica. The type of vegetation found in Jamaica and parts of South America favoured its development. The Jamaican hylas all deposit their eggs in bromeliads.

Bromeliads are relatives of the pineapple, but, unlike it, they are epiphytic; that is, they attack themselves to trees and obtain their sustenance from the air through specially modified aerial roots. The bromeliad leaves from a rosette, just as do those of the pineapple. In the very center of the rosette is a hollow that generally holds a considerable amount of water. It is here that the Jamaican hylas place their eggs, and the tadpoles develop in this water. Even in the driest of seasons, the bromeliad has a store of water in the center of its leaves. Other foreign frogs, such as the treefrogs of the family Rhacophoridae, found in Asia, deposit their eggs in trees overhanging pools, and upon hatching, the tadpoles fall into the water below and carry on their larval life there.

In order to protect the eggs from drying out, the mother first deposits some albuminous jelly without any eggs. This is whipped into a frothy mass by the hind legs of one or both parents until it is as full of air as beaten eggs whites. The jelly-covered eggs are then placed on top and are covered by an additional layer of jelly which is also well beaten. The outside of this nest soon hardens, while the inside liquefies, providing the eggs with a suitable medium in which to develop. The changes in the amphibian eggs from the moment it is laid and fertilized to the time it hatches are fascinating to watch. This is how most life begins and develops, from the mammal down to the lowly earthworm. If you go to the nearest pond and collect the largest eggs that you can find, you may watch the entire process with the naked eye or with an inexpensive magnifying glass. Suffice it to say here that the eggs changes—divides from one into many cells—lengthens out, and develops, and finally the tadpole or larva wriggles its way out of the confining jelly and hatches. The length of time between laying and hatching varies with the species and the temperature. It may be as short a time as 36 or 58 hours, as in some toads, or as long as 3 or 4 months in some salamanders.

In general, salamander eggs are slower in hatching than are those of the frogs and toads. By and large, amphibians lay their eggs and then desert them, devoting little thought as to the possible fate of the eggs. However, there are many exceptions to this rule, and some of them, especially the foreign ones, are most interesting. Our first exception is the male Hellbender, native to this country, who as we mentioned earlier, scoops out a nest on the river bottom, and welcomes the females as they enter to deposit their rosary-like strings. The male is a most faithful father, for he remains with the eggs, guarding the opening of the nest against all intruders, until the eggs hatch out 2½ or 3 months later, while the female goes her own way, her responsibility ended. The female three-toed Amphiuma, *Amphiuma means tridactylum*, who lays in the mud, remains with her eggs until they hatch. One wonders why the other subspecies, *Amphiuma means means*, who also lays in the mud, does not remain with hers. This habit of one or both parents remaining with the eggs laid on land is quite common. We find it true of the Marbled Salamander, the Dusky Salamander, the Oak Salamander, and the Redbacked Salamander. When this habit exists, you may be sure it is necessary, not so much for the protection of the eggs from enemies as from desiccation. The damp body of the parent helps in no small measure to keep the eggs themselves moist.

It has also been suggested that the skin secretions and cloacal excretions and/or secretions of the parents may serve as a deterrent to any fungus growths that might destroy the eggs capsule and egg. Recent research indicates, however, that mold develops only on dead eggs or embryos. The role that skin and cloacal secretions and the brooding habit play in increasing the number of young which survive has not been fully explained to date. When we come to the foreign species who care for their eggs, we see all manner of strange sights. The male Midwife Toad, *Alytes obstetricans*, found in western Europe, is tied down by his parental duties.

The female lays her jelly-covered eggs in long strings; the male then takes these, wraps them around his hind legs, and retires to a damp cavity. Here he remains during the day, coming out at night to bathe the eggs in dew or in a nearby pond. When the tadpoles are ready to emerge, the male goes to the water and the tadpoles hatch. One wonders how the male is able to get around at all without crushing the eggs. Perhaps a large number are destroyed, though we have never read anything to indicate it. But whether or not a large percentage of the eggs develop. Nature has amply provided for the survival of the Midwife Toad, for the female lays not once or twice annually, as do most amphibians, but three or four times each year. The male must be quite adept at carrying his burden of eggs, since he obviously gets plenty of practice. But if you think the Midwife Toad is badly off, you have never heard of *Rhinoderma darwii*, and others of the genus *Rhinoderma*.

The males carry the eggs and tadpoles until they are transformed into fully developed frogs in their mouths—or, properly speaking, in their vocal sacs, which are necessarily quite large and elastic. In the general *Phyllobates* and *Dendrobates*, found in South America, the male again cares for the young. In his case, he builds a mud dam to trap water for a suitable site for egg laying. Should this artificially made pool dry up before the tadpoles are transformed, their father takes them for a pickaback ride to a nearby stream where they can complete their development. They, in turn, have special suctorial lips with which they can fasten themselves firmly to his back so that they will not be dislodged on the overland journey. Naturally, if they should fall off their life is soon ended, for though they are able to withstand short trips out of water, they are unable, while tadpoles, to live away from an aquatic medium for any extended period.

14

EMBRYOLOGY

The frog is most common and easily available tailless amphibians. It lives chiefly in ponds and swamps, though some of its species may be also to exist in damp or shady places away from the water. Eggs of frog are easily obtained and may be studied in the laboratory from the moment of fertilization onward. This is not so easily true to the higher vertebrates. The amphibian embryos, also, are relatively straight, not coiled. This makes description easier. The stages of embryonic development differs in various chordates, yet the chief phases are basically similar in all. The differences are related primarily to the amount of yolk particles present in an egg. The yolk particles provide nourishment of the developing embryo.

MALE GENITAL ORGANS

As already noted the main male genital organ that produces and discharges sex cells in the testis. The testes are whitish or yellowish cylindrical or ovoid organs longer than broad. They lie ventral to the kidney attached to the outer edge of its anteriormost portion. They are suspended by a fold of peritoneum called *mesorchium*. From the testes arise a number of slender tubes, the *vasa efferentia*, and enter the kidney. Each tubule is slender tough-walled and interbranching duct lined with closely packed cuboidal cells. Each duct is connected directly with a number (8 to 12) of Malpighian corpuscles of the kidneys by way of the Bowman's capsules. These connection are permanent, so that the anterior uriniferous tubules of the kidney contain spermatozoa during breeding season. The testes produce male cells known as *spermatozoa* that pass through the substance of the kidney into the urinogenital duct. In some frogs the free portion of the urinogenital

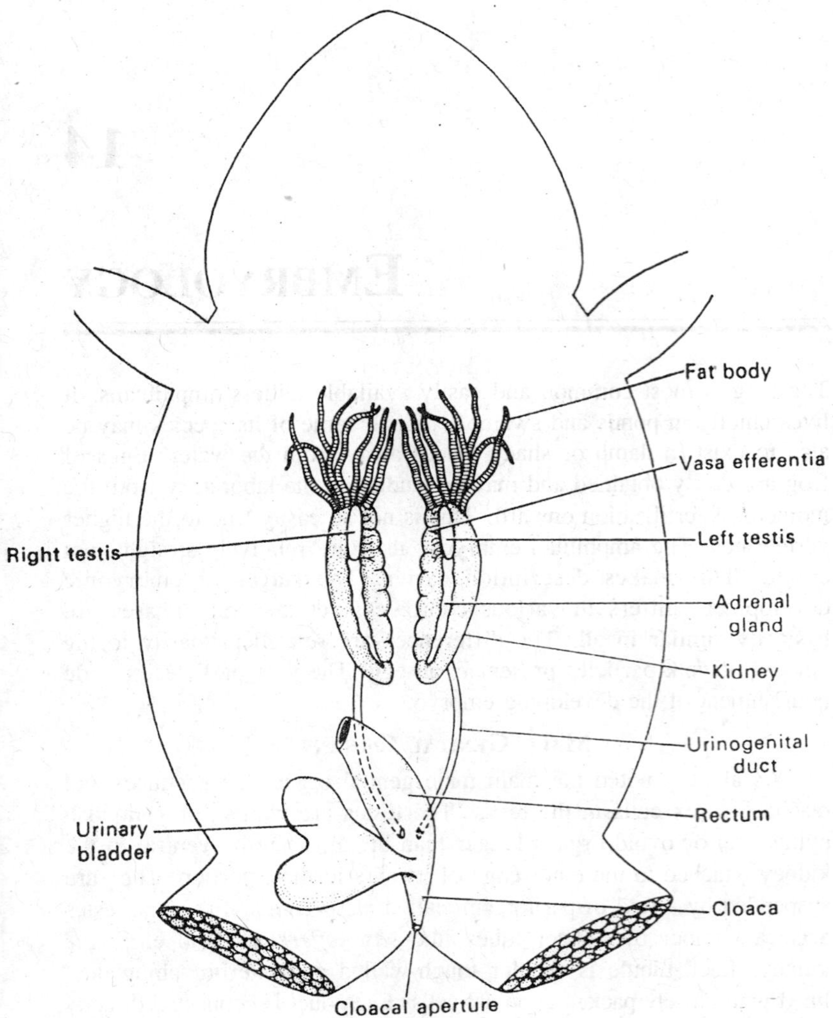

Fig. 14.1. Reproductive system of male frog (ventral view).

duct gets dilated forming the *seminal receptacle*. This store the spermatozoa till such as required. Each testis consists of a large number of closely packed, oval-shaped sac called *seminiferous tubules*.

The spermatozoa are produced within them. They are separated from each other by thin partition (*septula*) of connective tissue called *interstitial tissue*. It is likely that this tissue has some endocrine function. The thickness of this tissue is much reduced immediately after breeding or pituitary stimulation. The interstitial tissue is continuous with the covering of the testes called the *tunica albuginea*. Each seminiferous tubule is lined with germinal epithelium which produces sperm-mother cells or spermatogonia. They keep on multiplying for some time. Shortly after this each spermatogonium enters upon a period of rest, the mitotic activity stops but nuclear activity continues. In between the spermatogonia some large cells, the *primary spermatocytes*, may be found. They give off polar bodies and form *secondary spermatocytes*, which are smaller than the primaries and lie toward the lumen of the tubule.

Following another division the *spermatids* are produced which are even smaller and possess a condensed nucleus of irregular shape. Clusters of spermatids appear as clusters of granules. The spermatids metamorphose into *spermatozoa*. The mature *spermatozoon* is about .03 mm, in length, has an elongated solid staining head containing the nucleus, with an anterior acrosome. The short middle piece generally is not visible, but the tail appears as a gray filamentous extension into the lumen, about four or more times the length of the sperm head. In any trasverse section of he testis, bundles of sperm heads or tails may be cut at right angles or tangentially giving wrong suggestions about the structure. The mature sperm depends upon external sources of nutrition, so that it joins to form a congregation of 25 to 40 spermatozoa, the heads of all of which may converge into the cytoplasm of a relatively large columnar basal cell known as the *Sertoli cells*. Functionally this is a nurse-cell supplying nutriment to the cluster of mature spermatozoa till they are liberated.

Secondary Sexual Character

The mature male is generally smaller than the female. The male possesses a slender and *streamlined* body, a *darkened thumb* which changes its colour-intensity and thickness as the breeding season approaches, and the male produces distinct low *guttural croaking sound*. In many species males show other features such as brilliant colour on the ventral surface of legs (*R. sylvatica*), black chin (*B. fowleri*) or differentiation in the colour and size of tympanic membrane.

Fat-body

Attached to the anterior end of the testes may be seen a yellowish organ, the *fat-body* (*corpora adiposa*). Each consists of a number of finger-like out growth, whose number varies very greatly. Even in the same individual the number varies at different periods. The fat bodies represent stored nutrition for the long period of hibernation and for the pre-breeding season when food is scarce. The fat-bodies consist of clusters of vacuolated cells. It is believed that the fat-bodies as well as the gonad arise from the genital ridges of the early embryo. The fat-bodies tend to be reduced soon after the breeding season, only to be built up again as the period of hibernation comes near.

Female Genital Organs

The ovaries of the frog are paired, multilobed attached to the dorsal wall of the body by a double fold of peritoneum, the *mesovarium*, which surrounds the entire ovary as *theca externa*. Each lobe of the ovary is hollow and its cavity is continuous with that of other lobes. Each ovary may contain as many as 7 to 12 lobes. The ovaries occupy the same relative position in the body cavity as is occupied by the testes in the male. The size of the ovary varies with season. In the breeding season they may fill the body cavity and may even distend the abdomen. The number of egg varies from 2000 (*Rana pipiens*) to as many as 20000 (*Rana catesbiana*) each measuring about 1.75 mm. in diameter (*Rana pipiens*). The mature eggs are highly pigmented on the surface of the animal pole. It is for this reason that the mature ovary presents a speckled appearance of black pigment (animal pole) and white yolk (vegetal pole).

After the discharge of the ova the ovary becomes so small and pigmentless that it becomes difficult to locate it. It, however, contains oogonia for the following year. Within the outer peritoneum the *theca externa*, are suspended thousands of individual sacs, each of which consists of another membrane the *theca interna* (or *cyst*) which contains smooth muscle fibres. The theca interna surrounds the entire egg except for a small area bulging towards the body cavity. At this bulge the egg is covered only by the theca externa and it is this region that ruptures for ovulation. Within the theca interna and the ovum there are *follicle cells* derived from oogonia. The theca interna plus, the limited covering of the theca externa and the follicle cells form the *ovarian follicle*. In the ovary of the frog the ovarian connective tissue or *stroma* is limited very much, being represented by the two membranes the theca externa and the theca interna. Each egg is surrounded by a non-cellular transparent *vitelline membrane*, probably derived from the ovum as well as the follicle cells. This membrane develops during maturation. After the egg is fertilized a fluid-filled space appears within this membrane and is called the *perivitelline space*.

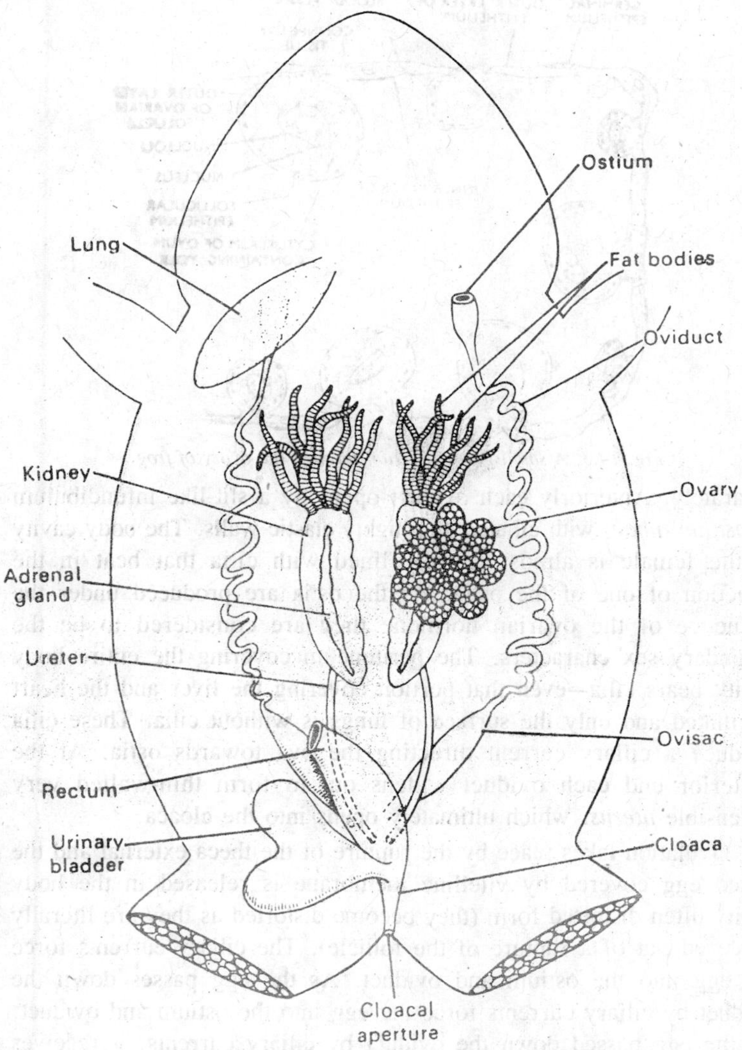

Fig. 14.2. Reproductive system of a female frog (ventral view).

Essentially the egg is just like large sac of yolk, the heavier and larger granules of which concentrate at the *vegetal pole*. On the side of each ovary lies greatly convoluted tube of white colour. These are the *oviducts* and are suspended by a double fold of peritoneum. The oviducts run from below the lungs to the cloaca where they open

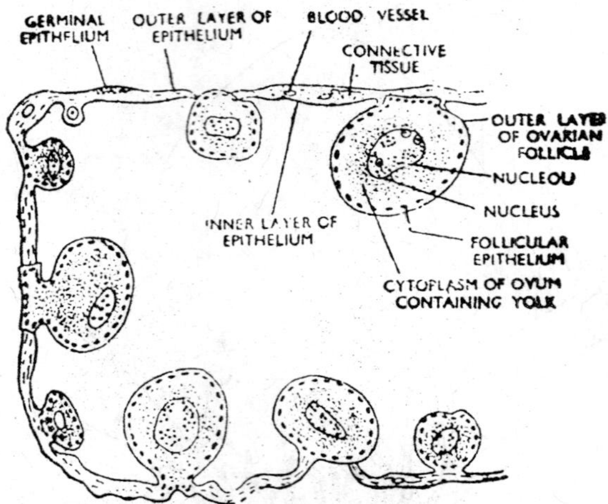

Fig. 14.3. A small part of section through the ovary of frog.

separately. Anteriorly each oviduct opens by a slit-like infundibulum or *ostium tubae*, with ciliated and highly elastic walls. The body cavity of the female is almost entirely lined with cilia that beat in the direction of one of the ostia. As the ostia are produced under the influence of the ovarian hormone they are considered to be the secondary sex characters. The peritoneum covering the entire body cavity bears cilia—even that portion covering the liver and the heart is ciliated and only the surface of lungs is without cilia. These cilia produce a ciliary current directing the ova towards ostia. At the posterior end each oviduct widens out to form thin-walled very distensible *uterus*, which ultimately opens into the cloaca.

Ovulation takes place by the rupture of the theca external and the naked egg covered by vitelline membrane is released in the body cavity often distorted form (they become distorted as they are literally squeezed out of a rupture of the follicle). The ciliary currents force the egg into the ostium and oviduct. As the egg passes down the oviduct by ciliary currents force the egg into the ostium and oviduct. As the egg passed down the oviduct by ciliary currents, it receives coating of *albumen* (jelly). The initial coat of jelly is thin but of heavy consistency. It is closely applied to the vitelline membrane. As the egg moves down more coatings are added. These layers are transparent and of uniform thickness, the jelly becomes very thick on coming in contact with water when the egg is laid. The thickness may

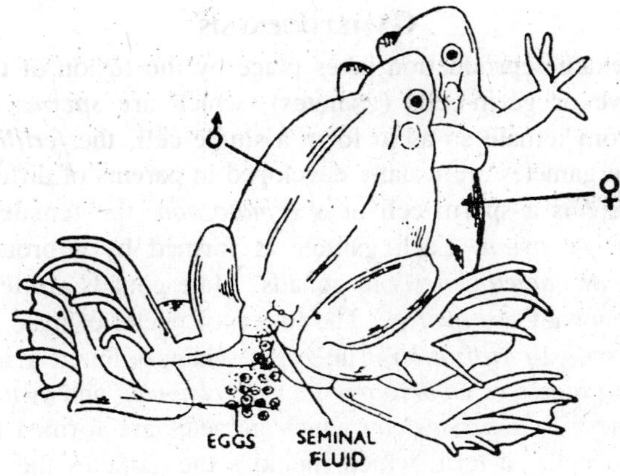

Fig. 14.4. False copulation in frog.

exceed the diameter of the egg. The egg are discharged only in the presence of the male.

Secondary Sexual Characters

The mature female frog is generally larger than the male of the same species at the same age. In the female the dark thumb pad of the male is lacking and they are not able to produce the chick pouches as do the males. Their abdomen is usually flabby and distended and the peritoneum lining the body cavity is ciliated. In some cases during the breeding season dermal papillae occur on the skin evidently to aid clasping by the male.

Fat-bodies

The *fat-bodies* (*corpora adiposa*) are found in female also, but in these it is less closely attached to the gonad than in the male. In other detail the fat-bodies in both the sexes are similar and function as store-house of nutriment.

Secondary Axis

According to them the ovum itself has a fixed inner chemical organization for the axis pattern. This secondary axis is also called dorsoventrally axis, the region of grey crescent becoming the dorsal side of the future animal and the opposite area the ventral side. This axis establishes the bilateral symmetry of the developing ovum. A perpendicular plane conciding with the primary and secondary axes divides the ovum upto right and left half, representing the right and left side respectively of the future animal. This plane coincides with first cleavages.

GAMETOGENESIS

The sexual reproduction takes place by the fusion of two mature reproductive or germ-cells (*gametes*), which are *sperms* from male and *ova* from female so as to form a single cell, the *fertilized egg* or *zygote*. The gametes are usually developed in parents of different sexes. Male gamete is a sperm cell or *spermatozoon*, the female gamete is an egg-cell or *ovum*. Each gamete is formed by a process, called maturation or *gametogenesis* in gonads. Male gonads are testes, while the female gonads are *ovaries*. The fusion of nuclei of male and female gametes is called *fertilization*. The zygote changes into a mature animal through the process of *embryology* i.e., *development* and *metamorphosis*. Gametogenesis is a *process* by which gametes are formed from cells, called germ-cells, a term which includes the gametes themselves.

Earliest germ-cells are called *primordial germ cells*, which can be recognised very early in the life of an animal as found in *Ascaris*, where germ cells are formed from one of the four cells of the embryo. However, in majority of the other animals, they are distinguished at early stages, before the gonads appear. By mitotic cell-divisions *primordial germcells* give rise to *gonia* or additional germ-cells. The development of sperm cells from the primordial germ-cells is known as *spermatogenesis*, while the development of egg-cells from primordial germ-cells is known as *oogenesis*. In both these processes, the number of chromosomes is reduced to *half*. This reduced number is called *haploid* and, at the time of fertilization, *diploid* number of chromosomes is formed. Thus, the *constant number* of chromosomes characteristics of a given species is maintained.

Spermatogenesis

A typical testis is made up of thousands of cylindrical sperm-tubules, in each of which numerous sperms develop. The walls of the sperm-tubules consist of spermatogonia or unspecialised germ-cells. For the growth of the testis, the *spermatogona* divide the mitosis so as to give rise to additional spermatogonia. When the sexual maturity is attained, some spermatogonia undergo spermatogenesis, which consists of two *meiotic divisions*, followed by cellular changes resulting in maturation of sperms. Other spermatogonia still continue dividing so as to form additional spermatogonia for spermatogenesis at a later period. During breeding season, testis increases in size and spermatogenesis takes place. During spermatogenesis, spermatogonia grow into larger cells, called *primary spermatocytes*.

Each *primary spermatocyte* divides by first meiotic division so as to form two cells of equal size, called *secondary spermatocytes*. These divide by second meiotic division so as to form four *spermatids*. Secondary spermatocytes are known as *diploid* (2x) i.e., the number of chromosomes is the same as is present in most of the cells of the animal body. But in each spermatid, there is a bit of cytoplasm and the number of chromosomes is haploid (x). Now, no division takes place but spermatid changes into *spermatozoon*. The changes include sprinking of nucleus in size, and formation of head. Most of the cytoplasm is shed off.

Some of the *Golgi bodies* aggregate at the anterior end of the sperm so as to form a point, which is said to be helpful for puncturing the cell-membrane of the egg during fertilization. A bit of cytoplasm also develops a *tail*, which helps movements of the sperm. The *mitochondria* lie at the junction of the head and tail so as to form a *middle piece*, which gives energy for the beating of the tail. The structure of mature spermatozoa is different species. Sperm of *Ascaris* is tailless. The sperm of *crabs* and *lobsters* are curious having pointed projections on the head.

Table 14.1. Comparison of Spermatogenesis and Oogenesis.

Spermatogenesis	*Oogenesis*
1. Spermatogenesis takes place in testes.	1. Ogenesis take place in ovaries
2. As a result, sperms are formed, which are typically much smaller and motile adapted to swim activity to the egg by lashing movements of the tail.	2. Egg is typically large, non-motile and contain yolk, which supplied nourishment to the developing embryo.
3. Secondary spermatocytes divide by second meiotic division to form four spermatids	3. Secondary oocyte divides by second meiotic division to form a large ootid and a small second polocyte.
4. Four sperms are produced from each primary spermatocyte.	4. Only a single ovum is produced from each primary oocyte
5. Polocytes are not formed.	5. Polocytes are formed so that mature egg may have haploid number of chromosomes.

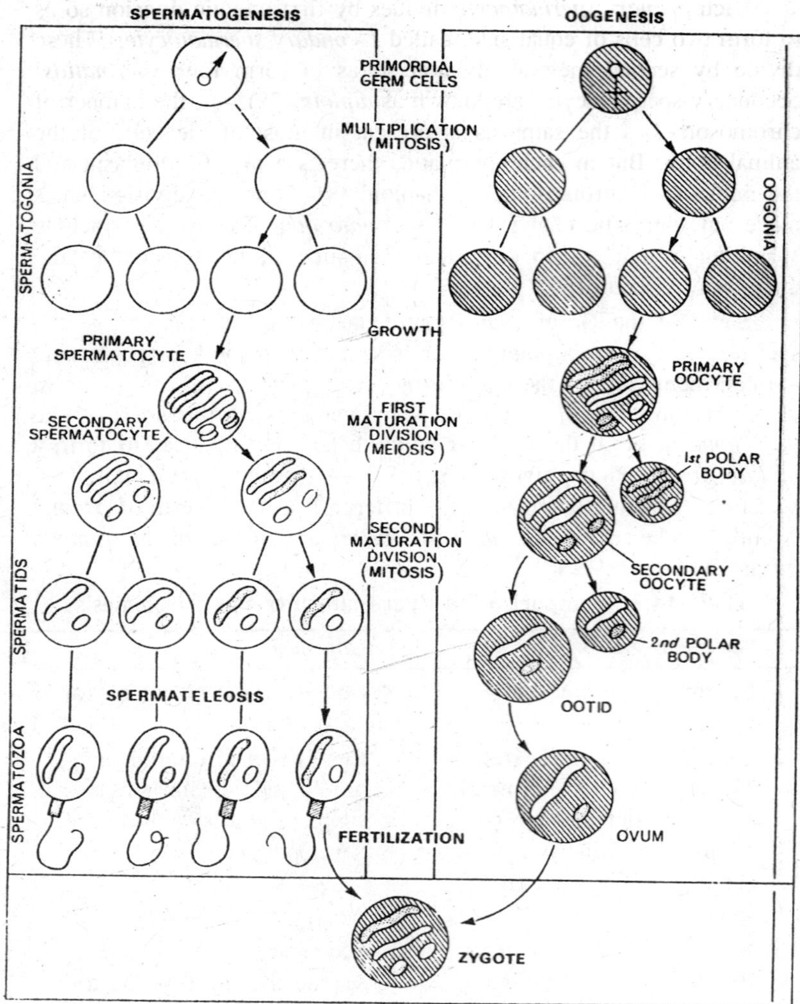

Fig. 14.5. Stages of gametogenesis (spermatogenesis and oogenesis).

Oogenesis

Immature sex-cells of the ovary are called oogonia, which, as a result of mitotic division, from additional *oogonia*. When the individual attains sexual maturity, oogonia develop into large *primary oocytes*. They are much larger in the size than the primary spermatocytes, as *yolk* is found in them. After growth, the primary oocyte divides by first meiotic division; as a result, two daughter-cells are formed. However, these daughter-cells are not of equal size. One daughter-cell is called the *secondary oocyte*, which gets all type cytoplasm and yolk, while the other is called the *polocyte* or *polar body* and is simply a nucleus and may divide into two additional *polocyte*. The secondary oocyte divides by second meiotic division again with an unequal division of cytoplasm so as to form a large *ootid* with yolk and cytoplasm and a small second polocyte.

Further cell-division does not take place and ootid changes and the result is formation of a mature *ovum* or *egg*. Polocytes disintegrated and disappear, thus, each primary oocyte forms a single ovum. Polocytes are formed so as to get rid of the excess chromosomes, so that the mature egg gets enough cytoplasm and yolk to survive. Egg are of different size, which vary due to the presence of the amount of yolk. Eggs with little or no yolk are small. Human egg is about 0.1 mm in diameter. Frog's egg is about 2 mm, yolk of hen's egg is about 30 mm. in diameter.

Structure of an Ovum

The ovum of the frog is a large cell, about the size of 1.6 mm. The *upper hemisphere* of the egg is black due to the presence of many pigments granules. The dark colour of the egg is an adaptation for absorbing and retaining heat as eggs are laid in water in the early

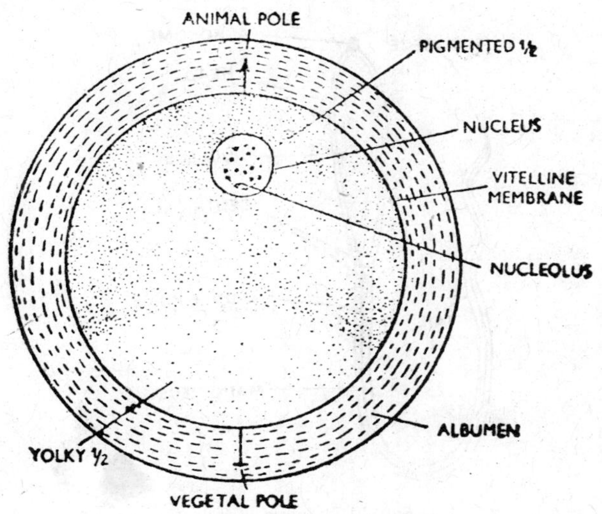

Fig. 14.6. Unfertilized egg of frog.

spring, when ponds and streams are very cold and there is the need for the egg to absorb sun's heat. Around the ovum, there is delicate *vitelline membrane*, which is enclosed by a gelatinous investment, that swells up in the water. The jelly serves to have a protective functions so as to safeguard it form the attacks to bacteria and to help it to retain the heat for further development. Roughly, one half of its surface is deep blackishbrown in colour due to the presence of a superficial layer of pigment, which is called *animal hemisphere* or *animal pole*. The other half is lighter in colour provided with yolk; this is called *vegetal pole* or *yolk pole*. The nucleus or the *germinal vesicle* is situated in the upper pigmented part.

Structure of a Spermatozoon

The structure of the spermatozoon is similar to that of the other vertebrates. It is a minute, elongated body made up of a narrow, pointed head having nucleus and a middle piece just behind the head and a long, very slender tail. *Spermatozoa*, which are liberated in the seminal fluid, are very active and swim here and there with the movement of the tail and enter through the *jelly* of egg mass until one spermatozoon comes in contact with an egg. The spermatozoon slowly enters the egg.

As a result of some changes in the substances of the egg, other spermatozoa cannot enter in them. Thus, only one *egg* is fertilized by one *sperm cell*. The sperm-nucleus is called the *male pro-nucleus*. The

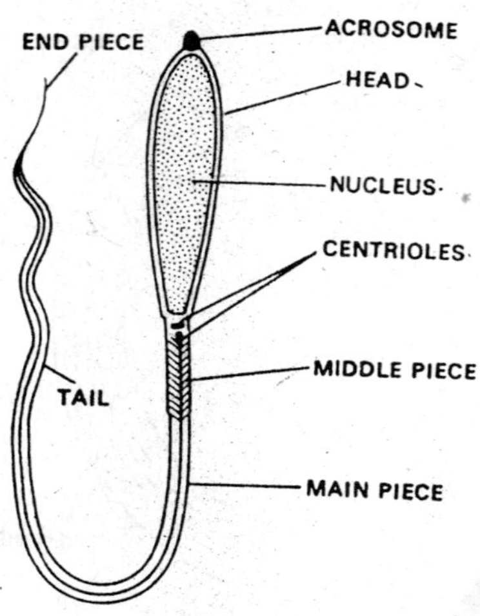

Fig. 14.7. Sperm of frog.

first polar body is already round as a minute globular cell under the vitelline membrane. The *second polar body* is soon budded off as the sperm-head penetrates into the egg. The nucleus of the ovum after the extrusion of the second polar body reaches the *male pro-nucleus* and the middle piece of thc sperm having *centrosome* and *mitochondria* becomes apparent. The male and female pro-nuclei move towards each other and ultimately form *zygote nucleus*. After fertilization, *perivitelline fluid* is formed between the egg and vitelline membrane. Due to the accumulation of this fluid, which is the derivative of egg itself, egg becomes free to rotate and the dark pole, which has a specific gravity less than the light coloured yolk-laden region, soon comes to lie the uppermost.

Ovulation

Ovulation is the process in which eggs are released from the ovary to the body cavity. This is done by a sex stimulating harmone from the anterior part of the pituitary gland. Ovulation can be seen artificially by injecting crushed pituitary harmone to female having ripe eggs. Pituitary from female frogs are more effective than those from males. In normal way endocrine is effected by light temperature and food etc., to prepare frogs for breeding. The process of ovulation is completed after the rupture and emergence of eggs from their individual follicles. Non vascular tissues which are theca externa separates the surface of the egg from the body cavity. During ovulation the area of theca externa is effected and dissolved by the pituitary harmone resulting the emergence of eggs. During the emergence of eggs smooth muscle fibres of the theca interna are stimulated which completes the liberation or ovulation.

Spawning

Each frog egg, as it comes from the cloaca of female, is spherical, unicellular, 1.75 mm in diameter, mesolecithal and telolecithal. It is surrounded by the vitelline membrane (secondary egg membrane) and many layers called tertiary egg membranes of adhesive jelly which were added as it passed through the oviduct. The outermost jelly layer is absorptive and tends to hold the eggs together. A cluster of frog eggs which remain stick together is called *spawn*. The spawn is laid, when during pseudocopulation or amplexus, male clasps the female and sheds a suspension of sperm over the eggs as they leave the cloaca. On reaching water, the jelly layers of fertilized egg absorb water, begins to well and lose much of their stickness. These jelly layers of egg serve various important functions—1. They serve to attach

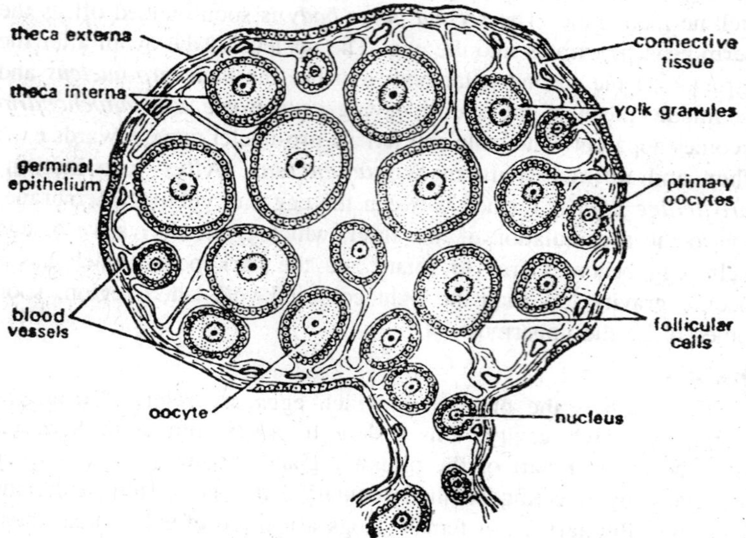

Fig. 14.8. Frog. T.S. of a part of ovary.

them to each other and to debris, so that they are not readily washed out. They protected the eggs from the mechanical injury and make the eggs unedible or distasteful to water snails and other aquatic creatures. 3. The jelly layers serve as insulators to protect the egg from the drastic effects of radiant energy and heat of the sun rays.

FERTILIZATION AND EGG SYMMETRY

Fertilization

Fertilization is external and only one spermatozoon penetrates the ovum. A sequence of events set in to complete the whole process of fertilization.

(i) The penetration of the sperm and the formation of fertilization membrane

The egg is laid at the secondary oocyte stage at a time when the second maturation division has already reached half-way. Spermatic fluid is shed almost simultaneously with oviposition and no time is lost in bringing about fertilization. *Monospermy* is a rule in mesolecithal or moderately telolecithal eggs, although several sperms pierce the outer jelly, but usually one of them is slightly in advance of its fellows and thus arrives first at the surface of the egg itself. The moment it has started to enter, cortical granules burst open into the perivitelline space and their contents become adhere to the vitelline membrane which now becomes thick to form the *fertilization membrane*. This membrane checks the entry of other sperms into the egg. Polyspermy is thus an abnormal phenomenon in the frog. The entrance of the sperm always occurs in the animal hemisphere of the egg, and usually about 40 degree from the pole. The sperm breaks through the egg-membranes with the help of an enzyme, *hyaluronidase*. Further movements of the sperm and its way inside the ovum is probably not an active part of the sperm. The ovum forms a small crator-like projection, the *cone of reception* or of *fertilization*, which engulfs the sperm by a complex process resembling to phagocytosis.

(ii) The Perivitelline space and rotation of egg

The penetration of ovum by the sperm has caused some cortical reactions which give up certain amount of its fluid. This fluid collects between the vitelline membrane and the egg plasma membrane and the space containing this fluid is as usual termed the *perivitelline space*. Its formation releases the egg from the grip of its covering, permits the egg to rotate freely within them. So that the heavier vegetal hemisphere comes to lie downward and the darker and lighter animal hemisphere comes to be upward. Such rotation is called "*rotation of orientation*."

(iii) Penetration and Copulation path

In case of frog, the whole spermatozoon enters the ovum in the animal hemisphere. It usually takes a minute or two to get entirely inside. The tail then disintegrates; the head and middle piece travel steadily carrying a trail of black pigment behind it which results an area opposite the sperm entrance point, to have a slightly decreased pigment content. This lighter pigmented area is called the *gray crescent*. The straight path followed by the head and middle piece inside the egg is called the *penetration* or *entrance path*. But in case of an excentrically placed female pronucleus, the path followed by the male pronucleus is first straight, then inclined at an angle. The inclined path is called the *copulation path*, and like the penetration path, it is marked by a trail of pigment. But when the female pronucleus is centrally placed and the sperm nucleus does not change its path, then the *penetration* and the *copulation path* remain the same. As the head and mid-piece of the sperm move along penetration path, the usual rotation of these parts occurs, thus placing the mid-piece, in the head.

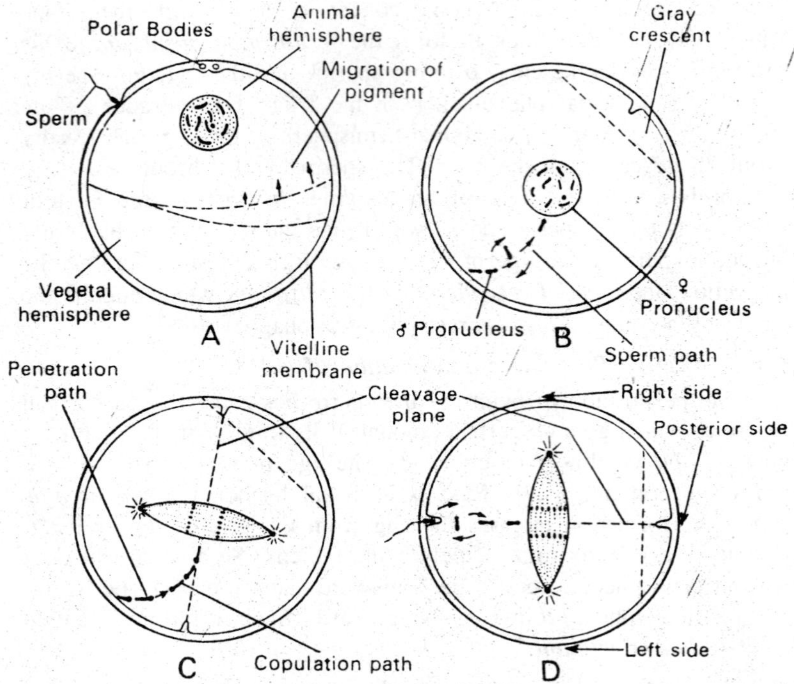

Fig. 14.9. Fertilization and fixation of symmetry of egg and cleavage plane in frog; A—lateral view; B to D—dorsal views.

(iv) Completion of second maturation division

Meanwhile the stimulus of the entrance of the sperm has incited the completion of second maturation division of the egg nucleus which was postponed in the metaphase second. After throwing off the second polar body, the egg nucleus withdraws from the surface of the ovum, usually to a position in the egg axis, and becomes vesicular to be called as *egg pronucleus*.

(v) Amphimixis

The sperm nucleus also becomes vesicular and is called *sperm pronucleus*. The sperm pronucleus along with its centriole moves towards the egg pronucleus and ultimately both pronuclei fuse together to form *zygotic nucleus*.

Egg Symmetrization

The process of bilateral symmetrization of an unfertilized radially symmetrical egg occurs during fertilization and remains intact during the entire embryogenic development. Therefore, the bilateral symmetry of the adult animal originates during fertilization process. During fertilization, certain very significant changes occur in the egg cortex to establish the bilateral symmetry. There is a temporary shifting of the pigments resulting from a raising and concentration of the cortical layer toward the animal pole, but the components soon returning to their original place.

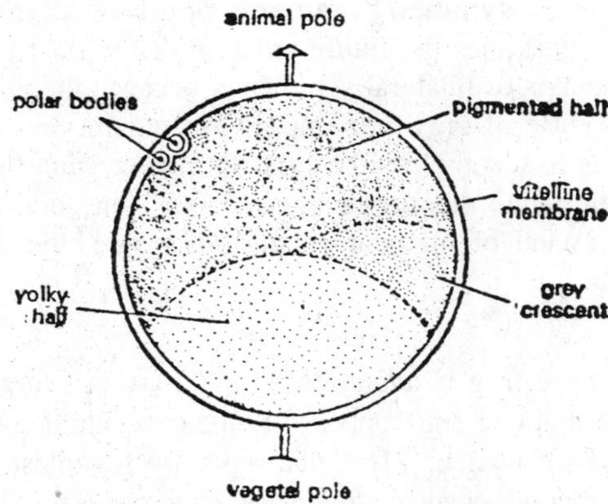

Fig. 14.10. A fertilized egg of frog.

Thereafter, a displacement of the egg contents brought about by a rotation of the cortical layer around the underlying layers, with relative concentration of the pigmented layer towards a particular meridian. This is accompanied by a corresponding thinning of the cortex on the opposite side of the egg, thereby producing a crescent-shaped surface known as the *gray crescent*. The plane passing through the centre of the gray crescent and the animal pole defines the median plane of bilateral symmetry in each case, i.e., it coincides with the embryonic axis, and is the only plane which separates the egg into two equivalent parts, each containing half of the crescent material. In frog, the gray crescent represents material which is no longer covered by the pigmented cortical layer, and it is gray because it lacks reflective yolk platelets.

Thus, due to slight rotation of pigmented cortex around the endoplasmic mass, the median axis of the bilateral symmetry is established and becomes irreversibly fixed. Such a cortical movement which fixes the median axis of the bilateral symmetry is called the "*rotation of symmetrization.*" The cortical pigment of the frog's egg not makes visible the shift of the cortex relative to the underlying endoplasmic layers but permits determination of the path taken by the sperm head within the egg cytoplasm. As the sperm nucleus and centriole move inward towards female pronucleus from the surface, a train of pigment granules is drawn along. This is the *sperm track*, and it serves as a record of both point of penetration and path taken.

It has been found that the sperm track always occurs in the median plane of bilateral symmetry and the point of entrance of the spermatozoon determines the future position of the gray crescent and, therefore of the axis of bilateral symmetry, because the point of sperm entry and the centre of gray crescent always remain on opposite sides of the egg. This is accomplished by the sperms causing the descent of pigment on one side, the future *ventral side*, the side on which it enters and an ascent of pigment on the other side, the future *dorsal side* of the egg.

Cleavage

The egg of the frog is *telolecithal* i.e., there is a large amount of yolk concentrated at one pole, opposite to the concentration of cytoplasm and the location of nucleus. The cleavages are holoblastic, i.e., total and after the second cleavage they are unequal. The first cleavage appears about 2½ to 3 hours after fertilization. It begins as a small depression near the centre of the animal hemisphere. It appears as if some internal force is drawing the surface of the egg towards its centre. This small inverted fold gradually extends in the form of a groove until it surrounds the egg. This furrow is shallow in the beginning, but becomes deeper ultimately dividing the fertilized egg into two halves called the blastomeres. Internally the division is mitotic, thus, each daughter cell contains a nucleus derived from the copulation nucleus of the fertilized egg. This cleavage is *vertical* or *meridional*. The two cells are identical in respect of cytoplasm, pigment and yolk. The second cleavage appears about an hour after the first. The furrow of this cleavage begins at the centre of the animal hemisphere, is at right angles to the first and is vertical. This divides the egg into four blastomeres. The four blastomeres so produced are not qualitatively identical, because of these only two contain the material from the gray crescent. The cleavage begins about thirty minutes after the second is completed or four hours after fertilization. The cleavage plane of the third furrow is horizontal and is slightly above the equator.

Thus the four upper cells are a little smaller than the four lower cells. The smaller blastomeres are called *micromeres* and the larger

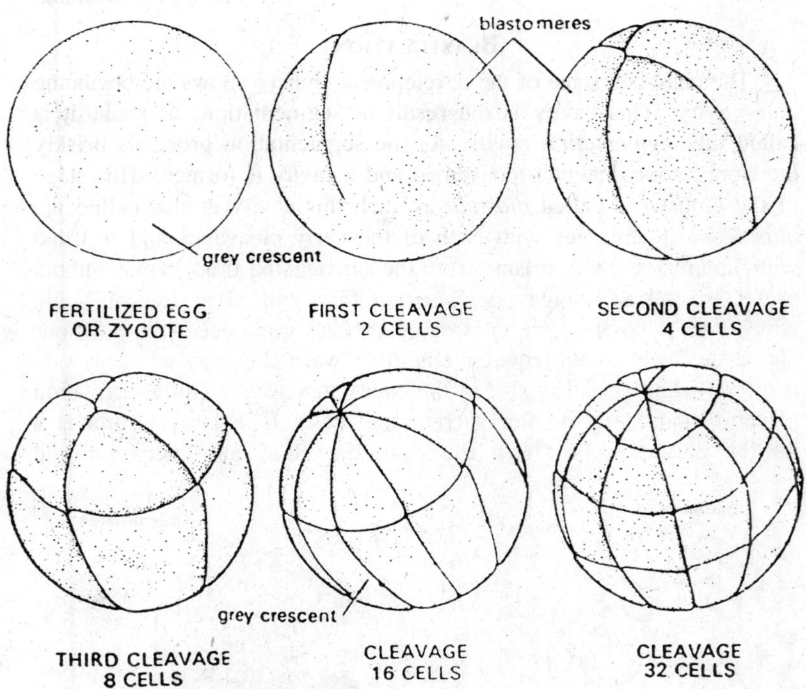

Fig. 14.11. Cleavage in fertilized egg or zygote.

blastomeres are called *macromeres*. The fourth cleavage follows 20 minutes after the third and tends to be vertical. This is usually a double furrow. The cleavage rate is accelerated with each of the early divisions and since the blastomeres are of unequal size and have varying amounts of cytoplasm and yolk, synchronous cleavage is lost and there is an obvious overlapping of the divisions. The uppermost cells divide more rapidly than the lowermost cells. From this point onward perfect symmetry in cleavage and in blastomeres is very rare, although the embryo develops perfectly. The *fifth cleavage* is also double, appearing first in the upper hemisphere and then in the lower. The cleavages thus far follow the rule that each cleavage plane comes in at right angles to the previous one. The subsequent divisions become so irregular that it is impossible to trace out any plan of procedure. The segmentation continues more rapidly in the pigmented region, since at that place the protoplasm is most dense, whereas, yolk which is very abundant in the vegetal side delays cell division. The multicellular embryo at this stage is called *morula* by some writers.

BLASTULATION

The eight-cell stage of the development embryo shows the beginning of a cavity. This cavity is the result of segmentation, as such, it is called the *segmentation cavity*. As the segmentation proceeds briskly the stored nourishment is consumed and a cavity is formed. This stage of the embryo is called *blastula* as such this cavity is also called the *blastocoel*. It enlarges with each of the early cleavages and is filled with albuminous fluid, arising from the surrounding fluid. Some authors prefer to call this stage *coeloblastula* (as opposed to stereoblastula which has no cavity). The blastocoel appears in an eccentric position, above the level of the equator slightly toward the gray crescent side of the dividing egg. The size of the cavity increases with the formation of more and more smaller surrounding cells. The late blastula is a hollow ball of cells. Now the rhythmic cleavages disappear and

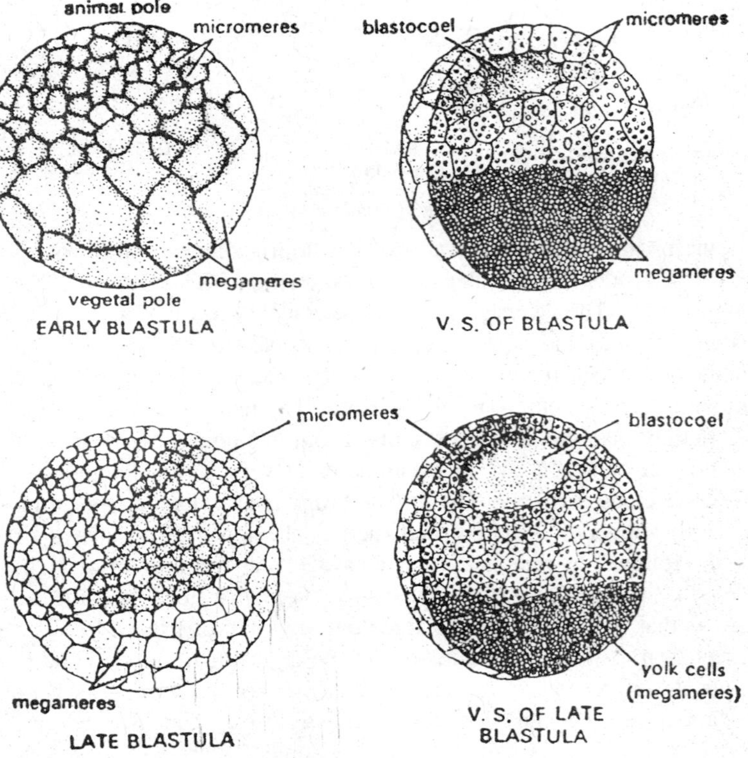

Fig. 14.12. Cleavage in fertilized egg or zygote.

multiplication takes irregular course. The small pigmented an male pole cells tend to spread their activity in downward direction toward the vegetal pole in such a way that there seems to be a regular migration of pigmented cell toward the vegetal pole.

Further the horizontal cleavages in the very small animal pole cells give the blastula a double or multilayered roof. The single outer layer of cell contains most of the superficial pigment forming the epidermal layer which forms epithelium of he integument or lining of the nervous system. The inner layers of the cells of the roof of the blastula give rise largely to the neuroblasts of the nervous system as such as designated as the nervous layer. At the junction of the animal and vegetal pole the encroaching cells of the animal hemisphere become thickened forming an equatorial band or the *marginal zone*, where yolk is most actively being transformed into cytoplasm. Such a marginal region of activity is found in *Amphioxus* and chick, and probably in all vertebrates. The cells of the blastula do not present much distinction in their make-up except in the degree of pigmentation, size and yolk content. There is no apparent indication as to what part the cells will play in the formation of the future embryo.

Many scientists have succeeded in demonstrating the ultimate fate of different cells even at the blastula stage. They have used many experimental devices, such as the vital staining, and constructed maps (*fate-maps*) showing the ultimate fate of various areas. This is called the *prospective* or *presumptive fate* of the different regions. Thus a fate-map is simply a topographical surface mapping of the blastula. The blastula is stained with various harmless dyes and the coloured areas are watched. For instance, a small area of cell stained by Nile blue sulphate (a vital dye) was picked up an watched. These cells moved from the marginal zone (between the animal and vegetal hemispheres) over the dorsal lip of the early blastopore into the embryo and became pharyngeal endoderm. The egg of the frog is very dark as such the effect of vital dyes is not clear in its case. But the fate-maps of various amphibians are basically alike (not sufficiently alike in detail), as such fate-maps of closely related species have been worked out, and giving due consideration to specific differences and the behaviour of the frog's egg in development, a reasonably accurate map of the frog blastula has been prepared.

Under normal conditions the ectoderm of the frog is derived from the cells of the animal hemisphere and in part from intermediate or equatorial plate cells. These regions are designated as *presumptive ectoderm* on the fate-map. The endoderm arises partly from the intermediate zone but largely from the vegetal hemisphere area which is called the *presumptive endoderm*. The mesoderm and the notochord arise between ectoderm and endoderm, largely from the region known as the lip of the blastopore. It must be clearly understood that the limits of any particular region are not sharply defined and the fate-maps show the mean results of a number of marking experiments. It is also worth noting that with the exception of cells of the gray crescent the fate of other areas is not fully fixed or determined.

Fate Maps of Blastula of Frog

In the blastula, the blastomeres which have to form different germinal layers and different organs of the adult frog, have their representation at the external surface of the blastula. The fate of each type of blastomeres has been observed by artificial-vital staining methods of *Vogt* (1925) and prospective organ region maps or fate maps have been prepared for the blastulae of frog. According to the fate map studies, the whole surface of the blastula of frog can be divided into following three areas :

Prospective Ectoderm Area

The entire pigmented area on and around the animal pole is the prospective *ectoderm*. The *neural ectoderm*, which develops into the central nervous system, occurs largely on the future dorsal side of the blastula, while the *epidermal ectoderm*, which develops into the skin epidermii of the embryo occupies the anteroventral side of the blastula. The material for the sense organs is also contained in these two areas. Inside the neural ectoderm area occurs a small sub-area which develops into the eye of the embryo. Likewise, inside the epidermal ectoderm area occurs the sub areas for nose, sucker, ears and ectodermal part of mouth. *Gilchrist* (1968) has called the zone of prospective ectoderm as "*zone of expansion*", because, during gastrulation, this zone expands downward and then, converges toward the vegetal pole. At the end of gastrulation, the ectodermal zone gives rise to the entire outer layer of gastrula.

Prospective Notochord and Mesoderm Area

The pigmented animal hemisphere (the prospective ectodermal area) is followed by a crescentic area of gray colour, the *marginal zone*, which goes all along the equator of the blastula, and has blastomeres for the formation of *notochord* and *mesoderm* of the frog embryo. The notochordal cells occupy the large area of dorsal side of the gray crescent. Beneath the notochordal area, and toward the vegetal pole,

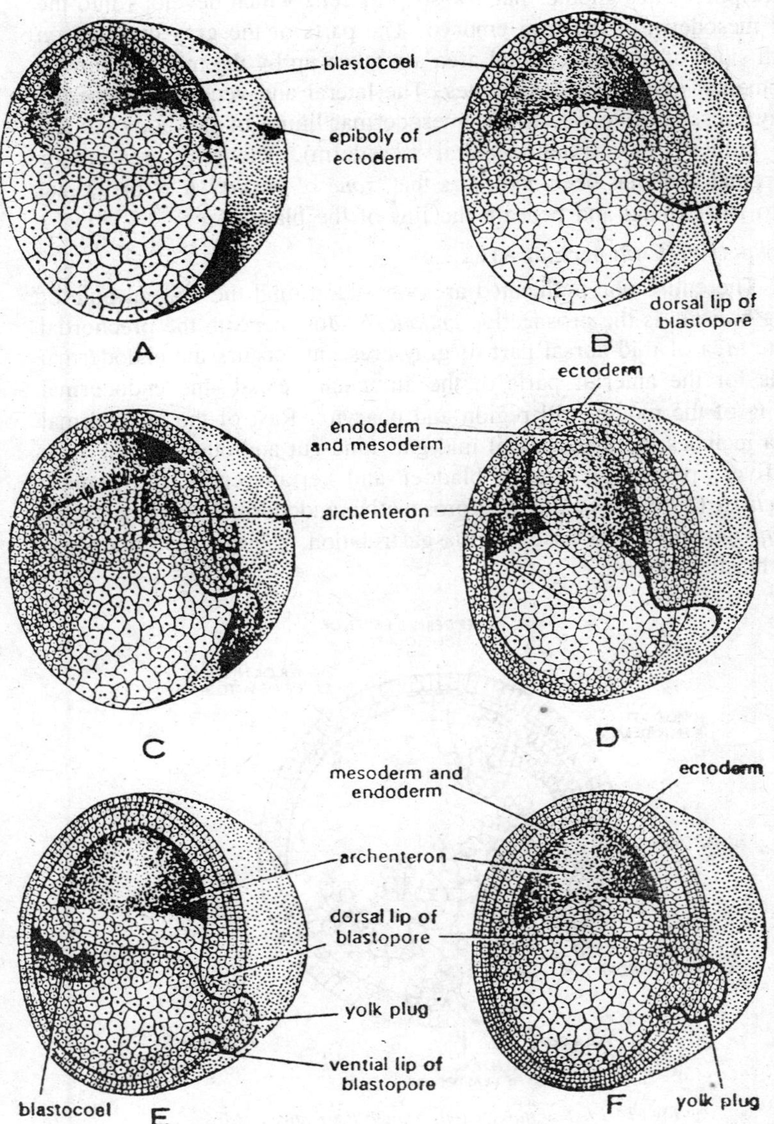

Fig. 14.13. The blastula (A) and its transformation into gastrula (B-F).

lie a narrow strip of cells which from the pre-chordal connective tissues of the embryo and hence, this strip is called prechordal plate. At the upper side of notochordal area and toward the animal hemisphere, lies another narrow strip of cells which develops into the tail mesoderm of the frog embryo. The parts of the gray crescent on both sides of the notochordal area are taken up by the material for the segmental muscles, i.e., somites. The lateral and ventral parts of the gray crescent give rise to the mesodermal lining of the body cavity, the kidneys, etc., (ventro-lateral mesoderm). The gray crescent or marginal zone has been named as the "*zone of involution*", for its fate is to roll around and beneath the lips of the blastopore.

Prospective Endodermal Area

The entire non-pigmented area on and around the vegetal pole of frog blastula is the prospective *endoderm*. Just beneath the prechordal plate area of mid-dorsal part of gray crescent, occurs the endodermal cells for the anterior parts of the alimentary canal—the endodermal lining of the mouth, gill region and pharynx. Rest of the endodermal area includes the materials of mid-gut, hind-gut and other organs such as liver, pancreas, urinary bladder and certain endocrine glands. *Gilchrist* (1968) has called the prospective endodermal zone the "*zone of invagination*", because, during gastrulation, it sinks and glides into the blastula of frog.

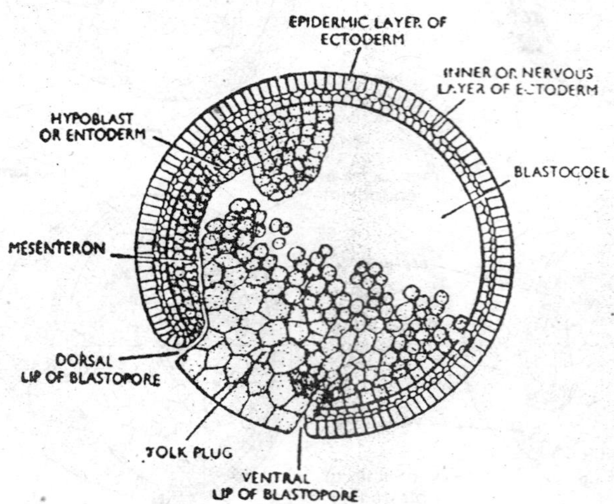

Fig. 14.14. Sagittal section through the embryo of frog.

GASTRULATION

Blastulation is followed by the complex process of gastrulation. In frog the gastrulation involves the following processes.

1. Epiboly

The pigmented cells of blastula have the tendency to overgrow the yolk cells. This process which is called epiboly begins slowly in the late blastula and accelerates with advancing gastrulation. The cells of the animal pole proliferate and grow downward on all sides except in the region where the dorsal lip of the blastopore has been formed.

2. Convergence

Vogt (1928) and Goerttler (1925) suggested that there is a streaming of the material of the dorsal and lateral surfaces of the early gastrula towards the blastopore. At the same time, their is a shifting of the lateral regions towards the mid-line. This combined type of movement is now generally described as *convergence*. Thus like amphioxus, here too the lips of blastopore do not actually constitute the sides of the embryo. Holtfreter (1943) has tried to give physiochemical explanations in this movement. He suggested that an unfolding of denatured protein molecules is partly responsible. This unfolding, he thought causes a spreading of the superficial cells over a substrate with appropriate adsorption properties. Hence epiboly and convergence are due to this spreading tendency, which is apparently augmented by a lowering of surface tension in parts of the spreading cells.

3. Rotation

The epiboly continues until the dorsal lip has passed over to an are a some what greater than 90°, and the area of white (blastopore) is reduced to small circle. This area will be situated rather beyond the vegetal pole. Side by side, the entire developing gastrula has been rotating about a horizontal axis lying at right angles to the original median plane of the egg. It means that the direction of rotation is such that the dorsal lip is in a sense carried backward in one direction as fast or faster than epiboly moves it forward in the other. The result will be that the blastopore, formed at approximately the vegetal pole, is posterior, and the morphologically dorsal and ventral lips are actually dorsal and ventral. This also makes it clear that the animal pole of the egg is to form the antero-ventral side of the future embryo while the region marked by the grey crescent is to form the dorsal side.

4. Invagination

Side by side to the above described external processes, a small cleft-like invagination appears half way between the equator and the vegetal pole where the cells of animal portion and vegetal portion

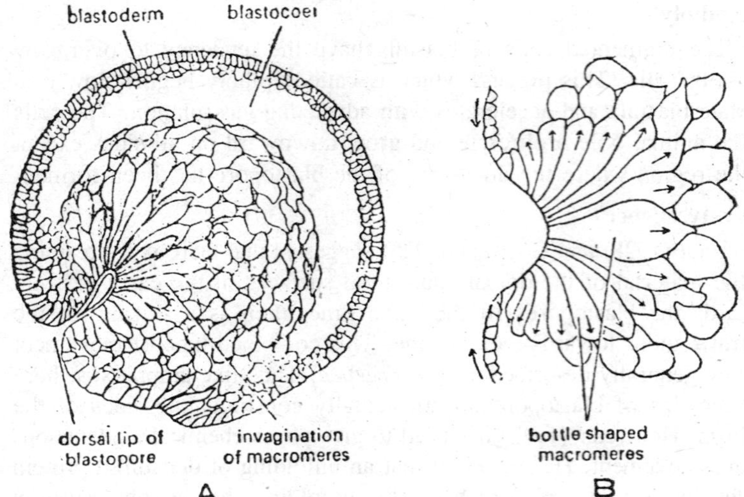

Fig. 14.15. Invagination of endodermal cells in an amphibian egg. A—Slightly schematized section through an advanced gastrula. B—The mechanics of gastrulation, showing active streaming of endodermal cells by botteling process.

merge. The upper or dorsal edge of the cleft forms the dorsal lip of the blastopore. This input swing begins on the dorsal side nearest the dorsal lip and spreads part way around he margins of the blastocoel in company with the external extension of the lateral lips. This modified invagination continues until the blastocoel cavity has been vertually eliminated, except for the narrow slit separating the outer layer of cells (epiblast) from the inner yolk cells (hypoblast). The new cavity formed is called the archenteron.

5. Involution

In addition to above processes a distinct *involution* also occurs at the blastoporal margin. This is most active at the median dorsal lip and progressively less so as one passes around either side until at the ventral lip where there is always more at all. In this process, cells located along the upper margin of the blastoporal lip move over the lip to the inside edge of the lip. There involuted cells are deposited on the inside of the embryo along the inner margin of the blastopore.

This means the roof of archenteron is composed of involuted cells (*presumptive choda-mesoderm*) and beyond this the outer layer (ectoderm). The floor and sides contain largely the endodermal cells, which have been derived from larger yolk cells, formerly located in

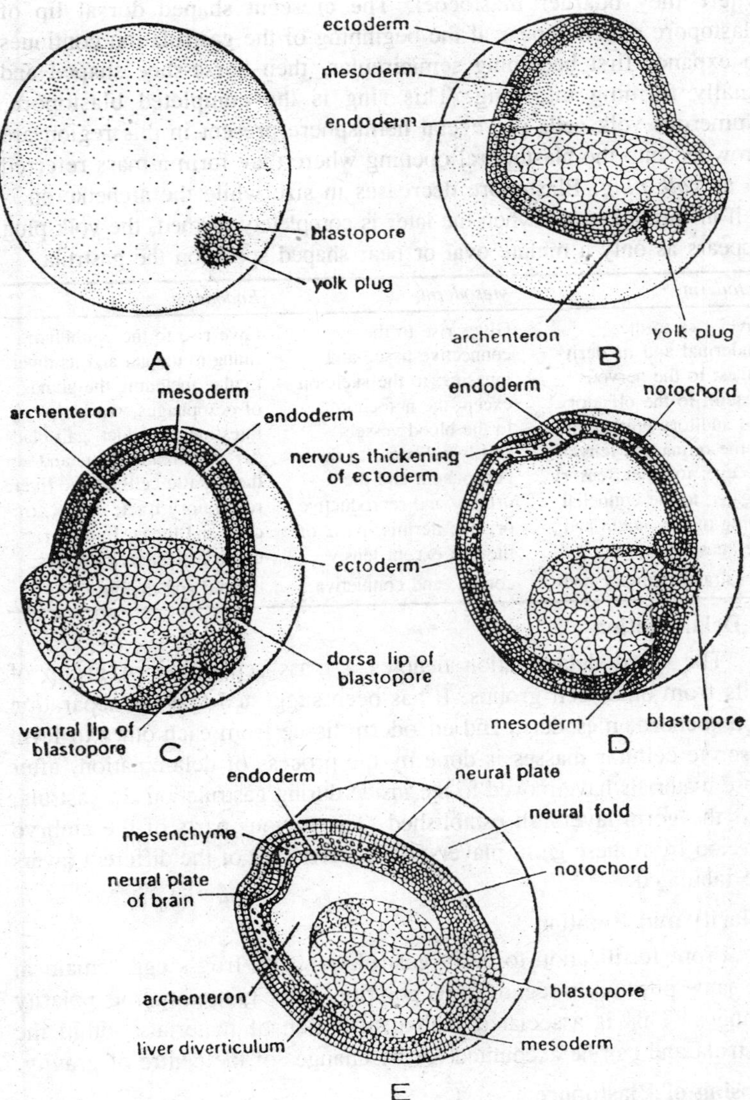

Fig. 14.16. A—External appearance of gastrula; B-C—Partial sections of early gastrula showing delaminations of endoderm and mesoderm; D, E—Sections of late gastrula showing the three germ layers.

the vegetal hemesphere of the blastula. In the later stages of the gastrula, the cells forming the floor of archenteron thin out considerably where they boarder blastocoel. The crescent shaped dorsal lip of blastopore which appear at the beginning of the gastrulation continues to expand, first becoming semicircular, then horse-shoe shaped and finally forming into ring. This ring is the completed blastopore. Numerous yolk cells of vegetal hemisphere present in this region are crowded into the blastoporal opening where they form a mass referred to as *yolk plug*. Blastopore decreases in size while the archenteron is still in formation and when the later is completely formed, the yolk plug appears as only a minute oval or pear shaped speck on the gastrula.

Ectoderm	*Mesoderm*	*Endoderm*
Gives rise to the epidermal and its derivatives; to the nervous system; to the olfactory and auditory epithelium; to the retina and lens of the eye; to other sensory organs; to the epithelial lining of the mouth and the anus; and to the pineal and pituitary body.	Gives rise to the connective tissue and muscles; to the skeleton except the notochord : to the blood vessels and lymphatics; to the peritoneum and the urinary and reproductive organs; dermis; parts of the eye except lens, cornea, and conjuctiva.	Give rise to the epithelial lining to the gut and its diverticulae including the glands of oesophagus, stomach, intestine, bile duct, gall bladder, pencreatic duct, and the hepatic cells of the liver; respiratory tract; larynx, trachea and lungs; lining of urinary bladder; pancreas; thyroid and thymus.

6. Delamination

The word delamination denotes a mass separation of group of cells from other cell groups. It has been suggested that the separation of notochord, mesoderm and endoderm tissue from each other to form descrete cellular masses is done by the process of delamination, after these materials have moved to the inside during gastrulation. In gastrula, thus the germ layers all established. The various parts of the embryo develop from these germ players. The derivatives of the different layers are tabulated.

Polarity and Rotation

From fertilization to early gastrulation, the frog's egg remain in the same position in regard to its polarity. After gastrulation its polarity changes. This is associated with distribution of materials within the gastrula and can be accounted for by changes of the centre of gravity.

Closing of Blastopore

When the gastrula of formed, the blastopore appears as a small round opening filled with yolk plug. As it continues to diminish in size, it assumes a pear-shaped outline through the mutual approach of its lateral lips. Finally these fuse completely to form a longitudinal groove, the *primitive streak*, ending dorsally and ventrally in a small

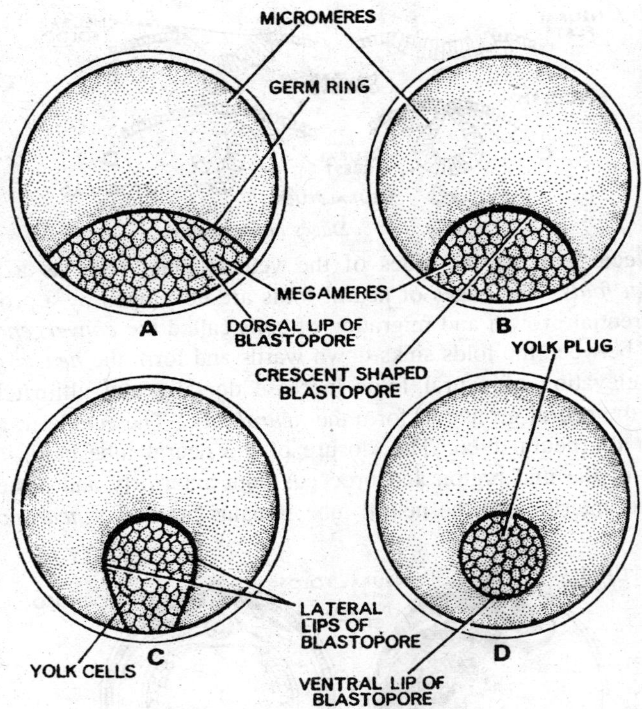

Fig. 14.17. Diagrams showing expansion of the rim of blastopore.

aperture. Soon the lower aperture close, leaving only a dipression called the *anal pit*. The upper ones remain open for some long time and closes for later when the neural groove is laid down thus initiating the next figure' the *neurala*.

Post Gastrulation or Organogenesis

During pregastrulation, all material for different organs disappear from the surface of blastula and come inside to take their final position in the embryo where different organs are developed from their respective prospective areas. So *organogenesis* concerns with the formation and differentiation of various organs. Organogenesis converts an embryo into free swimming larva.

The Formation of the Neural Tube

Neural tube is developed from prospective neurectoderm which during pregastrular processes takes its position some where in the mid dorsal region of the gastrula. Pear shaped *neural plate* or *medullary plate* is developed simply by the thickening of neurectodermal cells.

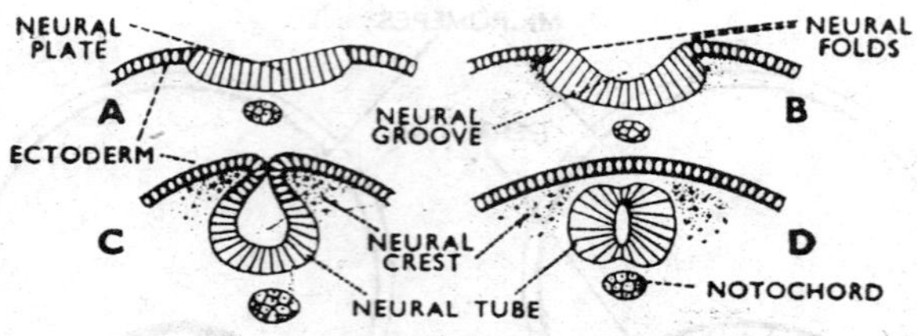

Fig. 14.18. Stages in neurulation.

The ectoderm on the sides of the ventral plate rises as a pair of *neural folds*. The edges of neural folds are developed by a process of differential growth and migration of cells called the *convergence*. The plate between the folds sinks down wards and form the *neural groove*. The elevation of neural folds increase dorsally and ultimately fuse over the neural groove to form the *neural* tube. The ectoderm reforms about the neural tube. The closure of the neural tube begins just in front of the mid region and proceeds both anteriorly and posteriorly. At the anterior end the neural tube remains open as a *neuropore*.

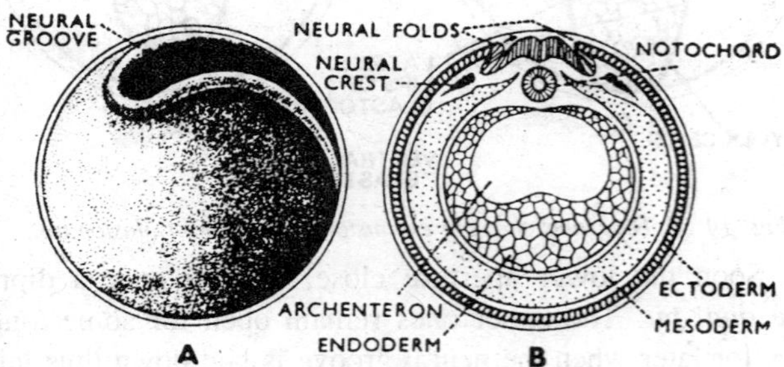

Fig. 14.19. Early neurula (A) and its transverse section (B).

The neuropore persists only for some time and closes ultimately. At the posterior end, the neural folds extend on the sides of the blastopore and meet over it. As a result of this enclosure of the blastopore, the neural canal is in continuity with the archenteron by a passage called the *neurenteric canal*. The yolk plug has been withdrawn. Both brain and spinal cord are differentiated from neural tube. Anterior wider end of this tube forms the *brain* while the remaining posterior part becomes the *spinal cord*. The anterior wider portion to neural tube forming the brain is further differentiated into three cerebral vesicles by three dialation. Of the three vesicles, the first is called the *fore brain*, the middle one is called the *mid brain* and the posterior one forms the *hind brain*.

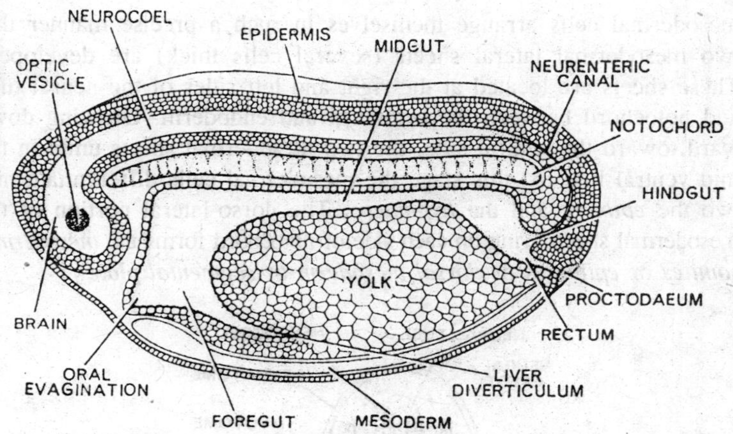

Fig. 14.20. Sagittal section of fully developed neurula.

Various parts of the brain are developed by the thickening and foldings of three vesicles. During the differentiation of brain, there develops a liner band of cells forming *neural crest* on the side of the neural tube. This linear band of cells then divides into many small parts forming the *dorsal root ganglia* of cranial and spinal nerves. The arrangement of the ganglia is metameric and their positions correspond with those of the somites. Visceral arch skeleton and sympathetic ganglia are also developed by the proliferated cells of neural crest.

Formation of Notochord

The *notochord* develops from the chorda-mesoderm, lying in the mid dorsal line of the gastrula. In the begining, the cells forming the notochord are indistinguishable from the mesoderm but later on these cells separate from mesoderm by a narrow cleft, by a process of *delamination*. Then these separated chorda cells very quickly expand and arrange them selves in the form of a cylindrical rod and at this stage they become vacuolated. Even the intercellular material of the notochord becomes vacuolated. An elastic and fibrous sheath is developed around these cells. This sheath is called the *notochordal sheath*.

Mesoderm and its Derivatives

As already told that during invagination chorda-mesodermal cells migrate from the outer surface to the interior where the mesodermal cells become separated from the chorda cells which from the notochord. In the late gastrula after getting separation from chorda cells the mesodermal cells arrange themselves in such a precise manner that two mesodermal lateral sheets (several cells thick) are developed. These sheets are located at the right and left sides of the neural tube and notochord between the ectoderm and endoderm. Growing down ward toward the ventral side, the two mesodermal sheets unite in the mid ventral line. At this stage the mesodermal cells differentiate into two the *epimere* and the *hypomere*. The dorso-lateral portion of the mesodermal sheets lying on each side of notochord forms the *mesodermal somites* or *epimeres* or *dorsal mesoderm* or *segmental plate*.

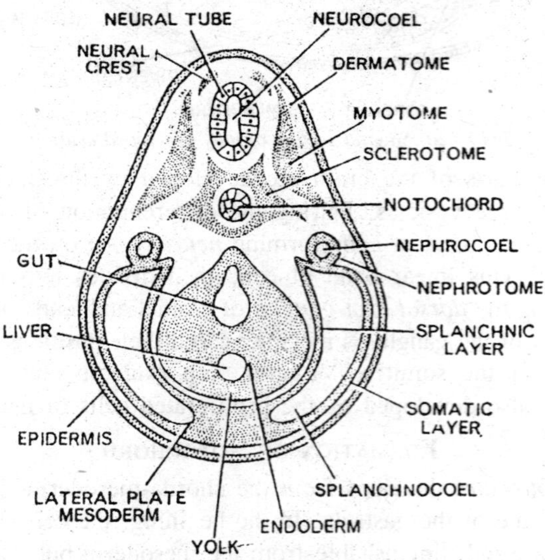

Fig. 14.21. Transverse section of completed neurula.

The remaining ventro-lateral portion of mesodermal sheets, is not affected by dorsal segmentation and is reffered to as the *lateral plate mesoderm* or *lateral mesoderm*. Mesodermal somites which increase in number as the embryo grows larger, are arranged segmentally and from each somite three embryonic structures, the *myotomes*; the *dermatomes*; and the *sclerotomes* are developed in the following ways— After the formation of mesodermal somites from mesodermal sheets, each somite splits in two halves. That portion of this splitted somite lying against the body wall or ectoderm forms the dermatome while the remaining portion of mesodermal somite constitutes the myotome.

At the same time a number of cells are proliferated from mesodermal somites that constitute the sclerotome from which the axial skeleton is developed. From dermatomes, the dermis of the skin is developed and from myotomes most of the body muscles are developed. In the due course of time, the lateral mesoderm is separated into two layers, an outer one of *somatic mesoderm* which lines the ectoderm of he body and an inner of *splanchnic mesoderm* which lines the gut. Both the layers are separated from each other except at their free ends, by a cavity which forms the *coelom* of the organism. When the two lateral mesodermal plates unite in the *midventral line*, the lateral coelomic cavities also unite to form a single continuous coelom. This coelom is also called the perivisceral cavity or splanchnocoel. From this splanchnocoel, the *pericardial cavity* is developed. Inside the pericardial cavity *heart* develops.

Kidney is developed from lateral mesodermal plate thus enclosing a part of body coelom. The only region where the coelom does not become continuous is the area dorsal to the gut or ventral to the notochord. Here at this region the splanchnic mesoderm of right and left side come together and form the double membranous suspensory mesentery for the gut; the *dorsal mesentery*.

Development of Heart

Heart is mesodermal in origin specially from splanchnic mesoderm. From splanchnic mesoderm and probably from endoderm even, a number of cells are proliferated. These cells arrange themselves to form a tubular structure. This tubular structure is the *endocardial tube*. The cells of the endocardial tube constitute the innermost layer of the heart. This tube is covered in the begining from lateral sides and

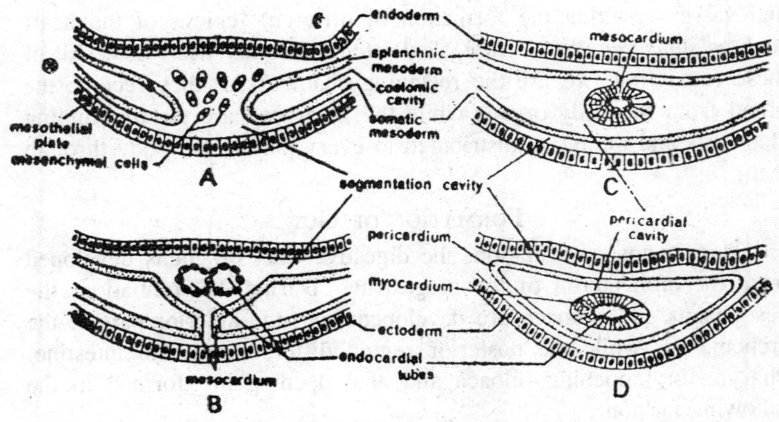

Fig. 14.22. Development of heart.

later on from all the sides by *epimyocardium*. From this epimyocardium both second and third or middle and outer most layers are *epicardium* and the *myocardium*. The myocardium thicknes to form the muscular wall of the heart.

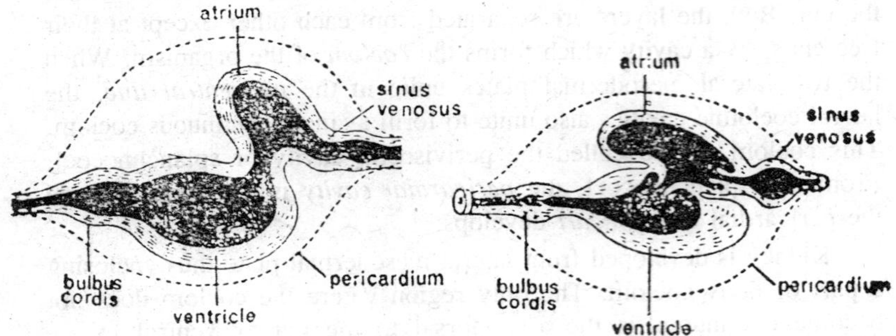

Fig. 14.23. Differentiation of various regions of heart.

In the early stages the heart is suspended in the pericardial cavity by a mesentery the *dorsal mesocardium*. The heart is located free in the coelomic cavity, formed above the splanchnic and below and laterally by the somatic mesoderm. This space is the *pericardial* cavity and the enveloper of this cavity is the *pericardium*. The heart thus developed, is in the form of a simple tube which in the latter stages of development becomes S shaped in the pericardial cavity. The lumen of the heart is divided into chambers by the development of *valves* which are originated by the reduplication of the living membrane. Externally divisions of heart is due to the formation of some constrictions and valves resulting the formation of different regions of the heart such as *sinus venosus*, *atrium*, *ventricle*, and *conus arteriosus*. Out of these four, first two are the receiving chambers as they receive the blood from the body organs while the latter two are the distributing chambers and the blood distributed to every part of the body through them.

FORMATION OF GUT

During post gastrulation, the digestive tract or gut is developed from the archenteron of the pregastrula. During differentiation, the oesophagus and stomach are developed from the anterior part of the archenteron, while the posterior region differentiates into intestine. Pharynx oral opening, cloaca and anal opening are formed in the following fashion :

Two processes; invagination and evagination, controll the formation of pharynx mouth, rectum etc. At the time of hatching of larva an ectodermal pocket is developed by a process of invagination and at the same time another pocket or pouch is developed from endoderm lining the archenteron by a process of evagination. Them both ectodermal pouch and endodermal pouch meet to form *oral plate* which generally ruptures to form the *mouth*. *Pharynx* is developed by the evaginated portion of endoderm.

Anal region of hind gut is also developed by the same fashion. The endoderm ventral to the closed blastopore evaginates to fuse with the invaginating proctodael ectoderm to form the *anal plate*. This plate ruptures to form the *anus*. *Rectum* is developed by the evagianted portion of endoderm. The portion between rectum and anus forms *cloaca* which is lined by ectodermal cells. Shortly behind the stomach, the yolky floor of the gut forms the *liver diverticulum* and its connection persists in the adult as the *bile duct*.

Development of Eye

Eye is developed from three different sources they are; the brain; the mesenchyme cells of the head; and the epidermis of the embryo. During the development *optic vesicles* are developed on either side of the brain simply by the out pushing of the ventro-lateral part of the fore brain. The connection between optic vesicle and brain is the *optic stalk*. In the due course of time *lens* is developed by the thickening of epidermis on the side of the embryo, opposite the optic vesicle. Actually this is done by invagination. Then a portion of the layer of optic vesicle invaginates to form two layered *optic cup* in which the lens comes to lie in. The portion of the layer of optic vesicle invaginates, constitutes the *nervous layer* of the retina while the portion of optic vesicle not invaginated constitutes the *pigmented layer* of retina. After some time both nervous layer and pigmented layer of retina come to lie in contact with each other. The retina and lens become enveloped by the mesenchyme cells of the head regions. From these mesenchyme cells *choroid*, *scleroid* and *cornea* are developed. A set of six eye muscles is originated from the mesodermal somites of head. Further differentiation of pupil choroid fissure, optic nerve and optic chiasma etc. take place before metamorphosis.

Embryo of Frog

In vertebrates, the earliest stage in development maybe modified by the amount of yolk that is present in an egg and also by the presence or absence of foetal membranes. The embryos of all

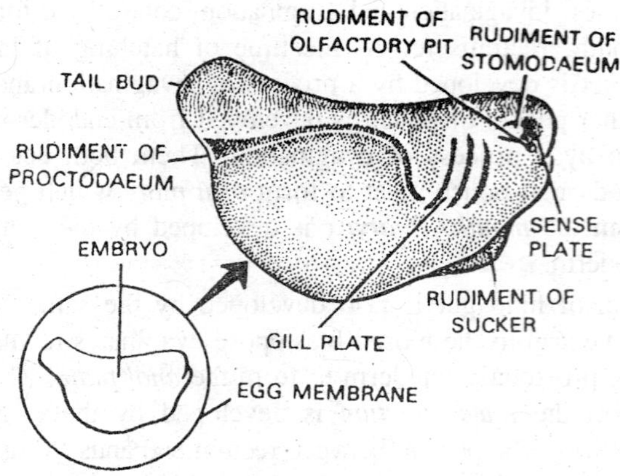

Fig. 14.24. Frog's tadpole before hatching.

vertebrates, particularly air-breathing, are surprisingly similar by the time of neurulation. This fact remains true for the embryos even if their habitat is water (amphibian), air (reptiles and birds), or the mother uterus (mammals). Moreover, the late neurula stage iş of great importance in the evolution of the vertebrates. The time needed for the development of an embryo depends upon (i) the temperature and (ii) the amount of oxygen present in the surrounding water. The early is well-recognised by its large size of structures.

The head is well-formed and quite distinct while the tail has just started. By that time, the heart starts beating and a circulation of blood is maintained in the gills. The embryo is also provided with a conspicous ventral sucker to be used for attachment by the hatched embryo. With all these structures developed, the embryo comes out of the jelly which is dissolved by the enzyme secreted by the embryo itself. The stomodeum is formed and mouth starts developing. The embryo is capable of contracting its muscle when stimulated. The young embryo is tubular and consists of following elements.

1. Outer epidermal covering.
2. Dorsal neural tube.
3. Notochord.
4. Endodermal canal or gut.
5. Mesodermal plate of right side.
6. Mesodermal plate of left side.

Besides these, there are also some loose cells in between these structures, which are called the mesenchyme cells. The mesenchyme, no doubt, is developed from mesoderm, yet it has some contribution from ectoderm also. These cells form the supporting tissues, or skeletal system, musculature, blood vessel and the blood. At this stage, the embryo of frog resembles the organisation of typical vertebrate embryo.

Elongation of the Embryo

The embryo, till the gastrulation is completed, is almost spherical or oval but after this, it undergoes elongation which is typically in anterio-posterior direction. With the so-called elongation of body beyond the anal aperture, tail formed and in this tail region the notochord, nerve cord, somatic mesoderm mesenchyme (excluding endoderm) are given in. This post-anal is oftenly called as the *tail bud*. But this term is not proper as this position is formed by the rearrangement of tissues present in and near the hinder end of gastrula. There is no proliferation of indifferent cells around the margin of almost closed blastopore. The significance of formation of tail is that their stretching of posterior past of neural plate and neural folds along the longitudinal direction. And this is recently proved that somatic mesoderm of the tail develops from the posterior part of neural plate which may be called as prospective tail somites. Then thus formed somites of tail are enclosed within by the hinder part of neural folds and the elongates to a great extent. This hinder end develops a dorsal skin fold or fin and a ventral fin and thus the tail is fully formed.

External Gills

Just after the embryo is hatched, the skin present on the dorsal part of the first and second and later on that of third branchial arch (i.e., visceral arches third, fourth and fifth) grows out to form three pairs of feathery *external gills*. These are the first formed gills and function as the respiratory organs.

Mucous Glands

When the external gills are developing, at the same time, on the other hand, ectoderm developes a deep pit lined with glandular cells. These pits are present on the ventral end of each mandibular arch which is the tissue between the hyomandibular pouch and the stomodeal invagination. These pits are called the *mucous glands* which help the embryo in attaching to some substratum, after hatching.

A. Formation of Tadpole Larvae

About four days after fertilization, the embryo is said to be a the *tail bud* stage of development when it is about 3 mm. long, The tail bud larva undergoes a series of changes in its external and internal

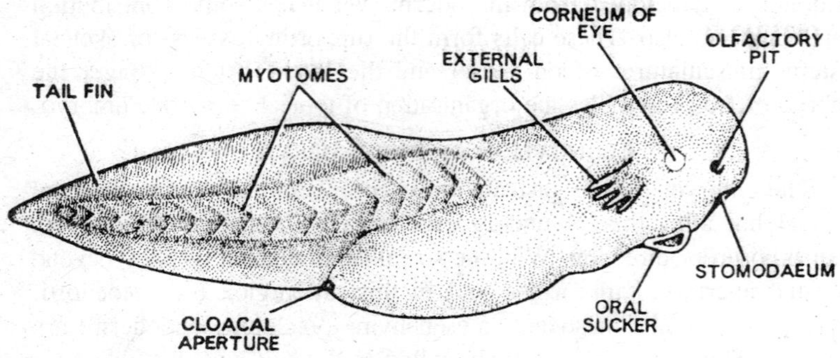

Fig. 14.25. Lateral view of early tadpole.

structure within the egg membranes to become the tadpole larva. The larva elongates in an antero-posterior direction and shows distinct differentiation into *head* and *trunk*. At the hind end appears a small outgrowth or *tail bud* indicating the beginning of the tail. Small swellings on each lateral side of the head indicate the position of the internally developing *eyes*. In front of each bulge is present a small *nasal pit*, and behind the nasal pit of each side lies the rudiment of the *ear*. A 'V'-shaped glandular pit, the *adhesive organ* or *cement gland* develops just below the prospective *stomodaeum* on the ventral surface of the head. In the anterior part of the trunk, three *pharyngeal pouches* are formed as lateral outpushings of the fore-gut. These are the rudiments of first three gill clefts. The trunk bulges ventrally and laterally due to the presence of a large amount of yolk beneath and at the sides of

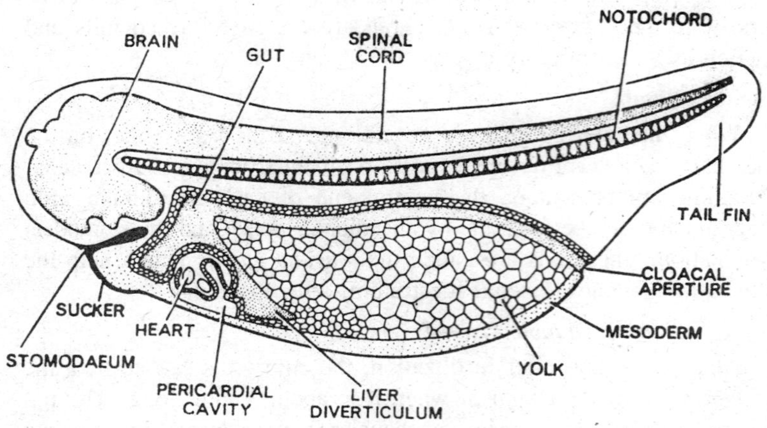

Fig. 14.26. Longitudinal section of early tadpole.

the developing gut. A small depression, the *proctodaeum* lies between the tail and the trunk on the ventral side. It will form the *cloaca* and the cloacal aperture. The larva also has a well-differentiated brain.

B. Hatching

When the embryo is about 12 days old, it dissolves the egg membranes by *hatching enzymes* produced in the skin and emerges from the jelly envelope. This process is known as *hatching* and the hatched larva is called a *young tadpole*. It is a blackish, fish-like creature, about 7 mm long. The newly hatched larva swims for a short while, but soon fixes itself to some floating vegetation or moss-covered walls of the ponds by means of a sticky secretion of mucous glands present in the adhesive organ. It does not take any food due to the absence of mouth and is nourished from the yolk present in the ventral bulge of the trunk. During the larval life of tadpole, following developments occur:

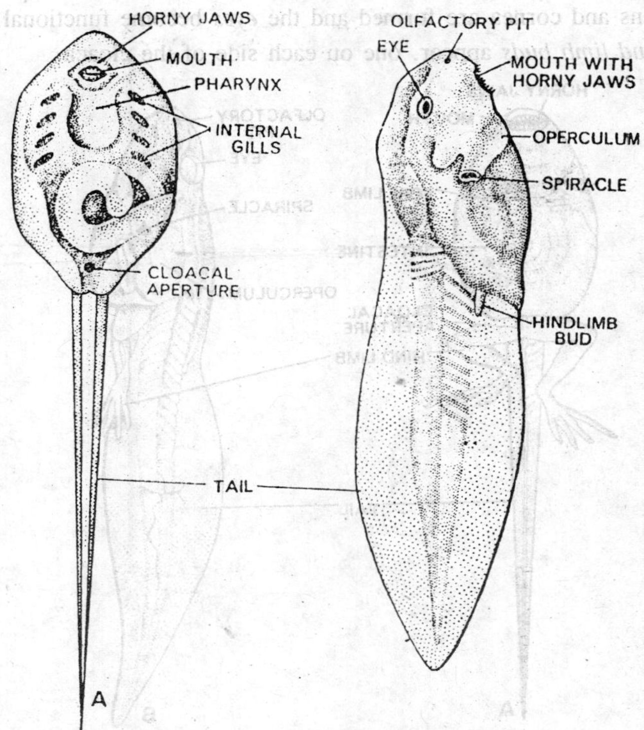

Fig. 14.27. Internal gill stage of tadpole: ventral (A) and lateral (B) views.

1. The *mouth* forms.
2. A pair of *horny jaws*, bearing *papillae* or *honey teeth*, developed around the mouth. The larva now begins to feed upon algae and other vegetable matter by the rasping action of the teeth.
3. Three pairs of finger-like *external gills* project from the sides of head in the branchial region.
4. The external gills gradually disappear under an opercular fold of the skin. Meanwhile, *true* or *internal gills* have been gradually developing by perforation of the pharyngeal gill pouches. The *opercular chamber* opens to the outside by an aperture called *spiracle* on the left side. The space between the body wall and operculum is called branchial or gill chamber.
5. The *tail* elongates and becomes a powerful organ of locomotion. It acquires a dorsal and a ventral *tail fin*.
6. Adhesive organ begins to degenerate.
7. Lens and cornea are formed and the *eyes* become functional.
8. *Hind limb buds* appear, one on each side of the cloaca.

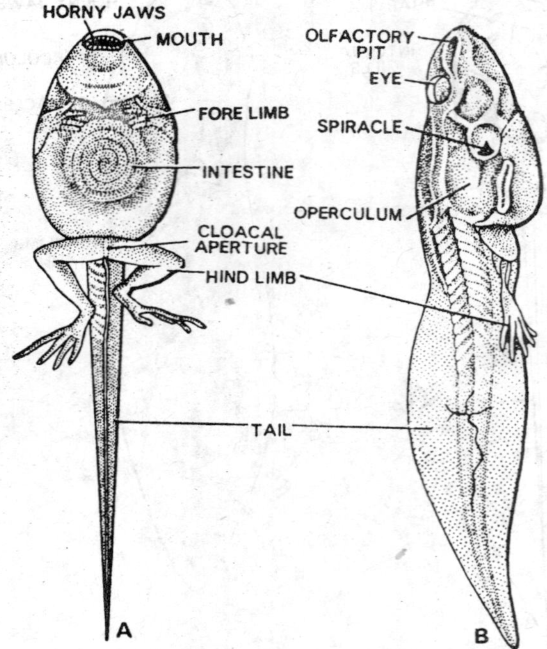

Fig. 14.28. Stage of limb-development from ventral (A) and lateral (B) views.

The embryo is now a free-swimming, actively freeding *tadpole larva*, which has no resemblance with the fish-like tailed semiterrestrial adult frogs.

Metamorphosis

The term *metamorphosis* is derived from a Greek word *metamorphoun*, which means "to transform". In embryology it is defined as transformation of the larva into adult during which many structural as well as physiological changes take place. Larval forms and accompanying metamorphosis are found in most groups of animal kingdom. When about 8 weeks old, the tadpole larva undergoes a series of rapid changes, which transform it into a young frog. This is called *metamorphosis*. The metamorphosis changes in frog tadpole may be grouped into three categories : (A) changes in habits and habitat, (B) changes in morphology, and (C) changes in physiology.

A. Changes in Habits and Habitat

1. In frog, metamorphosis is associated with a transition from an *aquatic* to a *terrestrial* mode of life. The metamorphosing larva comes frequently to the surface of water to breath air. Later, it begins to take short trips to land, i.e., it becomes *amphibious*.
2. The change of environment during metamorphosis is associated with a change in feeding. The tadpole are *harbivorous* feeding upon algae and other vegetable matter—which they scrape off with the help of the horney teeth surrounding their mouths. Adult frogs, on the other hand, are *carnivorous*—living on insects, worms etc. Sometimes, they also devour larger prey, such as smaller frogs and even little birds and rodents which they overpower and swallow.

B. Changes in Morphology

(i) Regressive or destructive changes

These include the reduction or complete disappearance of those organs or structures which are necessary during larval life, but not essential in the adults. The important changes of this nature are as follows:

1. The long *tail* of the tadpole along with the fin folds is absorbed again during metamorphosis and disappears altogether at the end of the metamorphosis.

   The reduction of the tail is affected by *autolysis* of the component tissues with active participation of amoeboid *macrophages*. These are rich in lysosomes that contain a proteolytic enzyme *cathepsin*. The macrophage phagocytose the debris of the disintegrating cells.

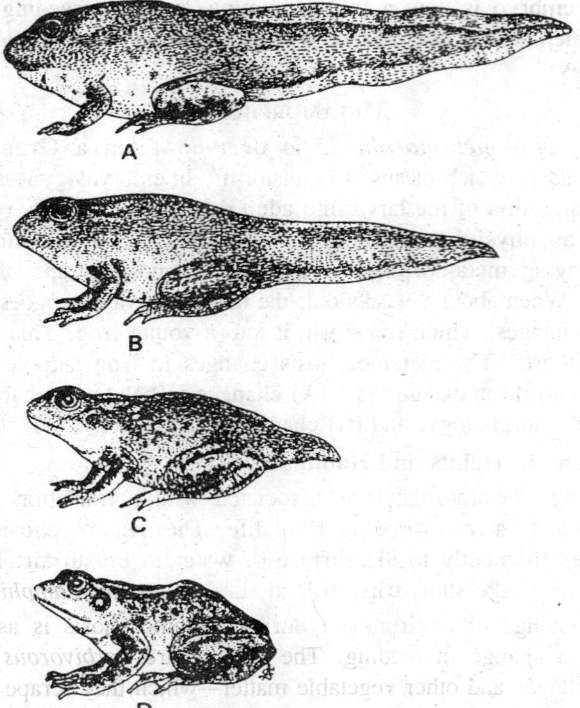

Fig. 14.29. Stages in metamorphosis of frog's tadpole.

During metamorphosis, however, the tadpole is nourished by the substance of its tail.

2. The *gills* are resorbed, the *gill clefts* are closed and the *branchial cavities* disappear. The resorption of gill also takes place by autolysis.
3. The *horny teeth* of the perioral disc as well as the *horny lining* of the jaws are shed.
4. The *lateral line sense organs* present in the skin of tadpoles, disappear during metamorphosis.
5. The *cloacal tube* becomes shortened and reduced.
6. Some *blood vessels*, including parts of the aortic arches, are reduced.

(ii) Progressive or constructive changes

These include the development of some organs which become functional only during and other metamorphosis.

1. There is a progressive development of the *limbs* which increase in size and differentiation. The *forelimbs*, which in the tadpole develop under cover of opercular membrane, break through to the exterior. At the same time there is a great increase in the length and strength of the *hindlimbs*. Joints develop in them and the toes appear.
2. The *middle ear* develops in connection with the first pharyngeal pouch. The tympanic membrane develops. It is supported by the circular tympanic cartilage and enables the frog to receive air-borne vibrations.
3. The *eyes* bulge up on the dorsal surface of the head and develop nictitating membrane.
4. There is development of the *tongue* and the *vomerine teeth*.

(iii) Remodelling of some structures

Some organs which function both in the larva and the adult, change their differentiation during metamorphosis so as to meet the requirements of the adult mode of life.

1. The *skin* of the tadpole is covered with a double-layered epidermis. The number of layers of cells in the *epidermis* increases during metamorphosis. Surface layers become cornified. Multicellular *mucous* and *serous glands* develop in the skin. The *pigmentation* of the skin is changed; new patterns and colours appear.
2. There is a widening of the *mouth* gap due to rotation of the quadrate cartilage backwards and the *true jaws* become functional.
3. The *tongue* grows much larger and more muscular.
4. The *stomach* and the liver enlarge.
5. The *eyes* grow more prominent.
6. In tadpole, the intestine is very long and wound up into a spiral like a watch spring. The intensive becomes greatly elongated due to a *harbivorous* habit because the vegetable diet contains less nourishment than the animal diet. Hence, a longer gut is required to extract most of the nourishment. During metamorphosis, the intestine is greatly shortened and most of the coils which it forms in the tadpole become straightened out. This change is apparently correlated with the *carnivorous* habits assumed by the adult.
7. Two chambered *heart* of frog is transformed into a three-chambered one.
8. *Lungs* do not undergo major changes. They develop very gradually and become fully functional.

9. The cartilaginous *skeleton* of tadpole undergoes ossification.
10. The larval *pronephric kidney* is transformed into a *mesonephros*.
11. The *blood vascular system* is modified with the disappearance of some aortic arches and increase of blood supply to the lungs and the skin.

C. Changes in Physiology

1. A marked change takes place in the excretory function. In the tadpole, the end product of nitrogen metabolism is ammonia which is easily disposed off by diffusion in water. After metamorphosis, however, the frogs excrete most of their nitrogen in the form of urea and only small amounts as ammonia. The latter is highly toxic and if accumulation in a terrestrial animal, it could become dangerous. This change-over, occurring in the late stages of metamorphosis is due to a changed function of the liver, which begins to synthesize urea.
2. The endocrine function of the pancreas starts at metamorphosis as it begins to secrete *insulin* and *glucagon* hormones. This is related to the increased role of the liver in the turnover of glycogen.

Physiological Control of Metamorphosis in Frog (Anura)

Metamorphosis is the anurans (frogs and toads) is initiated by the *anterior lobe of the pituitary* when it reaches a certain degree of differentiation and becomes capable of producing the *thyrotropic hormone* (TTH). The thyrotropic hormone activates the *thyroid gland*. Which builds up and releases the thyroid hormones. The most important of them are *tri-iodothyronine* (I_3Thn) and *tetra-iodothyronine* (I_4Thn) or *thyroxine*. In both of these compounds two residues of the amino acid *tyrosine* are joined together, and to these are attached 3 (in I_3Thn) or 4 (in I_4Thn) atoms of iodine. The structural formulae of these hormones are as follows: Of the two compounds, thyroxine is produced in much greater quantities and is thus the main action substance released by the thyroid gland, although the activity of tri-iodothyronine is three to five times as high as that of thyroxine. The thyroid hormone is carried to the various body organs through the blood stream and affects the tissues directly. While it causes the degeneration and necrosis of some cells, it stimulates the growth and differentiation of others, thus causing the metamorphosis.

15

Transformation of Larva

A remarkable series of changes takes place when an aquatic fishlike tadepole is transformed into a land-dwelling frog. Throughout embryonic and larval life the development of the frog has been almost entirely a gradual, unfolding process of growth and differentiation, but during the metamorphic period occur simultaneously with dramatic suddenness. The term "metamorphosis" has broad zoological usage. It is used to designate any group of developmental changes that are completed within a period of time which is brief in comparison with the full developmental period of the individual concerned. A variety of animals undergo metamorphoses of diverse sorts. Each of the events of amphibian metamorphosis contributes toward the preparation of the developing individual for life on land, but one cannot say that preparation is confined exclusively to the metamorphoic period, since, after all, the limbs and the lungs have been differentiating during most of the larval period.

Staging of Anuran Development

Development is, of course, a continuous process, but in order to make critical comparisons between different species or between different individuals of the same species it is desirable that the developmental period be divided into discrete steps or stages and that they be described in a staging table. A number of such tables have been prepared for different amphibians. The ones for anurans with aquatic larvae are quite similar to one another. Gosner (1960) has combined two of the best of them—those of Limbaugh and Volpe (1957) and Taylor and Kollros (1946)—into a simplified and generalized series that should be readily adapted for use with any frog or toad having free-swimming larvae. The larval and metamorphic stages of Gosner's series are presented here because they illustrate the sequence of external changes that are most easily observed during typical anuran metamorphosis. That stages 23, 24 and 25 are marked by the formation of the operculum and the consequent covering over of the external gills by its anterior component which grows backward from the hyoid arch.

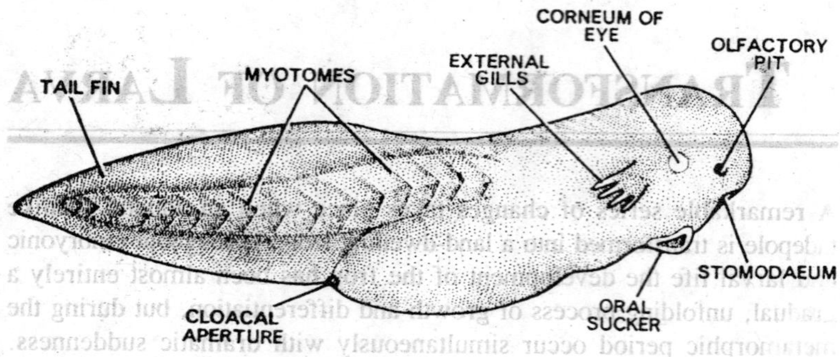

Fig. 15.1. Lateral view of early tadpole.

By stage 25 the anterior and posterior components of the operculum fuse except for a small funnel-shaped opening to the outside, the spiracle. At about stage 23 the oral disc and rows of keratinized labial teeth and horny beaks begin to form. In stages 23 to 25 the initial formation of pigmentary patterns takes place, chromatophores of several types appearing at about stages 23 and 24. In distinguishing between families the diagnostic feature is the form of the oral disc. Its essential peculiarities are present by about stage 26, although changes occur subsequently in the number and form of oral papillae. The tooth rows develop gradually.

Although the "mature" tooth rows formula of a species is usually established in the early larval stages, the relative proportions and numbers of the rows sometimes change during ontogeny. Independent feeding commences sometime between stages 25 and 26. Primarily on this basis stage 26 is arbitrarily designated as the first larval stage. Identification of stages 26 to 40 is made by examination of the hind limbs. Stages 26 to 30 are distinguished by changes in the ratio of length to diameter in the hind limb primordium. At stage 31 the "foot" is paddled-shaped, and in subsequent stages the configuration of the paddle is altered as individual toes are differentiated. The distinguishing criteria of stages 23 to 40 are proportional changes in the length of

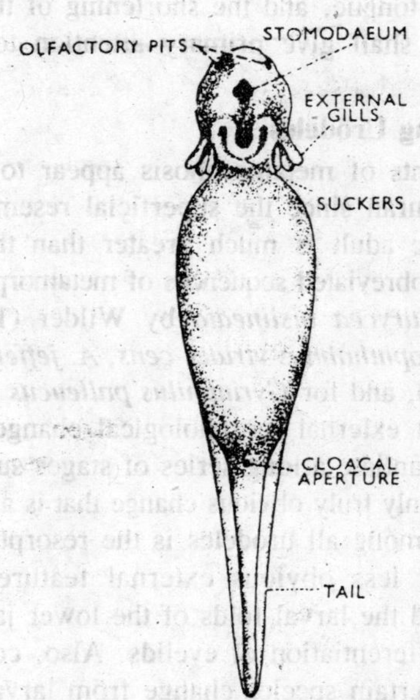

Fig. 15.2. Ventral view of early tadpole.

individual toes and in the appearance of metatarsal of metatarsal and subarticular tubercles. Etkin (1932, 1955 and 1964) regards anuran metamorphosis as consisting of two phases, prometamorphosis and metamorphic climax. Prometamorphosis begins at stage 36 and is marked by the rapid elongation of the hind limbs. During stage 40 the cloacal tail piece is resorbed, shifting the position of the anus. The completion of this shift marks the beginning of stage 41. The operculum thins and becomes translucent over the developing forelimbs, forming a "skin window" through which the forelimbs erupt. The appearance of the forelimbs marks the beginning of stage 42, the first stage of metamorphic climax.

Stages 42 to 46 are characterized by the remodelling of the head and are distinguished primarily by changes in the mouth. Larval mouthparts are lost. The tympanum is formed. At stage 46 metamorphosis is essentially complete. Newly transformed young may or may not resemble the adults sufficiently to permit positive identification. The extensive alterations in external features noted are accompanied by equally profound changes in internal anatomy such as the adaptation of the aortic arches and the hyoid cartilages and muscles for pulmonary respiration, the loss of the lateral line system, the development of a muscular tongue, and the shortening of the gut. In this chapter, however, we shall give primary attention to external morphology.

Typical Development among Urodeles

In the urodele the events of metamorphosis appear to be much less striking than in the anuran since the superficial resemblance of the salamander larva to the adult is much greater than that of the tadpole to the frog. Rather abbreviated sequences of metamorphic stages have been described for *Eurycea bislineata* by Wilder (1925), for *Ambystoma maculatum*, *Notophthalmus virides cens*, *A. jeffersonianum* by Grant (1930a and 1930b), and for *Gyrinphilus palleucus* by a Debt and Kirby-smith (1963) but external morphological changes are not sufficiently marked to substantiate a long series of stages such as that described for anurans. The only truly obvious change that is a consistent feature of metamorphosis among all urodeles is the resorption of the gills. Other consistent but less obvious external features are the resorption of the tail fin and the larval folds of the lower jaw, fusion of the gill slits, and the differentiation of eyelids. Also, colours and colour patterns typical of certain species change from larval to adult shades and configurations.

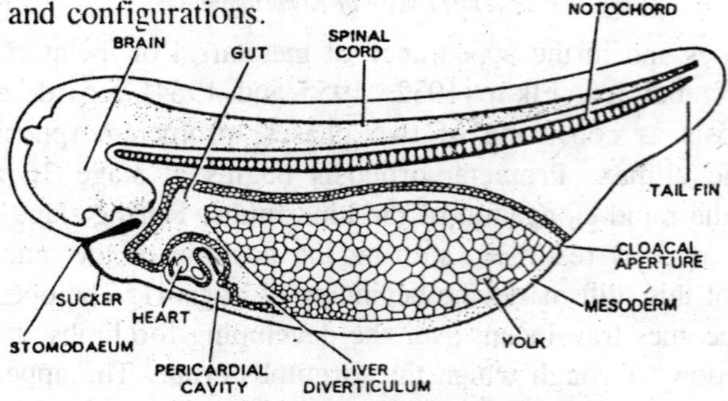

Fig. 15.3. External gill stage of tadpole.

In addition to the events common to all urodeles, there are numerous changes in features peculiar to certain groups. For example, the Plethodontidae usually develop naso-labial folds during the latter states of metamorphosis. The internal changes of urodele metamorphosis are most extensive in the head region. Valves develop around the nares to prevent the entrance of water if the adult becomes submerged. Alterations take place in the skull and in the hyobranchial apparatus. The skeletal changes are accompanied by alterations in associated musculature. In the integument there is a loss of large glandlike cells, the Leydig cells, and the formation of a keratinized stratum corneum.

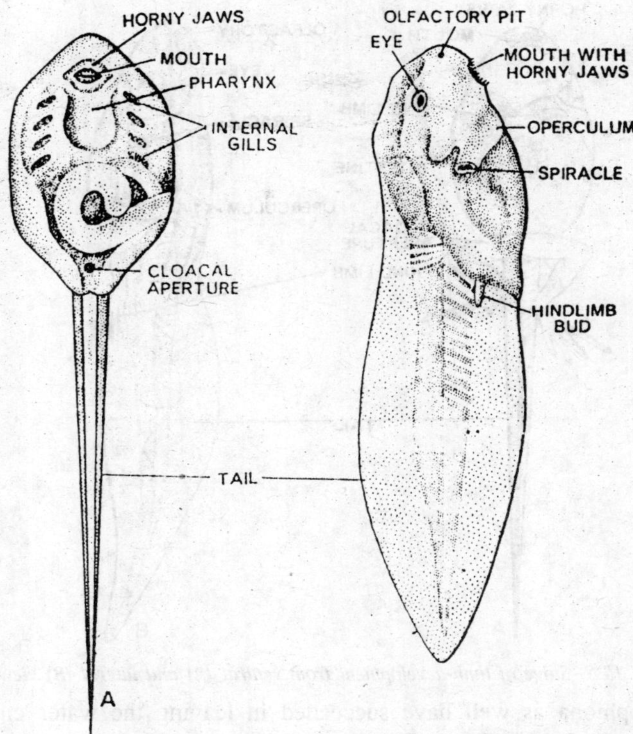

Fig. 15.4. Internal gill stage of tadpole: ventral (A) and lateral (B) views.

Variation in Amphibian Life Histories

Although there are many typical anurans and urodeles whose life histories follow the patterns that have just been described, one does not venture far into amphibian systematics and biology without discovering that there are wide variations upon the basic theme. In the evolution of the vertebrates the amphibians provided the link between the fishes and the reptiles. One tends to think of the amphibians being in the process of withdrawing from the water for life on land. A survey of the various taxonomic groups discloses that there is a fairly complete gradation in a broad range between amphibians that live completely in the water and those that live exclusively on land. The various levels of this gradation, however, neither represent a single orderly sequence nor links in a single evolutionary chain.

Instead, it is apparent that on several occasions not only frog and salamanders of diverse lineage but also the legless members of

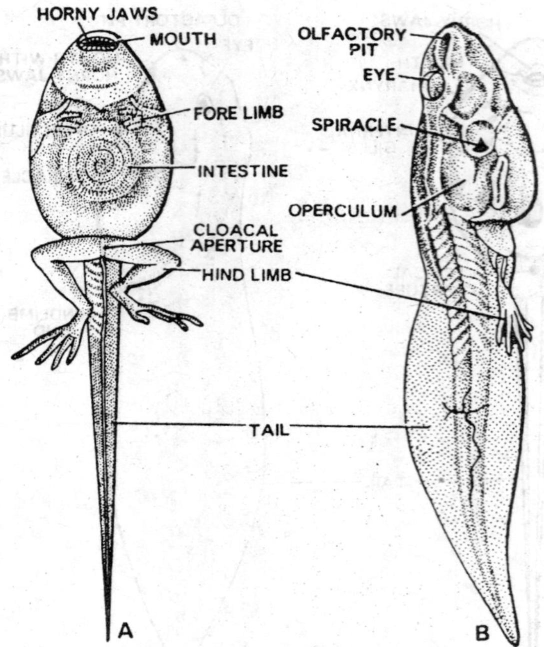

Fig. 15.5. Stage of limb-development from ventral (A) and lateral (B) views.

Gymnophiona as well have succeeded in leaving the water entirely (Orton, 1949 and 1951). Conversely, other species have returned to water or, possibly, have never left it. Withdrawal from or return to the water are not single phylogenetic trends but rather are general tendencies that have evolved repeatedly and independently in unrelated stocks. Among the 19 general of the order Gymnophiona one finds some with aquatic larvae, others with direct development, and still others that are ovoviviparous. Yet they are, in effect, small and degenerate groups, and it has been decided not to given them further attention in this chapter.

Completely Aquatic Amphibians

The completely aquatic amphibians may be divided into two categories: (1) those with aquatic larvae that complete metamorphosis in the water and then remain in it as adults, and (2) those that are neotenic. None of the urodeles quite fit into the first category. The spotted newt, *notophthalmus viridescens*, is completely aquatic as a larva and as an adult but following metamorphosis typically spends one to three years on land as a red eft. The eft has a tongue and a terrestrial integument lacking lateral line organs, but passes through a second metamorphosis before returning to the water during which the tongue is lost and lateral line organs are reacquired.

Leurognathus maramorata of the lungless Plethodontidae is rarely found outside the water but has no special aquatic adaptations as an adult. Among the Anura the Pipidae, such as the South African clawed toad *Xenopus laevis*, never leave the water. Both larvae and adults are tongueless and have lateral line organs. Occasional aquatic genera are also found among other families such a *Batrachophrynus* and *Pseudobufo* of the Bufonidae and *Telmatobius* of the Leptodactylidae.

Neoteny

There is some confusion in the use of the terms "neoteny" and "paedogenesis." Neoteny is a condition in which larval characters are retained for prolonged periods of time. Paedogenesis is the process by

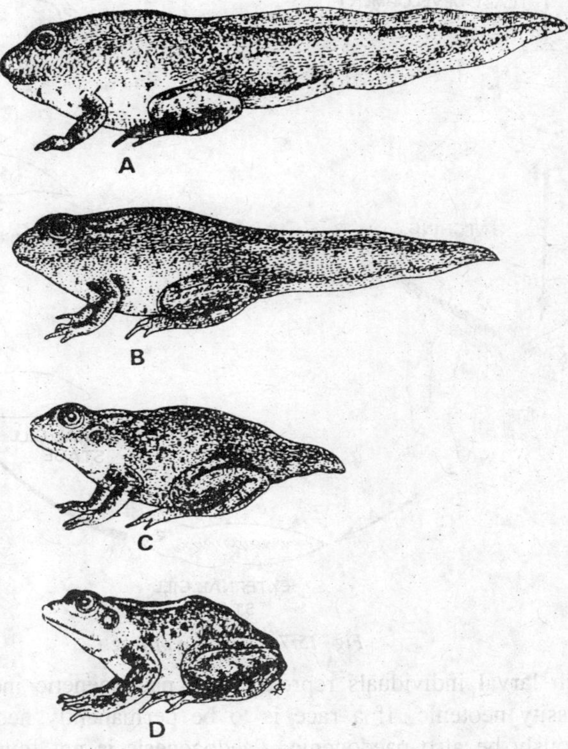

Fig. 15.6. Stages in metamorphosis of frog's tadpole.

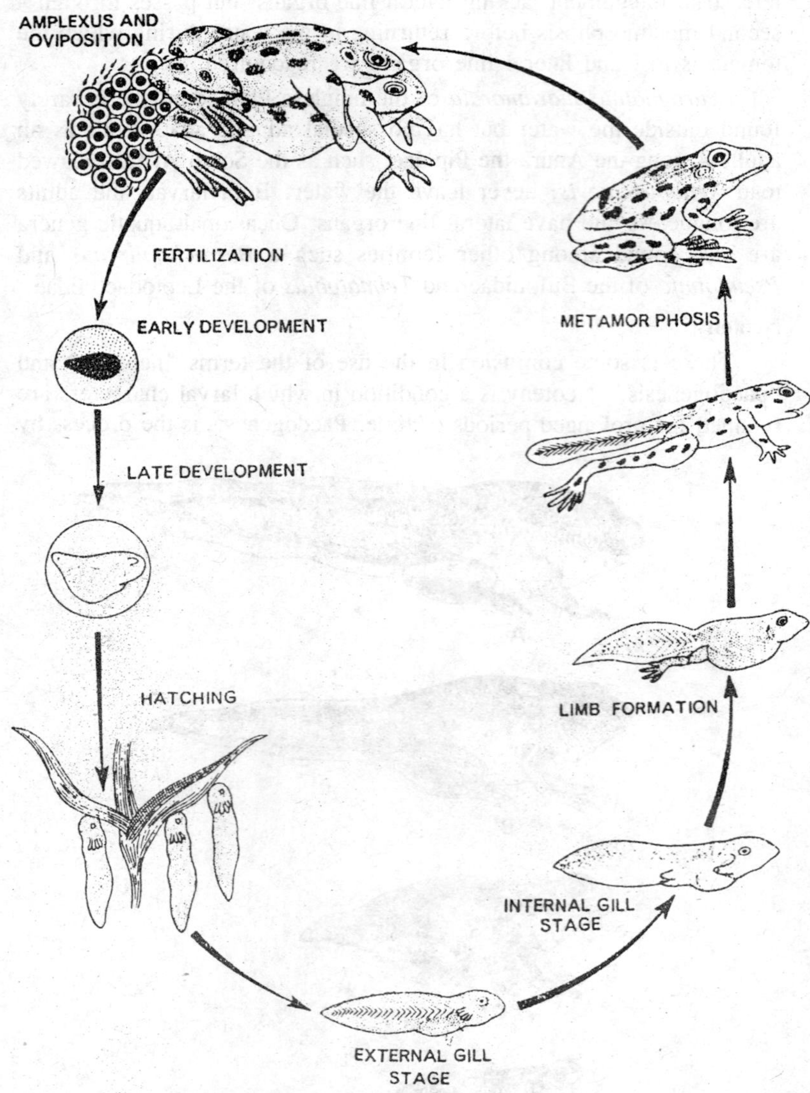

Fig. 15.7. Life cycle of frog.

which larval individuals reproduce. A paedogenetic individual is of necessity neotenic. If a race is to be permanently neotenic it must obviously be also paedogenic. Paedogenesis is not found among the Anura, although naturally occuring, non-metamorphosing anuran larvae have been reported (Jurand, 1955; Saxen, 1957; Boschwitz, 1957). In some instances the larval periods of certain species or individuals have been prolonged sufficiently for them to be classed as neotenic. Of the genera listed as permanent larvae, Noble (1931)—upon consideration of such metamorphic features as the development of limbs, maxillary bones, loss of gills, and reduction of branchial arches—proposed that *Siren* and *Pseudobranchus* be considered as forms that cease to differentiate at a very early stage of larval life; *Proteus* and *Necturus* as forms that reach a later stage of urodele ontogeny; *Cryptobranchus* as a from that had begun its metamorphosis; and *Megalobatrachus* and *Amphiuma* as forms that have nearly completed their metamorphosis.

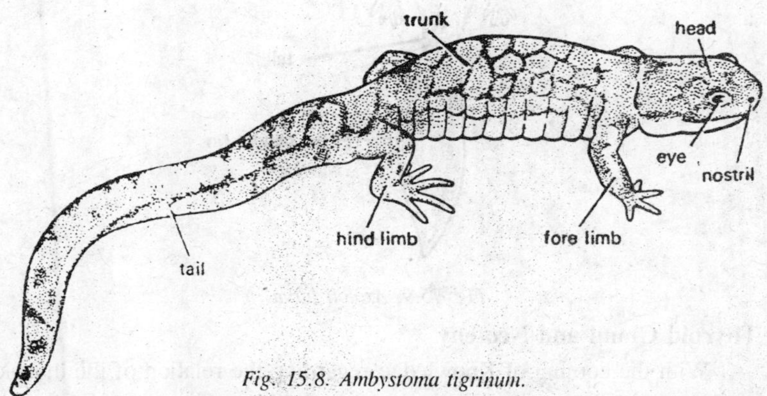

Fig. 15.8. Ambystoma tigrinum.

The outflow of water from the mouths of urodele larvae through the internal nares (choanae) is usually prevented by the action of a simple flap of connective tissue that serves as a valve. As metamorphosis a more elaborate valve activated by smooth muscle is developed. *Necturus* and *Proteus* have choanal valves and skin of the larval type (with Leydig cell and without stratum corneum), two pairs of gill slits, and three pairs of external gills (no urodele has internal gills). Moving in the adult direction, *Cryptobranchus* and *Amphiuma* each have one pair of gills slits. *Megalobatrachus* has none. These animals have not developed eyelids. The larval character of *Siredon* and *Pseudobranchus* is indicated by the absence of hind limbs and the presence of choanal valves, external gills (three pairs in *Siren* but only one in *Pseudobranchus*), and open gill slits. The skin of *Siren* has adult structure, but that of *Pseudobranchus* is larval.

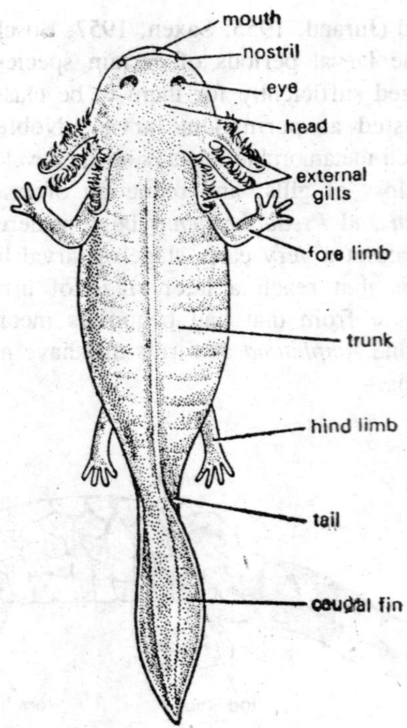

Fig. 15.9. Axolotl larva.

Thyroid Gland and Neoteny

With the coming of Knowledge regarding the relation of the thyroid gland to metamorphosis, questions were raised regarding neotenic forms. Do they fail to metamorphose because of deficient thyroid function or have the tissues hat ordinarily respond to the thyroid lost their sensitivity to it? In general, the second hormone explanation seems to be most nearly correct, *Necturus* filed more to respond to treatment with thyroxin alone, although it was in combination with adrenaline. The treatment was lethal, and no other metamorphic changes were noted. Etkin (1955) ascribes these results to generalizes effects of unfavourable conditions. Gill reduction without other changes was produced in *Siren* and *Pseudobranchus* by treatment with iodothyrine. This, also, may have been a generalized rather than a specific response.

Immersion in solutions of desiccated thyroid gland hastened the shedding of skin but has on further effect on *Cryptobranchus*. Equivalent results were obtained in *Amphiuma* and thyroxin. Although *Proteus* give no metamorphic response to treatments with thyroxin, its skin undergoes changes in the adult direction upon transplantation to metamorphosing salamanders—(*Triton*, *Salamandra*, or *Ambystoma*). The placement of the caverniculous salamanders *Typhlomolge* and *Haideotriton* in category 2 is perhaps questionable. In 1957 Dundee immersed a single small (33 mm snout-vent length) specimen of *Typhlomolge* in a 1:1,000,000 aqueous solution of racemic thyroxin. When no results were noted the concentration was increased first to 1,500,000 and then to 1:100,000. Immediately the labial folds, gills, and fins began to undergo gradual atrophy. Maxillae were developed and the beginnings of maxillary dentition were seen. Before the animal died at the end of 19 days, the fins and labial folds were resorbed, and the gills were reduced to mere stubs. The skin was unchanged, and no further metamorphic events were noted.

It is possible that if the experiment were repeated using larger animals metamorphosis might be brought to completion. It is well-known that distortion of the metamorphic pattern results when very young larvae are treated with thyroxin. The evidence favouring the inclusion of *Haideotriton* in category 2 is much stronger. Dundee (1961) immersed five specimens of *Haideotriton* in concentrations of thyroxin ranging between 1:100,000 and 1:,2000,000. All concentrations produced metamorphic changes with about equal effectiveness. In all instances, there was a loss of fins, labial folds, lateral line organs, gills, and the coronoid bone. There were no further changes in the skull, no changes in the skin, and the eyelids were not developed. All five animals died within 25 days of the beginning of the treatment.

Like the specimen of *Typhlomolge*, they were relatively small individuals, the larges being 27 mm in length. A mature specimen, the type specimen of *Haideotriton*, was reported by Carr (1939) to be 75.6 mm in length. Pylke and Warren (1958) found that an animal grown to be 43 mm long was still sexually immature. As with *Typhlomolge*, it seems likely that if larger specimens of *Haideotriton* were treated with thyroxin they might pass through a complete metamorphic sequence. It must be conceded, however, that the tissues of both *Typhlomolge* and *Haideotriton* are very resistant to the metamorphic action of the thyroid hormone and that they might be properly classed as permanent larvae. When immersed in a solution of thyroxin at a concentration of 1:500.000, the Tennessee cave salamander *Gyrinophilus palleucus* was seen after about tow weeks to begin metamorphosis with a regression of soft tissues at either side of the tip of the snout. The gills shrank too become mere stubs at the end of about three weeks.

By this time the labial folds and tail fin were resorbed, and parasphenoid teeth, maxillary bones and teeth were formed. The formation of eyelids took several more weeks, and the differentiation of the tongue was a protracted process that took up to five or six months for completion. Rather surprisingly, in view of the length of time required for complete metamorphosis at this rather high concentration of thyroxin, a few specimens underwent spontaneous metamorphosis in ordinary springwater. The factors that stimulated this spontaneous metamorphosis have not been discovered. Complete metamorphosis was brought about in *Eurycea tynerensis* and in *Eurycea neotines* within 18 days by immersion in a 1:500,000 solution thyroxin.

The other species of *Eurycea* listed in the second category have not been tested, but it seems reasonably safe to assume that because of their close relation to *E. tynerensis* and *E. neotines* they also are quire responsive to the thyroid hormone. The ease with which metamorphosis is induced in the axolotl *Siredon pisciformis* is common knowledge. The response of the species listed in category 3 to the thyroid hormone has not been tested in all instances, but it is safe to assume that they are all quite responsive.

Metamorphosis occurred quite readily in *Notophthalmus viridescens*, e.g., when neotenic specimens were exposed to quite low concentrations (1:2,000,000) of thyroxin. *Ambystoma gracile* completes metamorphosis at 18 days in a 1:500,000 solution of thyroxin. Thus, it is reasonably clear that the tissues of the animals in categories 2 and 3 of Table 9.1 are competent to respond to thyroid hormone. But what of their own endocrine glands? For many years it was generally believed that the thyroid gland was congenitally absent in *Typhlomolge rathbuni*, but eventually Gorbman (1957) mad a careful histological study and found in each of two specimens around 100 loosely arranged thyroid follicles. The thyroid of *Gyrinophilus palleucus* typically functions at a low level but produces sufficient hormone to induce metamorphosis when stimulated with exogenous thyroid-stimulating hormone. The thyroid of the axolotl induces metamorphosis when stimulated by pituitary grafts from its typically metamorphosing relative *Ambystoma tigrinum*, but *A. tigrinum* larvae retain their larval from when their pituitary glands are replaced with axolotl pituitaries. In general, these observations and others made on the thyroid glands of facultative neotines indicate that their metamorphic failure results from hypofunction of the pituitary gland.

Withdrawal from the Water

From the standpoint of evolution, several advantages have accrued to those amphibians that have taken up completely terrestrial existence. The survival value of the change is readily recognized when one becomes aware of the wide geographic distribution and the numbers of highly successful species that have acquired this mode of life. The developing terrestrial amphibian. On land, parental protection is feasible and often given. The hazards resulting from the drying up of temporary pools and the flooding of mountain streams are eliminated. Metamorphosis is apparently a traumatic process, as evidenced by the fact that in progenies of typical amphibians reared in protected laboratory situations the highest mortality rates are always observed during the metamorphic period.

The amphibians that have achieved direct development avoid the rigors that attend the extensive remodellings of typical amphibian metamorphosis. Adaptations that have resulted in various degrees of withdrawal from the water are found in both the Anura and the Urodela. The range of adaptation is broadest and most complete within the Anura, and we shall give first consideration to them. Gradations of withdrawal can be arranged in two series that are unsequential in the evolutionary sense. One illustration reaches its culmination with completely direct development in which strictly larval characters are almost entirely eliminated. The other, illustration reaches its peak when ovoviviparity is achieved.

Direct Development among Anurans

Rana pipiens is representative of all those anurans that have a typical life history in which jelly-encased eggs are laid in some body of water and in which typical aquatic larvae hatch from those eggs to eventually undergo a typical metamorphic sequence. The little West Indian tree frog *Hyla septentrionalis* is representative of this forms that lay their eggs in very small volumes of water such as those found in sections of bamboo or in the cavities created by the branching of leaves from the bases of bromeliads. Their development is morphologically typical but extremely rapid. Such small deposits of water would be likely to evaporate before slowly developing larve could metamorphose. The South American frogs *Hyla faber*, *H. pardalis*, and *H. rosenbergi* lay their eggs in small nests or puddles of water that are very close to or walled off from larger pounds or pools. *H. rosenbergi* has an additional adaptation. Its gills are modified for clinging to the surface film of the water in which it develops. A large number of species from several genera and families attach themselves to vegetation that overhangs some body of water.

In the South American *Hyperolius tuberillinquis* the eggs are contained within a stiff mass of jelly. Oftentimes, as in *Rhacophorus leucomvstax* from the Philippines, the nest is composed of froth beaten up from an albuminous secretion of the female by the limbs of one or both parents. Sometimes the nest remains relatively soft and sometimes it hardens superficially into a crust that resists dehydration. The female of *Chiromantis xerampolina* remains with and moistens her nest until the larvae within it are hatched, otherwise the crust becomes too hard for the larvae to break when the time comes for them to escape. Other forms such as *H. pusillus* attach their eggs individually to leaves around the eggs, as do the banana frogs of the genus *Hoplophyryne*.

In all of these species the larva upon hatching falls into the water and takes up the existence of a typical tadpole. The majority of the 50-odd species of the genus *Leptodactylus* lay their eggs in froth-filled nests near small bodies of water that fluctuate in depth with rainfall. The nest sites are from time to time flooded and the larvae are routinely washed into the body of water before metamorphosis takes place. *Hylambates natalensis* and a few species from other families are known to build quite similar nests. With her snout, the female of *Hemisus marmoratum* digs an underground nest chamber and remains with her eggs until they hatch. She then tunnels to the water and is followed by the larvae which, even though they become aquatic, do not develop external gills.

Thoropa petropolitana is representative of a few species of South American frog such as *Cyclorampus pinderi* whose tadpoles are only semiaquatic in that they spend their larval period upon moist rocks and not actually within the water. Their mouths are especially adapted for clinging to the rock, and their tails are modified somewhat as respiratory organs, although gills do develop. *Leiopelma archeyi* from New Zealand is occasionally semiaquatic throughout the larval period. The eggs are laid in a damp situation, but not in the water, and the young swim only if the next happens to be flooded before metamorphosis is complete. It has four pairs of open gill slits, but no gills form. As pointed out earlier, the anuran operculum is derived principally from a transverse dermal fold that grows backward from the hyoid region, but also from a low, ridge like fold across the anterior part of the belly just behind the forelimb buds. In aquatic larvae these two components are fused together prior to the formation of the forelimb bud. In *Leiopelma* fusion does not take place although the forelimb buds are covered by the anterior components. The tongue forms before hatching occurs.

In South American *Leptodactylus* represents a group of several species such as *Zachaenus parvulus* whose larvae are semi-aquatic, being laid and left throughout metamorphosis either in froth or in some moist situation some distance from water. In these animals essentially all larval features are retained and the metamorphic pattern is essentially unchanged. The South American frog *Arthroleptella lightfooti* has a semiaquatic larva but is considerably adapted in the direction of direct development with the loss of gills, gill slits, and larval mouth parts. It does retain the larval operculum. Its tail shows no special adaptations. Frogs of the genus *Cornufer* have essentially direct development. The operculum is the only larval structure that is retained. The embryos are distinctive in that their abdominal walls bulge into huge saclike structures that are obviously adapted for respiration.

Eleutherodactylus nubicola is representative of the some 200 different species of over 15 other genera in which the aquatic larval period and most of the larval characters are eliminated. The young are encased within gelatinous membranes throughout the developmental period to a stage equivalent to the termination of metamorphosis. Information is fragmentary concerning most of these species, but a reasonable amount of detail is available concerning a few such as Arthroleptella rattrayi as well as six species of *Eleutherodactylus*. In these forms gills have either disappeared completely or are present in vestigial form. Gill slits never develop, and without then neither do internal gills.

Among the species of *Eleutherodactylus* that have been studied, *nubicola* is the most advanced. In this species a trace of the lower larval lip and a small vestige of the posterior opercular anlarge are the only larval characters that remain. Among the other species of *Eleutherodactylus*, *E. portoricensis*, *E. nasutus*, and *E. guentheri* have evolved to a stage nearly equivalent to that of *E. nubicola*, whereas *E. inoptatus* and *E. martinicensis* have transitory vestiges of external gills. In all of these animals the highly vascularized tail expands into a balloonlike structure and serves as an organ of respiration.

Parental Care among Anurans

Although they lack fetal membranes, anurans with direct development have otherwise reached a condition of oviparity equivalent to that typical of the birds and of many reptiles. They show a

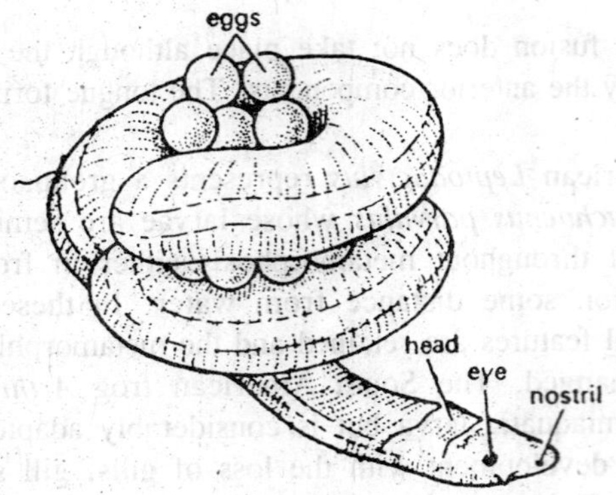

Fig. 15.10. Ichthyophis. Female coiling around the eggs.

progressive increase in parental care and handling that culminates in true ovoviviparity. The "brooding'' of eggs is not uncommon among those forms with direct or near direct development. A parent guards the eggs in several species of *Eleutherodactylus*, and the male of *Leiopelma* guards the eggs. It has been suggested by Jameson (1950) that the male of *Eleutherodactylus latrans* wets down the eggs that it guards with urine to prevent their drying out. Some wetting down of the encrusted forth nest of *Chicomantis* by a guarding parent has also been reported by Cochran (1961).

The attention of the female of *Hemisus marmoratum* to her young was noted in the preceding section. The male of *Alytes*, the midwife

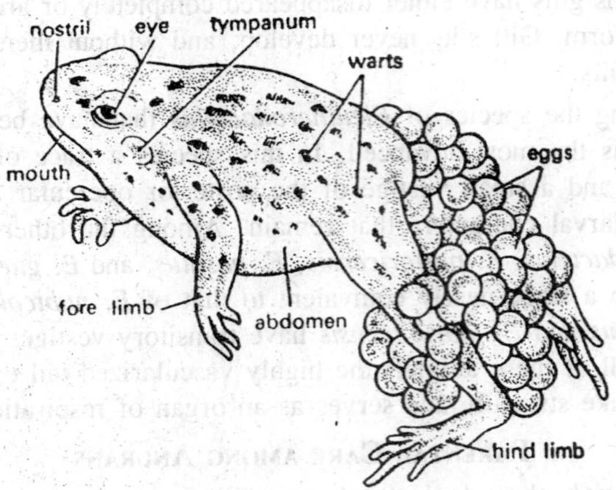

Fig. 15.11. Alytes.

toad, loops strings of eggs as they are laid around his hind legs and then carries them about, occasionally entering the water so that they are moistened, until finally just before they hatch the enters the water and remains there until the larvae emerge from their membranes. The males in some species of *Dendrobates* and also those of *Phyllobates* pick up eggs after they have been laid on land and transport them on their backs until some time after hatching takes place. The young of both these genera as well as those of *Cryptobatrachus*, in which the female provides the transportation, are deposited in the water, however, sometime prior to the metamorphosis, *Hemiphractus* and *Gastrotheca*, the "obstetrical frog," which is equipped with a pouch on its back, vary from species to species with regard to the amount of development that takes place while the young are being transported.

In the species cited, *H. divaricatus* and *G. oviferum*, transportation continues until metamorphosis is essentially complete but in some other species the larvae swim away before metamorphosis begins. There is also variation in the yolk content of the eggs from species to species, and those larvae with the greater amounts of yolk remain longer on the back of the female. The Surinam toad *Pipa pipa* is well-known for its adaptation by means of which embryos are transported in individual chambers on the back of the female. The chambers develop when skin thickens after the eggs have been pressed into position on the back of the female by the belly of the male during the mating process. There has been some loss of larval characters by these larvae. Horny mouthparts, for example, have been lost, but for the most part they

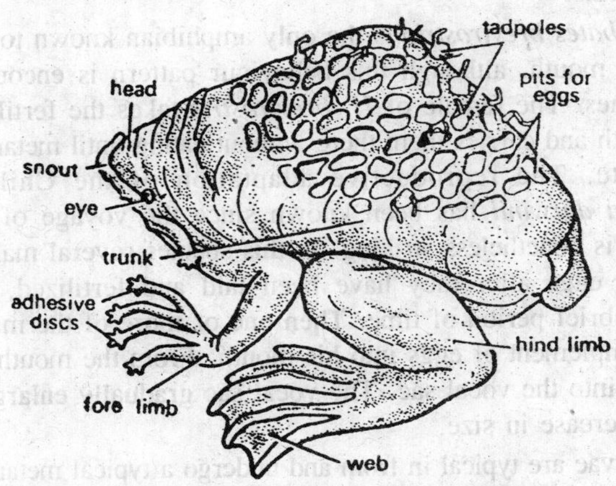

Fig. 15.12. Pipa pipa (female).

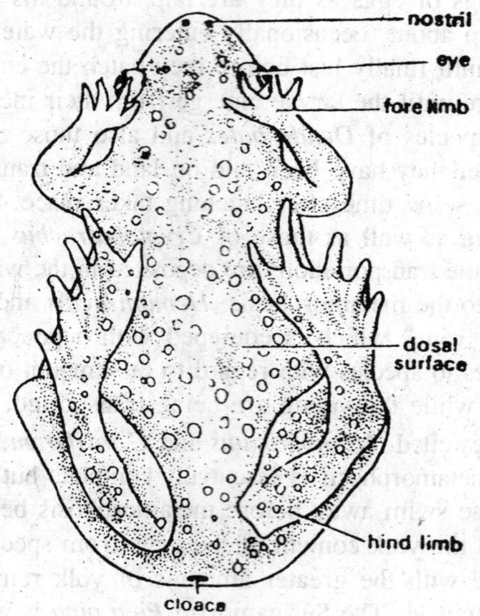

Fig. 15.13. Bombinator.

have the configuration of typical tadpoles and may actually feed on small organisms that come within their grasp during the last few weeks they are within their chambers. A similar pattern of life history is found in two other species of *Pipa* and in the species of the closely related genus *Protopipa*, except that the larvae the female prior to metamorphosis.

Hylambates brevirostris is the only amphibian known to brood its eggs in its mouth, although this behaviour pattern is encountered in several fishes. The female of *H. brevirostris* takes the fertilized eggs in her mouth and carries them there without eating until metamorphosis is complete. The reproductive adaptation of the Chilean frog *Rhinoderma darwinii* has been known since the voyage of H.M.S., Beagle but is nonetheless striking. In this species several males gather around the eggs after they have been laid and fertilized, guarding them for a brief period of time. Then one or more of the males takes a small complement of eggs into his mouth. From the mouth the eggs are moved into the vocal sac. The vocal sac gradually enlarges as the embryos increase in size.

The larvae are typical in from and undergo a typical metamorphosis with the exception that the horny teeth and beaks never harden. Although the embryos have a considerable amount of yolk, some feeding may take pace toward the end of the developmental period while they are still encased within the vocal sac. In the three species of *Nectophrynoides*, ovoviviparity is attained. The larvae are equipped with gill slits and suckers, but horny beaks and teeth are absent. The long, ratlike tail has very low fines. Although its surface are is not very great the tail is well vascularized and obviously has a respiratory function. This is quite apparent in *N. vivipara*. Here the uterus may contain as many as 100 larvae whose bodies tend to be located towards the central part of the uterus and whose elongate tails are arranged so as to have a maximal contact with the uterine wall. Angel and Lamotte (1944) report *N. occidentalis* to be without a spiracle. Orton (1949) states that a spiracle is present in *N. tornieri* but that it is quite small.

Withdrawal from the Water by Urodeles

The movement from the water in the Urodela is roughly parallel to that in the Aruna. Several species (including, oddly enough, the neotenic and otherwise aquatic *Amphiuma*) lay their eggs on land but near water that fluctuates in level so that the embryos are swept into it prior to hatching. Among the urodeles, direct development is accomplished only in the family Plethodontidae. The large, yolk-laden eggs are surrounded individually with heavy gelatinous membranes. In general, there is no loss of larval characters and no deviation from the typical metamorphic pattern except that the gills are larger and longer than those of aquatic larvae. In view of its neoteny it is somewhat surprising to find that *Proteus* is often ovoviviparous in the lower temperatures of its range.

The European newt *Salamandra salamandra* likewise retains embryos in utero until after they hatch, sometimes apparently stimulated by low temperatures, throughout metamorphosis, *Salamandra atra* is consistently ovoviviparious, as are the plethodontids, *Hydromantes* and *Oedipus*. As in the ovoviviparous urodeles, with direct development there is no loss of larval charactérs in ovoviviparious urodeles. The intrauterine larvae of *Salamandra*, for example, develop rudimentary balancers and lateral line organs. The elongate gills of these creatures functions in the exchange of respiratory gases between maternal and larval bloodstreams.

Role of the Thyroid Gland in Direct Development

In a preceding section to involvement of the thyroid gland in amphibian neoteny was commented upon. Its function in direct development is also a matter of considerable interest. As early as

1917 Hoskins and Hoskins showed that the course of typical metamorphosis was halted by removal of the thyroid gland. In 1925, Allen found some difference in *Bufo* and *Rana* in the degree to which limbs will grow in the absence of thyroid stimulation. In *Bufo* the hind limbs of thyroidectomized larvae may reach a length of 8 mm, whereas those of *Rana* remain as small buds. It is well-known that administration of thyroid hormone to normal tadpoles brings about precocious metamorphosis. It seems possible, then, that direct development may consist of a telescoping of larval stages resulting from a precocious metamorphosis brought about by an early and intensive functioning of the thyroid gland, by an usually high sensitivity of the tissue to a relatively low thyroidal activity, or by the emancipation of the tissues from the need for stimulation by the thyroid hormone.

Histological observations made by Lynn in 1936 indicated that in *Eleutherodactylus* the thyroid began to function early in the developmental period and therefore might be responsible for a precocious metamorphosis. Similar conclusions were reached by Brink in 1939 with regard to *Arthroleptella bicolor*, but Dent found that the thyroid differentiates late in the intraoval life of the terrestrial salamanders *Plethodon cinereus* (1942) and *Aneides anaeus* (1954). Embryos of neither *Plethodon* nor *Eleutherodactylus* Have survived attempts to remove the anlage of the thyroid by surgical procedures. In 1947, however, Lynn treated embryos of *Plethodon cinereus* with the goitrogen thiourea, blocking thyroid function by chemical means. Those embryos completed their development except for the resorption of the long and branched gills. Similarly, Lynn (1948) and (Lynn) and Peadon in 1955 reported results of treating with goitrogens *Eleutherodactylus ricordii* and *Eleutherodactylus martinicensis*.

Three metamorphic features were affected by the treatment. The pronephoroes did not degenerate, the tail was not resorbed, and the egg tooth the animals develops just prior to hatching did not drop off as it ordinarily does. Administration of thyroxin to *Eleutherodactylus martinicensis* caused precocious degeneration of the pronephroes and resorption of the tail. Thus, it seems clear that the final steps of direct development are under control of the thyroid but that in the course of evolution the greater part of the ontogenetic sequence has been freed from thyroid control in *Plethodon* and in *Eleutherodactylus*.

Environmental Factors in Metamorphosis

Dodd and Callan (1955) found sexually mature neotenic individuals within an otherwise typical population of *Triturus helveticus* congregated in a pond of the count of Fife in Scotland. Neoteny has been reported to occur occasionally in several species of *Triturus*. These specimens of *T. helveticus*, however, were unique in that their thyroid glands were goiterous and tremendously enlarged. This condition could hardly be attributed to a lack of iodine in the environment since the pond is situated within a few miles of the North Sea, and the iodine content of drinking water in the nearly town of Crail in 5 μg/l. Dodd and Callan noted that kale and turnips were cultivated in fields in the vicinity of the ponds.

The goitrogenic effects of the brassicas is well-known. Rabbits were frequently seen in the vicinity of the pond, and rabbit feces accumulated on the slopes from which water drained into the pond. Dodd and Callan concluded that this drainage water should contain the "brassica factor" at times when the rabbits were feeding on turnips and kale and that larvae overwintering in the pond might well, as a result, become goiterous and neotenic. Although it is interesting, it is unlikely that this condition described by Dodd and Callan for *T. helveticus* obtains generally among other occasionally neoteni forms. *Notophthalmus viridescens* is neotenic in the lowlands of Louisiana, on Cape Cod, and on Long Island; but the majority of the neotenous individuals among the species live in lakes located at altitudes of several thousand feet, as does *Siredon pisciformis*, the axolotl cited in category 2. Snyder (1956) has made comparative study of the Pacific Coast salamander *Ambystoma gracile* from ponds found at sea level and from other ponds at high elevations (4,300 to 5,500 ft.) in Mount Ranier National Park. At sea level a high proportion of larvae transform into immature adults at one year of age, and most of the remainder transform at two years of age as sexual maturity is attained. A small number persist as neotenic larvae.

Adult animals are found in the montane habitats, but metamorphosis appears to be rare and most of the mature animals are neotenic. Snyder states that among amphibians cold either inhibits the release of thyroid hormone or the ability is tissues to respond to its presence, and Huxley (1929) has demonstrated that below 50°F the tissues of tadpoles are not stimulated by thyroidal material. Frienden has studied this effects extensively. Snyder state further that younger larvae respond more readily than older animals to the metamorphic effect of the thyroid hormone. This contention is supported by observation of Lipchina (1929) made on the axolotl. At high elevations the annual period of activity and growth lasts only about three months, whereas at sea level larvae may be active and feeding throughout the year. Snyder suggests that since the montane larvae reach metamorphic size at a later age they are possibly less responsive to the thyroid hormone or perhaps produce less thyroid hormone and thus do not metamorphose.

The presence some neotenic animals at sea level is accounted for by Snyder on the grounds that eggs are laid over a period of five to six weeks and that perhaps the last eggs give rise to animals that overwinter as larvae and reach the age at which they are resistant to the thyroid hormone. Some brief experiments of a preliminary nature appeared to give support to this hypothesis, but further studies should be made. From an evolutionary point of view its seems reasonable that the selective pressures would be greater on the barren icy borders of a mountain lake than within its waters. Similarly, in the caves inhabited by most of the animals of category 2, it is usually quite apparent that food is more abundant and other conditions are more favourable to the salamander within the water than outside of it. Very likely it is these selective pressures that brought about the rise neoteny in caves and in mountain lakes.

Pedogenesis has never been found to occur among the Anura, but the larval period in some frogs varies greatly in length. Temperature appears to be the predominant factor involved. *Rana clamitans*, the green frog, is usually said to metamorphose after spending one winter as a larva, but Ting (1951) showed that tadpoles of *R. clamitans* metamorphosed in 92 days in the laboratory, and Martof (1952) found that great numbers of larvae metamorphosed in August from a pond that had been dry the preceding summer. He though that overwintering larvae come from eggs laid late in the season.

The bullfrog *Rana catesbeiana* also has a larval period of variable length, which is taken from a paper by Willis, Moyle and Baskett (1956). These figures indicate that the larval period increase with the length and severity of winters. The length of the larval period may also be increased in urodeles without the occurrence of paedogenesis. Many individuals of *Eurycea bislineata* metamorphose in their second year, but some (presumably those from eggs laid late in the first year) metamorphose in the third year. In southern Europe *Triturus vulgaris* metamorphose in the fall of its first summer, but in northern Russia it hibernates for one winter as a larva, no doubt because the summers in the latter region are not long enough or warm enough for it to complete larval development in one of them. In looking about for environmental features that may be responsible for neoteny in any given group of amphibians one always given consideration to the iodine content of the water.

Actually, one has yet reported finding neotenous larvae in water with an iodine content too low to support metamorphosis in other amphibian species. Elair (1961) reported that one of two specimens of neotenic *Gyrinophilus palleucus* metamorphosed after three months in a rather strong solution of sodium iodide, but the other was still uncharged after six months, and as was pointed out earlier Dent and Kirby-Smith (1965) found that specimens of this species occasionally metamorphose spontaneously in spring water under laboratory conditions. Although the induction of metamorphosis in the axolotl by the application of iodine has been reported, the amount given was relatively large. It seems safe to conclude that low iodine content of water is not ordinarily a cause of neoteny. It is to be expected that larvae deprived of food would soon stop growing and differentiating.

As far back as 1878 Yung reported that amphibian larvae are highly susceptible to nutritional deficiency and that their development is conditioned by the nature of the food. In 1887, Barfurth noted that under certain conditions the development of tadpoles was accelerated by starvation. This phenomenon was investigated in detail by D'Angelo, Gordon and Charipper (1941). They found that up to a critical period occurring in the early stages of hind-limb development metamorphic progress is slowed down and then halted by inanition so that tadpoles can be held for considerable periods of time in a "stasis" condition. Having passed the critical stage, the rate of metamorphosis indeed is increased by inanition. Possibly inanition accelerates the degenerative phases of metamorphosis. D' Angelo, Gordon, and Charipper observed that the thyroid and pituitary glands of starved animals underwent atrophic and degenerative changes.

Stasis tadpoles metamorphose readily after either immersion in thyroxin solutions or injection with pituitary material. The thyroid glands of stasis animals that are metamorphosing after treatment with thyroxin are histologically inactive, but those of pituitary-injected tadpoles give histological evidence of marked activity. On the basis of these observations it was concluded that the failure of starved animals to metamorphose is directly related to a decreased production and release of thyrotropic hormone from the anterior hypophysis. Any naturally occurring inanition would be likely to be associated with crowding, and crowding itself appears to have a direct effect on the growth of amphibian larvae.

It was Yung (1885), again, who first made observations on the relation of crowding to the delay in tadpole metamorphosis. His work was followed by that of others; notably, Adolph in 1931 and more recently Richards (1958 and 1962), Rose (1960) and Akin (1966). Adolph saw that uncrowded animals reaches a much larger size than crowded ones and that they metamorphosed in a much briefer period of time. Rose observed that whenever a group of *R. pipiens* embryos were put in a rather confined space all began to grow after hatching, but at different rates. Those that grew more rapidly at first and became larger than their bowl mates continued to grow if the water was changed daily, whereas those that lagged behind stopped growing and failed to eat even when abundant food was supplied to them. On the other hand, the stunted animals grew if they were removed and put into other containers, giving greater "lebensraum." Richards (1958) showed that if small tadpoles were put into culture water in which large tadpoles had been growing the growth of the small tadpoles was completely or almost completely inhibited, indicating that some sort of inhibitory product is thrown off into the medium by the larger tadpoles.

Further, the inhibitory effect could be removed by heating the culture water to 60°, by centrifugation, by sonification, or by filtration. Thus, the inhibitory product is a rather large particle. In an aquarium where both large and small animals were present and the small ones were growing poorly and dying out, the largest animals were just as large as large animals that had been growing in isolation (Rose, 1960). This indicates that the large animals were unaffected by anything produced by the small tadpoles or that the small ones were neither producing stimulatory nor inhibitory substances.

The growth of small tadpoles was supported, on the other hand, by culture water from large tadpoles after it had been treated with a proteinase. At least some of the inhibitory material, then, must be a protein. Richards (1962) reported that the inhibitory agent is associated with a type of algai cell in *R. pipiens*. Akin (1966) showed that the agent is elaborated by the pass through half of the gut in the growing tadpole. The algae pass through the gut and apparently transport the agent to the water Akin also confirmed earlier reports that the agent is largely species-specific. There are not reports of the inhibitory effects of crowding in nature, but such effects must exist since in the laboratory one rapidly growing large tadpole can appreciably retard the growth of small tadpoles inas much as 75 litres of water. There are not reports that various electrolytes have accelerating or inhibitory effect on metamorphosis.

Lynn and Wackowski (1951), however, are of the opinion that these apparent electrolytic effects are caused by changes in the hydrogen ion concentration of the medium rather than by any specific effects of the ion in question. Marzulli in 1941 confirmed earlier observations of Rosen (1938) to the effect that acidity below pH 4.8 and alkalinity above pH 11.0, respectively accelerate and retard the metamorphosing action of thyroxin. Marzulli further demonstrated that if thyroxin is injected its effects are not dependent on the pH of the culture medium. He concluded that the greater effectiveness of thyroxin dissolved in an acid culture medium results from the fact that it is taken up more actively by the tissues at acid rather than alkaline pH levels. For further discussion of the update of thyroxin. Disclos (1959) reported increased metamorphic rates among tadpoles of *Alytes obstetricans* given supplemental illumination, and Guyetant (1964) found that growth and metamorphosis were accelerated in larvae of *R. temporaria* maintained in constant illumination at 2,000 and 4,000 lux, whereas tadpoles kept in permanent darkness grew more slowly. It seems likely that the effects of illumination upon development will be found to be quite complex. Possibly they are mediated through the neuro-secretory pathways of the hypothalamus. This area of investigation is relatively untouched and may be the source of very interesting results in the future.

16

GENETIC CONTROL

Metamorphosis is one of the most dramatic development changes in late embryonic life. Some of the classical notion of functional maturation, as in the case of acquisition of urea formation in Amphibia, have emerged from a biochemical study of amphibian metamorphosis. The initiation and completion of metamorphosis in both amphibians and insects is under obligatory hormonal control which distinguishes it from other hormone-dependent late developmental changes. It is well known that an anuran tadpole or an insect larva will never turn into its adult form if the respective endogenous metamorphic hormones, thyroxine and ecdysone, were withdrawn or prevented from reaching the target tissues. Another feature that distinguishes both amphibian and insect metamorphosis from ordinary adaptational responses is that the process is begun and completed in anticipation of a change in environment.

The hormone serves to trigger a predetermined programme of developmental changes in order to prepare the embryo for an environment suitable for adult life. Since this chapter will deal with induction of new function and proteins, it is important to note that the hormone in no way directs cell to acquire differentiative characters but allows already differentiated but immature cells to acquire adult functions and structures. Ever since the discovery by Gudernatsch (1912) that exogenous thyroid hormone will cause the precocious onset of metamorphosis in tadpoles, hormonal induction of the process is a well-established procedure of studying sequential biochemical changes in a variety of amphibians and insects. In most studies little has been learned about the mechanism of action of the hormone, later usually serving as convenient tool to induce or delay developmental process in free-living embryos. It is in this context of a developmental tool that thyroid hormones will be treated in this article which is restricted to amphibian metamorphosis.

There are great similarities in the types of phenomena involved in insect metamorphosis which has been reviewed by many authors. The following two main facets of the regulation of protein synthesis leading to a well-ordered and sequential acquisition of new functions or structures during metamorphosis will be considered separately: (1) the regulation of RNA synthesis, formation of enzymes, and alterations in cellular structures in tissue, such as the hepatocyte, that undergo further development or functional maturation; (2) the importance of new or additional protein synthesis in tissues, such as the tail, that are programmed for death or regression. In accordance with the aims of this publication the account given below is based on personal interests and ideas and is not meant to be an exhaustive review of the subject. For the latter, the reader is referred to several reviews that have appeared in the last few years.

Role of Hormones in Amphibian Metamorphosis

Net long after the discovery by Gudernatsch (1912) that feeding tadpoles on mammalian thyroid tissue induced precocious metamorphosis, it was found that thyroidectomy of larvae (or feeding larvae on antithyroid drugs) prevented their development into adult frogs or toads. The dormant thyroid tissue of the developing larvae is activated into producing and secreting the two thyroid hormones, L-thyroxine and, 3, 3, 5-triiodi-L-thyronine (T_3), by thyrotropic hormone (TSH) produced by the anterior pituitary. The larval pituitary itself is under neural control. Thus eventually it is some environmental factor, such as illumination, temperature, salinity, that acts as the initial trigger for metamorphosis.

Although a hypothalamic tripeptide, thyrotropin-releasing factor (TRF), of the kind recently described in mammals has no yet been discovered in amphibian larvae, there is much indirect evidence to suggest that a similar neurochemical link may exist between external signals and the production of endogenous thyroid hormones. Of great interest in considering hormonal control of metamorphosis is the relatively recent discovery in amphibians of "*juvenilizing*" factors of the type known for insects. The level of insect juvenile hormone is well known to determine whether or not or how the larval target tissues will respond to the metamorphic stimulus of ecdysone. It now seems that prolactin obtained from mammalian pituitaries exerts a similar juvenilizing effect on tadpoles in that its administration at the same time of spontaneous or thyroid hormone-induced metamorphosis arrests, delays, or modifies the process. In salamanders and newts (urodeles), which do not undergo the same type of metamorphosis as in frogs and toads (anurans), prolactin causes the terrestrial forms to return to water, a phenomenon known as "*water drive*" or "*second metamorphosis.*"

Proteins Involved in Metamorphosis

Accompanying the dramatic visible morphological changes during metamorphosis tail (regression, limb emergence, positioning of eyes, etc.) are profound functional and biochemical changes in most tissues. Many of the changes or acquisition of new functions necessary for the organism for a terrestrial life are a consequence of the induction or preferential synthesis of proteins. The induction of the enzymes of urea cycle leading to the switch from ammonotelism to ureotelism during amphibian metamorphosis is new indeed a classic of the biochemical basis of development. First, virtually every type of cell in the embryonic tissue is subjected to the action of thyroid hormones, and none fails to respond in some important way. (This is not too surprising if one is declining with a rapid transition from an aquatic to a terrestrial life). Second, the hormone acts locally and directly on different cells, as for examples Kollros (1942) has shown that only the part of that tail to which thyroxine was applied underwent lysis, leaving the rest of the tail intact. In a series of elegant experiments, Wilt (1959; Ohtsu et al., 1964) had shown that only that eye to which thyroxine was applied developed rhodopsin, leaving the other with the larval visual pigment, porphyropsin.

There is thus no indirect systemic action of the hormone although an alteration in the metabolic pattern of one tissue is eventually bound to affect the activity of another via secondary adaptational routes. Third, the types of proteins whose synthesis is induced is determined by the nature of the cell, that is to say that there is qualitatively no hormone-determined protein synthesis. In some cases, the positioning of some types of cells determines the response. For example, epithelial cells of the tail synthesize collagenase while those on the skin accumulate collagen under the influence of thyroid hormones or that thyroxine may provoke regression as well as growth in adjacent cells of Mauthner's neurons. The last points is of considerable importance in the definition of the role of developmental hormones in the genetic control of protein synthesis. The multiplicity of responses to thyroxine in the same organisms makes it highly unlikely that the hormones has any inherent informational content to determine, say, via an interaction with genes or repressors, to induce the synthesis of a fixed number of proteins. When one extends the above list to be very different actions of thyroid hormones in fish, bird, and mammals, it is only reasonable to conclusion that once a cell has the receptor to recognize the hormone then it makes use of the hormone as a trigger mechanism to initiate processes determined, but not expressed, during early differentiation.

Regulation of Protein Synthesis During Metamorphosis

Current Concepts of Regulation of Protein Synthesis in Animal Cells

It is now widely accepted that, although some fundamental concepts of DNA transcription and messenger RNA translation are universal, regulation of protein synthesis in nucleated cells of higher organisms involves additional control steps not described in microorganisms. Several reviews have been devoted to this topic, and the reader's attention is drawn to two publications edited by San Pietro et al. (1968) and Wolstenholme and Knight (1970). The features mentioned briefly below are important for the interpretation of the work to be described later.

Translational control

The contrast between long-lived proteins synthesized on unstable bacterial messengers and some animal proteins of short half-life synthesized on relatively stable messengers has prompted many investigators to propose an extranuclear regulatory mechanism in higher organisms. Much of the earlier evidence for a translational control of protein synthesis was based on indirect manifestations of inhibitors of RNA and protein synthesis and did not establish the exact level at which such a control could be affected. The explanation of Wool et al. (1968) for the anabolic actions of insulin and of Korner (1970) for that of growth hormone are based on an almost direct modification of the functioning of ribosomal subunits. But the currently most favourable hypothesis of translational control is that put forward by Tomkins to account for the induction of tyrosine aminotransferase in rate hepatoma cells. Tomkins has proposed the existence of a cytoplasmic repressor whose function is to control stability or availability of messenger RNA, for translation. It is the synthesis of such a repressor that is thought to be under hormonal control. No such evidence is yet available for metamorphosis or similar late embryonic developmental systems, and direct translational control important as it may eventually turn out to be, will not be discussed in this article.

Nuclear restriction and breakdown of RNA

A substantial amount of RNA synthesized in the nucleus is rapidly turning over, heterodisperse and of high molecular weight. This RNA is not a precursor of rRNA, mRNA, or tRNA and DNA-RNA hybridization studies have shown that many of the species of RNA within the nucleus do not appear in the cytoplasm. Britten and Davidson (1969) have suggested that the rapidly turning over intranuclear RNA hybridizes very rapidly turning over intranuclear RNA hybridizes very rapidly with DNA and may be a product of repeating DNA sequences or "redundant" DNA. Although a precise role for this RNA is far from established, they and others have thought it may be somehow important in differentiation and growth.

Selective transfer of RNA from nucleus to cytoplasm

An intranuclear restriction of one whole class of RNA poses the question of what mechanisms control a selective transfer of RNA from the nucleus to the cytoplasm. The transfer of both ribosomal and messenger RNA seems to require the formation within the nucleus of ribonucleoprotein particles which are then incorporated into cytoplasmic polysomes. Several workers have now detected informosome-like particles both in the nucleus and the cytoplasm. The role of the protein associated with such informational particles in the polysomes may be an important factor in the availability or rate of mRNA translation. Very little is yet known about the nature of these proteins, their site of synthesis, how they react selectively with mRNA, etc. but any developmental process must depend on their ready availability.

Structural requirement for protein synthesis

This article deals in some detail with an important question which has not received much attention concerning regulation of protein synthesis in animal cells. It relates to the association between cellular structure and protein synthesis, particularly the attachment of ribosomes to membranes of the endoplasmic reticulum. Such an attachment is known to affect protein synthesis, but so far the main role for the structural association is thought to be that of secretion of proteins. However, the special role of membrane-bound ribosomes in bacteria and the presence of highly active membrane-bound ribosomes in predominantly non-protein secreting tissues suggests some other function for the membrane-ribosome attachment. It is of particular interest to note that the ribosomes and membranes to which they are attached are turning over continuously and that there exists a tight co-ordination between the proliferation of membranes and ribosomes when additional demands are made for protein synthesis during growth and development.

Regulation of Protein Synthesis in the Developing Tadpole Hepatocyte During Metamorphosis

The liver of the tadpole has been intensively studied in the laboratories of Frieden (1967; Friedsen and Just, 1970) and Cohen (1966, 1970) with respect to the synthesis of proteins that characterize amphibian metamorphosis, such as urea cycle enzymes, serum albumin, and adult haemoglobin. These and other workers had firmly established, especially in the bullfrog, *Rana catebeiana*, that administration of thyroid hormones to pre-metamorphic tadpoles causes a *de novo* synthesis of these proteins. Because of this firm biochemical background, the author's laboratory has over the last seven years investigated the formation and turnover of nuclear and cytoplasmic RNA in the pre-metamorphic bull frog hepatocyte at different stages after hormonal induction of metamorphosis in order to understand the nature of the process of induction. Cohen's group have also been extensively investigating the metabolism of RNA *in vivo* and in isolated preparations of tadpole liver.

Thyroid hormone induced formation of new or additional proteins

In bullfrog tadpoles (*Rana catesbeiana*) a rather long lag period of about 6 days elapses after exogenous thyroid hormone administration before new or additional proteins, characteristic of the metamorphic change in the function of the liver, could be detected. The increase of urea cycle enzymes or the appearance of serum albumin in the blood is preceded by a day or two by a rather abrupt increase in the rate of amino acid incorporation into protein *in vivo* per unit of ribosomal RNA. The decline in the rate of incorporation is only an apparent one caused by the progressive changes in levels of free amino acid as regression of organs like tail, intestine, and gills gets under way.

To some extent the lag period preceding the increase in protein synthetic rate represents the time for additional RNA to be synthesized and processed in the nucleus. Some of the RNA synthesized during the lag period is certainly important for the *de novo* synthesis of hepatic metamorphic proteins since actinomycin D, administered with or soon after thyroxine will prevent the rise in carbamyl phosphate

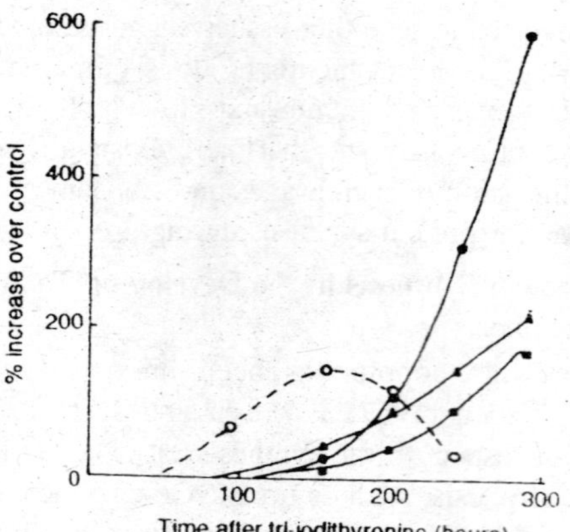

Fig. 16.1. Schematic representation of the lag period for the induction by triiodothyronine of de novo protein synthesis and the stimulation of amino acid incorporation into hepatic protein of Rana catesbeiana.

synthetase. However, hormonal control of transcription is not the exclusive mechanism for the induction of these proteins, and it is thought that thyroid hormones are also required for a sustained translation of messenger RNA or for some process of maturation of the protein molecules. Cohen and his colleagues have concluded that thyroxine exerts a dual action in regulating the rate of formation of the key urea cycle enzyme, carbamyl phosphate synthetase as well as for glutamate dehydrogenase.

It seems that part of the induction is due to control of transcription which involves all species of RNA and that almost simultaneously the hormone controls the activation of an inactive form of the enzyme. These results were based on the discrepancy between the detection of the enzyme by immunochemical methods and measurement of enzyme activity. Part of the inactive enzyme was preformed and part synthesized following hormone administration. Shambaugh et al. (1969) have further found that thyroxine also stimulates the synthesis of the inactive form of the messenger for carbamyl phosphate synthetase, but it is not certain whether or not it involves mechanism based on the inactivation by the hormone of a cytoplasmic translational repressor of the type described by Tomkins for explaining the induction of tyrosine amino transferase in cultured rat hepatoma cells by cortisol. How or at what rate-limiting step of translation the hormone exerts an effect on protein synthesis is not clear although a few observations may be relevant.

Unsworth and Cohen (1968) reported an enhanced activity of hepatic aminoacyl tRNA transferase during induced metamorphosis of bullfrog tadpoles, a phenomenon that is frequently observed in many rapidly developing systems. Tonoue et al. (1969) found an altered pattern of leucyl tRNA charged *in vivo* in a number of tadpole tissues during metamorphosis. It could also be argued that the enhanced uptake of amino acid under the influence of thyroid hormones is another way in which translational process is facilitated in a non-specific way. Perhaps it is not necessary to separate translational and transcriptional phenomena from one another, but to consider that the two processes are coupled and integrated into a well co-ordinated regulatory complex.

Nuclear RNA synthesis

It can be that there occurred, well within the latent period of 5-6 days for new proteins to be detected an acceleration of the rate of nuclear and cytoplasmic RNA synthesis in bullfrog tadpole liver following induction of metamorphosis with triidothyronine. Somewhat similar results in the same species of tadpole induced with thyroxin have also been observed in Conen's laboratory. The curves for rates of RNA synthesis have been derived from values of specific activity of RNA obtained after correction for changes in uptake of the radioactive precursor which occur earlier after triiodothyronine, as has also been bound by Eaton and Frieden.

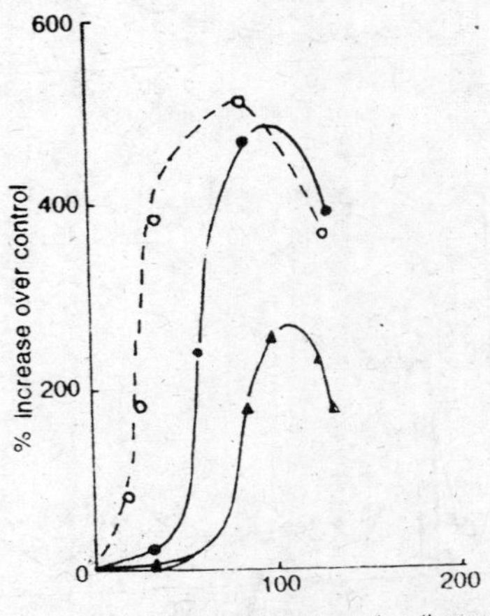

Fig. 16.2. Schematic representation of the stimulation of nuclear RNA synthesis and its turnover into the cytoplasm of liver of Rana catesbeiana tadpoles after induction of metamorphosis with triiodothyronine.

An early perturbation in the pool size or uptake of precursors of RNA and protein in target tissues has been observed with hormones known to affect protein synthesis. The question of pool sizes in radioactive labeling of constituents of non-regressing tissue is even more complicated when regression occurs in other tissues in the same organism during metamorphosis. Thus, the abrupt downward trend in the specific activity of nuclear and cytoplasmic RNA as was also noticed for amino acid incorporation, is not due to a sudden reversal of the accelerated rate of synthesis but merely reflects a dilution of radioactive precursors caused by the autolysis in regressing tissues, such as the tail, gut and gills. It would be tempting to suggest that the additional RNA made following hormone administration includes messengers for proteins like urea cycle enzymes, serum albumin, etc., especially as a sustained RNA synthesis is important for metamorphosis to occur.

Furthermore, an alteration in template activity of live chromatin from thyroxine-treated bullfrog tadpoles has been observed. In our laboratory, however, we failed to demonstrate, by base analysis, sucrose density gradient fractionation and DNA-RNA hybridization that a

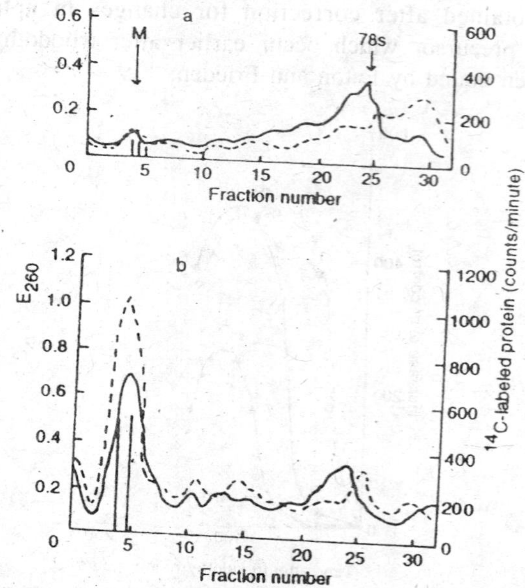

Fig. 16.3. Distribution of nascent protein synthesized in vivo on hepatic polysome obtained from (A) pre-metamorphic bullfrog tadpoles and (b) animals in which metamorphosis had been induced 7.0 days before the preparations were made.

significant part of the additional nuclear RNA synthesized *in vivo* at the onset of metamorphosis was messenger or even DNA-like RNA. Our DNA-RNA hybridization studies have shown that whereas there did occur a net increase in the amount of readily hybridizable nuclear RNA, there was also a drop in the overall "hybridization efficiency" of the RNA formed during metamorphosis. This paradoxical finding really suggest that very small increases in DNA-like RNA following hormone administration are accompanied by relatively massive increases in the rate of synthesis of ribosomal RNA. This is not to say that no changes occurred in the nature of genes transcribed under hormonal influence but that one does not yet have sensitive enough methods to detect a small change in a wide spectrum of nuclear RNA molecules with eventual messenger function. Much of the DNA-like RNA in the nucleus, which is also rapidly hybridizable, is not found in the cytoplasm, and although it may have a function in differentiation it is not thought to be direct precursor of cytoplasmic messenger RNA.

Cytoplasmic RNA

A consequence of the burst of nuclear RNA synthesis is the appearance of additional cytoplasmic RNA, mainly in the heavier polyribosomal aggregates with a concomitant decrease in the relative amount of monomeric ribosomes or ribosomal subunits. This build-up of polyribosomes occurs at a times just preceding the appearance of new proteins, as can be judged from a comparison. An increased rate of accumulation of newly synthesized ribosomes and polyribosomes coinciding with new or additional protein synthesis seems to be a common feature of regulation of protein synthesis during growth and

Table 16.1. Double-labelling or Ribosomal RNA Demonstrating that Both Breakdown and Synthesis of RNA are Accelerated at the onset of Metamorphosis

Day on which *orotic acid*^{14}*C administered*	*Metamorphosis induced*	*specific activity (cpm/mg of RNA)* ^{3}H	^{14}H	$3C^{14}C$ *raton*
0	–	17,350	4,600	3.77
	+	14,900	5,230	2.85
2	–	15,765	5,680	2.78
	+	8,010	7,825	1.02
4	–	15,500	3,650	4.24
	+	6.360	11,300	0.56
6	–	12,870	4,100	3.13
	+	5,085	9,420	0.54

development. However, unlike the situation of hormone-induced growth in many mammalian tissues, there is no appreciable accumulation of ribosomes per cell during the initial phase of induction of metamorphosis (6-10 days). This conclusion was borne out by the double isotope labeling studies shown in table above, which suggest that there is an accelerated turnover of cytoplasmic RNA and ribosomes as additional ribosomes appear after hormone administration. It seems that there is some mechanism within the developing cell that selectively breaks down "old" ribosomes and allows the preferential accumulation of "new" ribosomes formed after the metamorphic stimulus was applied.

Distribution of ribosomes in the cytoplasm and proliferation of Endoplasmic reticulum membranes

What is perhaps of utmost importance in studying the sequential events that occur between the initial burst of RNA synthesis following triiodothyronine administration and the appearance of newly synthesized proteins in a redistribution of ribosomes attached to membranes of the endoplasmic reticulum. That such a redistribution may have occurred was first suggested in experiments in which liver mitochondria free supernatant fractions from premetamorphic and induced tadpoles were titrated against Na deoxycholate. When the fraction of ribosomes remaining attaching to microsomal membranes was measured as more deoxycholate was added, those from metamorphosing tadpoles were found to be more tenaciously bound to endoplasmic reticulum. The next question was to find out whether there was a redistribution of all "old" and "new" ribosomes on pre-existing and stable membranes or whether there was also some alteration in the formation or proliferation of membranes.

It is known that cellular membranes, even in resting cells, are not metabolically stable but turning over quite rapidly. To study this problem we compared accumulation of newly formed ribosomes, their distribution on endoplasmic reticulum and the labeling of membrane phospholipids as an index of membrane proliferation. As shown in Figure there occurred an almost simultaneous increase in the synthesis of ribosomes and membranes of the endoplasmic reticulum, especially those to which ribosomes are found (rough endoplasmic reticulum). There were no obvious qualitative changes in the types of membrane phospholipids synthesized, but it is interesting to not that the simultaneous increase in the formation of the two rough endoplasmic reticulum components is accompanied by an enhanced rate of protein synthesis *in vivo*. Titration of mitochondria-free extracts with sodium

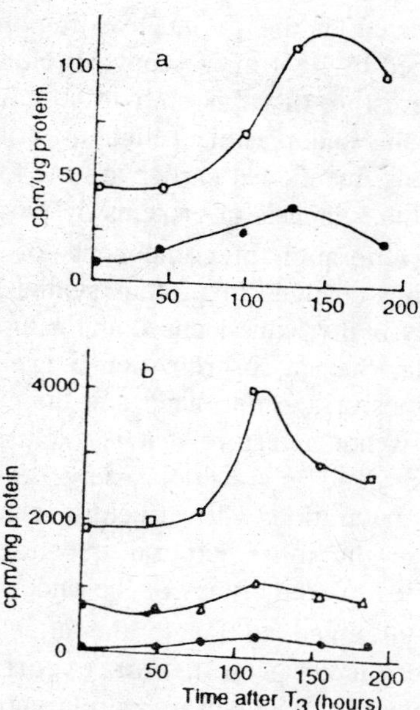

Fig. 16.4. Co-ordinated proliferation of endoplasmic reticulum and the increase in hepatic protein synthesis in vivo in the heavy rough membranes during triiodothyronine-induced metamorphosis of young bullfrog tadpoles.

deoxycholate showed that the newly formed ribosomes in induced animals were more tightly bound to membranes of the rough endoplasmic reticulum, i.e, they required a higher amount of detergent for their release. It can be seen that the "heavy rough membrane" fraction, i.e., the one in which ribosomes are more densely packed on membranes, was more active in protein synthesis *in vivo* than were free ribosomes or the "light rough membrane" fraction.

Although the evidence is too tenuous to warrant associating a distribution pattern in broken cell preparations with differential protein synthetic rates in the intact cell, it is tempting to suggest that the protein biosynthetic response to the hormone may be contained in the newly formed ribosomes. The above ribosomal redistribution changes are so marked that it is easy to corroborate them by electron microscopy. These studies revealed that induction of metamorphosis caused a shift in the distribution of ribosomes from around the simple vesicular membrane structures of the immature larvae to the more complex double lamellar structures more commonly seen in mature tissues. At the initial stage of metamorphosis, these new structures are predominantly located in the perinuclear region, an event that could by easily discerned by light microscopy of toluidine blue stained preparations. It is interesting that this shift in structural organization of polyribosomes on the endoplasmic reticulum coincided with the biochemical observations, mentioned earlier, on the formation of RNA and phospholipid and the synthesis of proteins by polyribosomes.

Rather similar proliferation of components of the endoplasmic reticulum accompanying changes in protein synthesis has also been described more recently in thyroxine-induced and natural metamorphosis of the bullfrog tadpole. Such a co-ordination between formation and distribution of ribosomes on the one hand and their protein synthetic activity on the other is not a feature that is exclusive to amphibian metamorphosis but is found in a wide variety of systems of rapid growth or functional maturation. Ultrastructural changes of the kind described above raise the more general question of the role of attachment of ribosomes to membranes of the endoplasmic reticulum.

The major role assigned so far to the attachment is that of facilitation of secretion of proteins for export. However, the proliferation of endoplasmic reticulum or a preferential shift from free to membrane-bound ribosomes is a characteristic of growth and development. This is true for both non-secretary and secretory cells in which relatively more protein for intracellular use is made during rapid growth. It is likely that the attachment of ribosomes to membranes serves some other role during growth and development. The significance of the coupling of structural alterations and biosynthetic function that is noticed in tadpole liver during metamorphosis, mentioned above, may be that: (a) a response to growth and developmental stimulus is contained in newly formed ribosomes, and (b) that there is a topographical segregation on membranes of the endoplasmic reticulum of differently precoded polyribosomes carrying out the synthesis of different groups of proteins.

There is, of course, no direct experimental evidence to verify either of these suggestions, but some recent work from our laboratory on the additive effects of growth and developmental hormones in mammalian organs provides an indirect support. In these experiments, when growth hormone, triiodothyronine and testosterone were administered in different combinations, the increased rates of formation of ribosomes and the membranes of the rough endoplasmic reticulum were co-ordinated with those of synthesis of different classes of proteins in the liver and seminal vesicles. Since these hormones have quite different latent periods of action, the simultaneous administrated in different combinations, the increased rates of formation of ribosomes and the membranes of the rough endoplasmic reticulum were co-ordinated with those of synthesis of different classes of proteins in the liver and seminal vesicles. Since these hormones have quite different latent periods of action, the simultaneous administration of any two caused stepwise increases in rates of proliferation of ribosomes and rough endoplasmic membranes, each burst corresponding in its time-course and magnitude to that induced by each individual hormone. It is assumed in the above suggestions, that an exchange between polysomes and membranes proceeds at a low rate relative to the lifetime of these structures.

An indirect approach to testing the idea of a topographical segregation would be to identify induced proteins on proliferating rough endoplasmic reticulum during development. We are now attempting by histochemical methods of detection and localization of ornithine and asparate carbamyl transferases as they increase during induced metamorphosis. The recent work of Pitot et. al. (1969) on the immunochemical location of serine dehydratase has shown that this inducible enzyme is almost exclusively synthesized on membrane-bound and not free ribosomes of the liver. In a different system, Ledue et al. (1968) have shown that antibody synthesized at the initial stages of stimulation of the lymphocyte by a given antigen is localized in the perinuclear rough endoplasmic reticulum structures. Even in reticulocytes it seems that different classes of proteins are synthesized on membrane-bound and free ribosomes. The suggestion of a segregation by membrane attachment of different population of ribosomes carrying out the synthesis of different classes of proteins may not be as far-fetched as it may seem.

Role of DNA Synthesis

The role of DNA synthesis is quite obvious in those tissues that are rapidly formed during metamorphosis, such as the limbs and lungs. However, the possible role of a restricted DNA synthesis is not clear in those tissues that do not grow rapidly but undergo a functional maturation, i.e., the tadpole liver and skin. That a relatively small burst of DNA synthesis during metamorphic maturation may be important and needs careful investigation is emphasized by the work of Topper's group on a limited DNA synthesis or cell division as a prerequisite for milk protein synthesis in prolactin-induced development of mammary gland in culture.

McGarry and Vanable (1969) have indeed found that cell division is a prerequisite for thyroxine-induced maturational changes in skin gland from *Xenopus* in organ culture. Ingram's work on the switch from fetal to adult haemoglobin during bullfrog metamorphosis suggests that DNA synthesis is important in that a different population of nucleated erythrocytes now takes over the synthesis of haemoglobin for terrestrial life. It is thought that the switching itself takes place in tadpole liver. A different kind of involvement of DNA synthesis may underlie the observations of A. M. Campbell et al. (1969) that triiodothyronine stimulated mitochondrial DNA polymerase in bullfrog tadpole liver. It is, however difficult to say whether an early alteration in cytoplasmic DNA metabolism is part of some general developmental process or they it merely is a first stage of an increase in mitochondrial size and number that is known to occur during metamorphosis.

Requirement of RNA and Protein Synthesis for Tissue Resorption

Tissue regression or cell death is an important and integral part of many embryonic developmental process. During amphibian metamorphosis, regression of organs like the tall gut, and gills accompany the maturation or formation of organs like the liver, limbs and eyes. A question that had not been answered until recently was whether or not regression or cell death was also based on a hormonal regulation of biosynthetic processes as we have seen above for maturation of the liver. We studied this question in our laboratory by using the technique of thyroid hormone-induced regression of the isolated tadpole tail maintained in organ culture.

The technique of tail organ culture was already successfully used by Weber (1963) and the induction of regression *in vitro* corresponds in magnitude and speed to that seen in the intact tadpole. Decrease in the size of the amputated tail *in vitro* was accompanied by an increase in the activity of enzymes involved in regression such as cathepsin, phosphatase, and deoxyribonuclease. Earlier work on the comparison of the properties of deoxyribonuclease and cathepsin in regressing and no-regressing tadpole tails had suggested that a part of the additional enzyme activity following hormone treatment may differ from the basal enzyme present in non-regressing tails. It is therefore interesting to note that there is a burst of both RNA and protein synthesis just when regression sets *in vitro*.

Recently, Tonoue and Frieden (1979) have found that *in vivo*, administration of triiodothyronine to bullfrog tadpoles rapidly (within 1-3 hours) lead to a decrease in the incorporation of radioactive leucine in proteins of the tail and other regressing tissues. This apparently opposite result from that found in organ cultures may be a manifestation of different properties of thyroid hormone. Besides the fact that, an enhanced rate of protein synthesis takes 1-2 days to be manifested, the rapid decrease in labeling of proteins may be due to changes in the precursor pool sizes although these authors seem to rule it out on the basis of indirect observations.

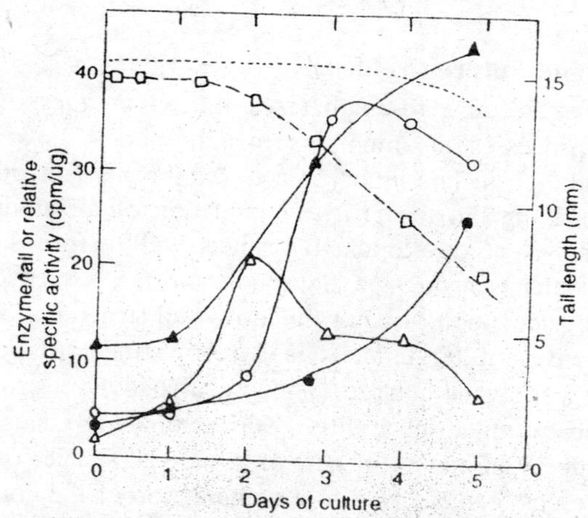

Fig. 16.5. Accumulation of cathepsin • and deoxyribonuclease Δ and burst of additional RNA and protein synthesis during regression of tadpole tails induced in organ culture with triiodothyronine added to the medium.

Not only thyroid hormones but all hormones, whether they affect growth and development or cause a rapid change in metabolic activity, are known to cause rapid alterations in pool size or uptake of sugars, amino acids and nucleotides in their target tissues. To determine whether any of the additional RNA and protein synthesized at the onset of regression induced in cultured tails was essential for the process itself, we turned to the use of inhibitors of RNA and protein synthesis. From these experiments, it was quite clear that inhibition of RNA synthesis with actinomycin D or of protein synthesis with puromycin or cycloheximide completely abolished the regression in tail organ cultures that was induced by triiodothyronine. It was also found that inhibition of protein synthesis abolished the hormone induced increase in hydrolytic enzyme activities.

Eeckhout (1966) has also noted, in a different species of the tadpole, that protein synthesis inhibitors (such as puromycin and cycloheximide) arrested tail regression in organ cultures, and Weber (1965) observed that in *Xenopus* tail regression was more sensitive than was limb generation to be administration of actinomycin D to the intact tadpole. It seems that tail regression, and perhaps even that of other tissues, is not brought about by a direct hormonal activation of lysosomal enzymes but by the formation of a new population of hydrolase molecules. Thus, not only cell growth and maturation, but also cell death during development may require a genetically determined synthesis of specific proteins.

Conclusions and Future Problems

The above account re-emphasizes the advantages of studying biochemical processes in amphibian metamorphosis as a model for studying regulation in late embryonic development. Not only are the embryos free-living but the artificial induction of metamorphosis by thyroid hormones at developmental stages well before spontaneous metamorphosis have made it possible to establish a sequence of events preceding the acquisition of adult functions and structures. Among the earliest responses of target cells is a readjustment of permeability barriers to a variety of nutrients and precursors of macromolecules. However, the eventual specificity of developmental changes is most likely to reside in the nature of new or additional species of RNA and protein molecules formed. Due to a combination of the complexity of nuclear RNA, the high degree of nuclear restriction of the rapidly turning over nuclear RNA and the inadequacies of the currently available analytical techniques it has not been possible to relate the burst of RNA synthesis with selective gene activation at the onset of metamorphosis. In the liver of the bullfrog tadpole a relatively long period of time elapses between an accelerated synthesis of RNA in the nucleus and the appearance of new proteins that characterize metamorphic changes. During this period there occurs a substantial increase in the turnover of ribosomes in the cytoplasm accompanied by their redistribution on membranes of the endoplasmic reticulum.

At the same time, there is a tight coordination in the rate of formation of ribosomes and the proliferation of membranes of the endoplasmic reticulum to which they are bound. This phenomenon of structural reorganization of the protein synthesizing machinery is common to many late embryonic developmental systems, and its significance may reside in a topographical segregation of precoded polysomes engaged in the synthesis of different classes of proteins. A prominent features of amphibian metamorphosis is the convenience of studying tissues regression or cell death. It seems that regression is not merely due to activation of existing lysosomes but that an activation of RNA and protein synthesis underlies the process of regression.

Experiments with inhibitors of RNA and protein synthesis in the resorption of tadpole tail in culture have shown that cell death during metamorphosis requires the formation of new proteins just as it is necessary for those cells that are programmed for further growth and development. Thus it is now possible to describe sequential phenomena concerning the synthesis of RNA and protein at the onset of metamorphic maturation or growth. But there are several questions still to be resolved. For example, the nature of RNA made, its transfer into the cytoplasm, and its role as a messenger will have to be clarified—a problem now facing almost all work on differentiation. Are new genes really transcribed or is there a switch in the selection and transfer to the cytoplasm of the types of RNA molecules which are constantly made since an early stage in differentiation?

In the cytoplasm, it will be most worthwhile exploring the possibility, by a combination of biochemical, histochemical, and immunochemical techniques, of identifying a separate class of ribosomes preferentially satisfying the demand for metamorphic proteins. The fact that cell growth and cell death proceed simultaneously within the same organism creates problems of a flux of precursor pools which is impossible to control. It is obvious that more emphasis will now have to be devoted to studying the late embryonic developmental process in culture. Organ cultures of tadpole tail, skin, and liver have already been studied, and it would be useful to extend these studies to dispersed cell cultures, providing that cultured cells retain the full developmental competence to respond to the hormonal stimulus. Finally, the question of developmental competence is related to that of the initial site of thyroid hormone action, i.e., the hormone receptors. Some recent experiments from our laboratory have shown that in *Xenopus* larvae metamorphic competence was acquire very early in development. Metamorphic competence was assessed biochemically by a change in the overall rate of synthesis of DNA, RNA, phospholipids and protein, water loss and altered permeability to anions like phosphate. These changes indicate that the cells (undetermined) which were initially unresponsive have suddenly become sensitive to thyroid hormones.

It should be realized, however, that in normal development there is a variation in the magnitude of sensitivity of different tissues as a function of the age of the tadpole. One interpretation of such

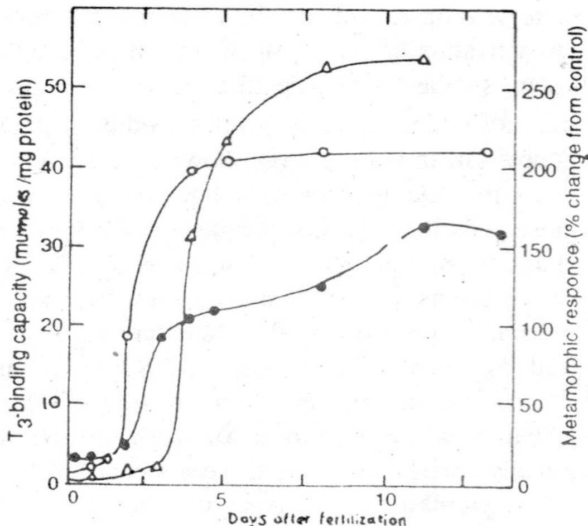

Fig. 16.6. Correlation between the appearance of temperature-sensitive binding capacity for triiodothyronine (•) and the acquisition of a metamorphic response to thyroid hormone by developing Xenopus larvae.

experiments was that the acquisition of sensitivity to the hormone was a consequence of the first appearance of hormone receptors. Thus when the capacity of *Xenopus* larvae to bind radioactive thyroid hormones was measured, there was an excellent correlation between high affinity (average Kd = 10^{-9} to 10^{-10}) thyroxine-binding sites and the rate at which the organism becomes hormone-sensitive. It has yet to be shown that one is dealing not with non-specific binding, but with true receptors whose interaction with the hormone would lead to the normal chain of events responsible for metamorphic changes. A selective rather than random distribution of hormone-binding sites has also to be determined. In the meantime, however, the study of co-ordinated acquisition of metamorphic competence and hormone-binding offers a developmental approach to the problem of hormone receptors which is now a key tissue in understanding the biochemical basis of hormonally regulated processes.

Metamorphosis and Gene Action

The improved resolution of starch gel electrophoresis and the use of enzyme histochemical techniques (3, 4) have greatly stimulated research on individual and developmental variation of proteins. Not surprisingly, this improved method of enzyme electrophoresis has been applied to amphibian metamorphosis. It was observed that appropriate one-third of the enzymes are specific to either the tadpole or the frog phase of the life cycle, whereas two-third of the major soluble proteins are specific to the different phases.

It was thus concluded that, although enzyme changes in the major soluble proteins, some of which may well be enzymes with unknown activities, are quantitatively more important. Thus many enzymes appear to be present in both phases of life cycle—e.g., are amphienzymes. In addition, a few proteins changes at times other than metamorphosis, and several proteins are present only during the metamorphic period. Two different approaches can be used to decide if certain structural genes are active in both phases of the life cycle, as well as to decide if others are expressed only in specific phases.

1. Isolation and characterization of enzymes with identical substrate specificities from tadpole and adult phases. In particular, if such enzymes are the products of different genetic loci, then they should differ in at least part of their amino acid sequence, such a difference being most readily demonstrated by peptide mapping ("fingerprinting"), although preferably being carried to the ultimate point of complete primary structure determination.
2. If two proteins are coded from different genetic loci, then genetic variation of one should not be paralleled by genetic variation of the other.

Each approach has its problems. The first is laborious. For race enzymes, large amounts of tissues are needed in order to obtain sufficient pure protein. This means that individuals must be pooled. The second approach is biochemically less sophisticated, if considerably simpler. As small amounts of tissue can be extracted and submitted to electrophoresis, many studies can be done without combining organs or individuals.

It has recently been realized that the pattern of genetic variation of many enzymes and other proteins is very easily determined by starch gel electrophoresis. For a number of proteins, heterozygotes for a polymorphism have the two zones characteristic of the corresponding homozygotes—e.g., serum albumin. For other proteins they may be multiple banding in homozygotes but heterozygotes have simply the sun of zones present in both homozygotes—e.g., transferrin. For still other proteins, heterozygotes are easily detected by the possession not only of the two zones corresponding to those of appropriate homozygotes but also of one or more hybrid zones of intermediate electrophoretic mobility. For example, polymorphism of 6-phosphogluconate dehydrogenase in man, the quail (*Coturnix coturnix*), snails *Cepaea nemoralis* and *C. hortensis*), and a sea anemone (*Actinia equina*) involves a single hybrid zone in heterozygotes. For a number of vertebrate lactate dehydrogenases, the pattern of individual variation involves basically three hybrid zones in heterozygotes. There are several complications to the approach either through biochemical characterization or through genetic variation in studying the proteins of different phases in a life cycle:

1. Appearance of a new proteins does not necessarily mean gene activity in the sense of messenger RNA formation followed immediately by protein synthesis. There can be a long delay between gene transcription and activation of the messenger RNA-ribosome complex.
2. Some proteins are the product of more than one genetic locus, e.g., the distinct α and β chains of most vertebrate haemoglobins. For a β chain mutation in man the polymorphism is confined to the major adult haemoglobin Hb A. For an α chain mutant there will be parallel variation of Hb A, Hb A_2, and Hb F.
3. Proteins can have different properties and still have been the product of the same genetic locus if there has been a subsequent difference in chemical modification. For example, transferrin conalbumin in galliform birds differ electrophoretically and yet always show parallel genetic variation. It has recently been shown that these proteins differ not in amino acid sequence but in the carbohydrate moiety attached to sequence.

The present paper includes additional data on individual variation of certain proteins in amphioxus and amphibians. Amphioxus is in some reports the most primitive known living chordate. It metamorphoses from a planktonic larva to a benthic adult phase which, though still motile, spends much of its time partly or completely buried in coarse sand. Although the matter is by no means settled, it has been suggested that the metamorphosis of amphioxus is controlled by thyroxine or some related iodine compound.

Materials and Methods

Amphioxus, *Branchiostoma lanceolatum*, were obtained from the English Channel in the vicinity of Plymouth, England. Larvae were collected in plankton tows during the summer and adults were collected by bottom trawling with appropriate fine-meshed netting. The amphibians sampled were edible frogs, *Rana esculenta*, common frogs, *Rana temporaria*, and newts, *Triturus vulgaris*. Treatment of samples, electrophoresis in vertical starch gel, and histochemical localization of enzymes are detailed in other papers. Amphioxus larvae are very small, as well as thin and elongate; thus it was necessary to pool 30 or 60 larvae at a time. Adult amphioxus were pooled for direct comparison with larvae. But, in addition, variation in a number of enzymes was studied in 40 individual-adult amphioxus. Separation of organs in adult amphioxus is difficult; the best that can be done with reasonable rapidly (so as to avoid autolytic modification) and with minimal contamination is the division of adults into body musculature (including notochord and nervous system), viscera (including the digestive system), and, depending on sex, ovary or testis.

RESULTS

Enzymes of Amphioxus, Branchiostoma Lanceolatum Esterases

Amphioxus esterases are resolved best in the pH Ferguson-Wallace discontinuous buffer, although good resolution of some esterases is obtained in pH 6.0 phosphate. The major amphioxus esterases are relatively non-specific in that almost identical patterns, differing only in relative activity in a few zones, are obtained whether the enzymes are localized by hydrolysis of α-naphthyl acetate, α-naphthyl butyrate,

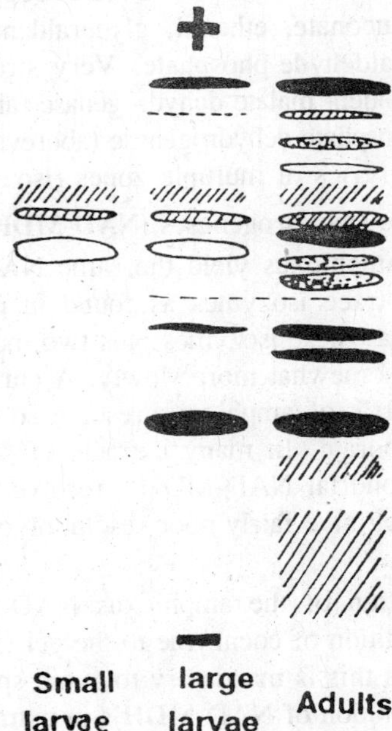

Fig. 16.7. Esterases of larval and adult amphioxus, Branchiostoma lanceolatum.

or 5-bromoindoxylacetate. None of these enzymes has activity on a-naphthyl phosphate, there being at least two distinct and active acid phosphatases in adult amphioxus, although no detectable phosphatase activity in extracts of larvae. The esterase pattern for small larvae (5 mm long), large larvae (8 mm long) and adults.

There is some individual variation in electrophoretic position of two of the more anodal esterases of adult amphioxus, though most of the esterase zones were identical in sample of 32 adults screened singly. It is clear that larval amphioxus lack many of the esterases present in adults. Only three zones of esterase activity are detectable in small larvae; the number increases to eight in large larvae. Seven of these eight zones are identical in position to zones found in extracts of adults. There is one conspicuous esterase of larval amphioxus which does not appear to be present, at least in anywhere near the same level of activity, in adults. In addition, adult amphioxus have at least seven esterases not present in larvae.

Dehydrogenases

No activity could be found in extracts of either larval or adult amphioxus to sliced starch gels at the completion of electrophoresis: *lactate*, *isocitrate* (neither NAD-nor NADP-dependent IDH's), glutamate, 6-phosphogluconate, ethanol, glyceraldehyde, α-glycerol-phosphate, or α-glyceraldehyde-phosphate. Very strong activity was observed for NAD-dependent malate dehydrogenase (abbreviated NAD-MDH) and glucose-6-phosphate dehydrogenase (abbreviated G-6-P DH), each of which yields a series of multiple zones (isozymes).

NAD-Dependent Malate Dehydrogenases (NAD-MDH's)

Extracts of larval amphioxus yield the same NAD-MDH pattern of two major and one trace isozymes as found in extracts of adult viscera. Adult muscle has these isozymes plus two more major NAD-MDH's which migrate somewhat more slowly. A curious observation is that these NAD-MDH's of amphioxus resolve so well in alkaline starch gels containing borate. In many extracts of other species the supernatant and mitochondrial NAD-MDH's resolve best in pH 6 to 7.5 phosphate gels, which give fairly poor resolution of the amphioxus NAD-MDH's.

In addition, resolution of the amphioxus NAD-MDH's is only slightly improved by addition of coenzyme to the gel (50 or 100 mg of NAD per liter), whereas this is mandatory to avoid spurious variation and excessively bad resolution of NAD-MDH's in extracts of a number of species of invertebrate and vertebrates. No individual variation of

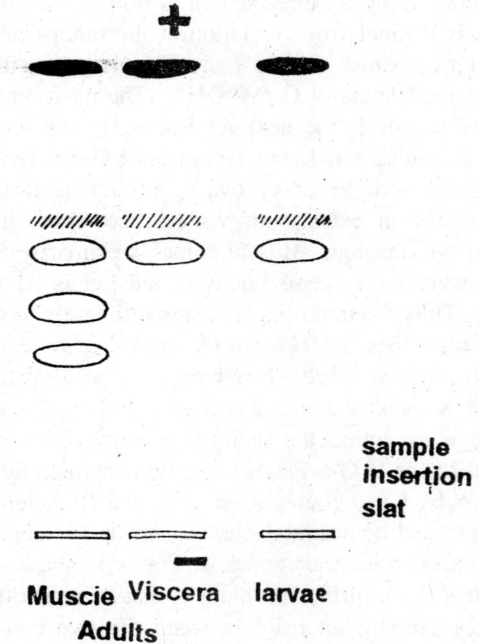

Fig. 16.8. NAD-dependent malate dehydrogenase isozymes of larval and adult amphioxus, Branchiostoma lanceolatum.

NAD-MDH's was observed in 40 adult amphioxus surveyed singly. The distinct NADP-dependent malate dehydrogenases ("malic enzymes"), which require Mg^{++} or MN^{++} for full activity, and which yield such nice sharp zones in electrophoresis, could not be localized after electrophoresis of amphioxus extracts even with the proper precaution of the addition of 20 mg of NADP per liter to both gel and the upper bridge buffers.

Glucose-6-Phosphate Dehydrogenases (G-6-P DH's)

These enzymes are extremely active after electrophoresis of amphioxus extracts, multiple zones appearing within 10 minutes after the staining solution is added to the sliced starch gel. The enzymes are specific for glucose-6 phosphate and no zones appear when gels are incubated for as long as 24 hours with glucose, glucose-1-phosphate, or 6-phosphogluconate. Magnesium ion added to incubation mixtures increases activity slightly. The multiple G-6-DH's have an obligatory requirement for NADP and, in addition, best resolution demands the presence of NADP in the gel and bridge buffers (20 mg NADP per liter). There are basically three sets of G-6-PDH isozymes. As variation at each set is distinct from variation at the others, the three sets are the product of distinct genetic loci. This is especially clear for the two rapidly anodal sets of G-6-P DH's. The most anodal set of zones is designated Locus I, the next set Locus II, and the slowly moving and poorly resorbing set, Locus III. Locus I G-6-P DH's are especially active in extracts of testes or ovary and Locus II G-6-P DH's are especially active in extracts of viscera; however, the specificity is relative and not absolute. Muscle contains primarily the Locus III G-6-P DH. Larvae have active Locus II and Locus III G-6 P DH's but no Locus I. Thus, metamorphosis leads ultimately to the activation of one of the three G-6-P- DH loci. The nature of individual variation in larvae was not studied; however, in adults many different electrophoretic patterns were observed.

With one exception the pattern of individual variation for both Locus I and Locus II G-6-P DH's can be explained by postulating five alleles at Locus I and four alleles at Locus II, heterozygotes having any two zones and homozygotes having one heavy zone. The exception is one individual with four zones at Locus I, although only a single zone at Locus II. A further problem is that there is often considerable difference in staining intensity between the two zones in postulated heterozygotes. Unfortunately, G-6-P DH, whether in sea anemones or in quail extracts, has a tendency to form satellite zones. Nevertheless, the pattern of variation of amphioxus G-6-P DH is unusual for a dehydrogenase in that there are no hybrid zones in heterozygotes. Except for sex-linked erythrocyte G-6-P DH in mammals, other G-6-P DH has been obtained; as for mammalian liver G-6-P DH and avian erythrocyte G-6-P DH's in vertebrates have only hybrid zone in

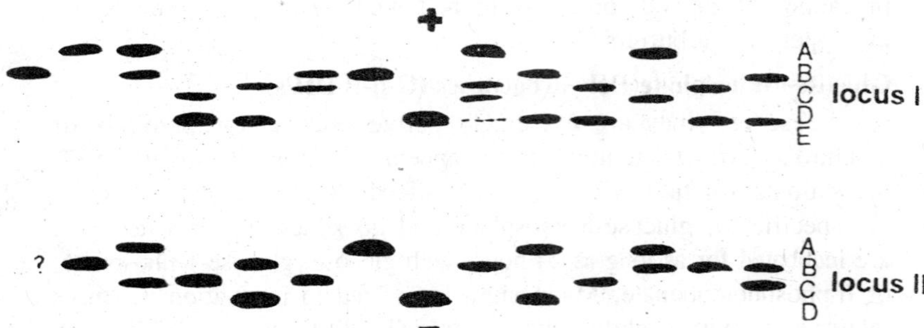

Fig. 16.9. Individual variation of Locus I and Locus II glucose-6-phosphate dehydrogenases of adult amphioxus, Branchiostoma lanceolatum.

heterozygotes–and in sea anemones the pattern resembles lactate dehydrogenase in that heterozygotes have five zones, three of which are hybrid isozymes (unpublished studies by the senior author on *Actinia equina*). No evidence for sex linkage of amphioxus G-6-P DH, the amphioxus G-6-P DH loci are autosomal. The absent of hybrid zones in heterozygotes in the variation of amphioxus Locus I and Locus II G-6-P DH's, together with the extremely rapid electrophoretic mobility in starch gel, are suggestive of a smaller and simpler and structure than other dehydrogenases. It is possible that, in contrast to other G-6-P DH's those of amphioxus lack a stable quaternary structure, a point which is compatible with the observed data but which can only be proven by careful biochemical characterization of the isolated enzymes.

Phosphoglucomutase (PGM)

Extracts of adult amphioxus also have a very high activity of phosphoglucomutase. Initially it was not realized that amphioxus PGM is completely inactivated by even brief freezing and thawing. Thus, PGM was studied in only seven individuals. However, the range of individual variation of PGM among those seven adults is considerable. In addition, there is a surprising similarity in the patterns observed in amphioxus PGM and those observed in studies on variation of human PGM. *Amphioxus* and man both have two or three minor, rapidly moving zones of PGM, whose variation is independent of the variation of the more active and more slowly moving zones of PGM. The minor zones are determined by a locus, designated PGM_2 in man, which is distinct from that determining the major zones, designated PGM_1 in man. For man, the existence of two distinct PGM loci has been confirmed by the observation of independent segregation in appropriate pedigrees;

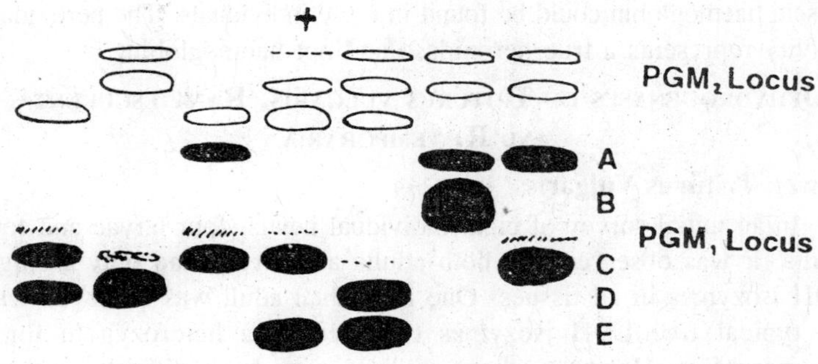

Fig. 16.10. Individual variation of seven adult amphioxus, Branchiostoma lanceolatum PGM_1 and PGM_2 phosphoglucomutases.

whether or not the two PGM loci in amphioxus are linked can only be determined by appropriate mating experiments, which have not been done.

As with variants of human PGM, the heterozygote is a simple sum of the zones of the corresponding homozygotes. Whereas in man two alleles at PGM1 are relatively common and other variants are rare: in only seven adult amphioxus, five different positions for the major zone are observed, suggesting five different reasonably common PGM_1 alleles. The PGM_1 locus appears to be active in most tissues, whereas PGM_2 is most readily observed is gonad and, irregularly, in some viscera extracts.

Aldolase

A single zone of fructose-1, 6-diphosphate aldolase is observed in extracts of adults. It is probable that this necessary enzymes occurs in larvae but, if so, its presence could not be detected after starch gel electrophoresis of extracts.

Peroxidase

Adult amphioxus but not larvae have from one to four zones of peroxidase, migrating fairly rapidly anodally in alkaline buffer (pH 8.0 discontinuous and pH 8.5 borate). There is extreme variation in the number of and relative activity of the peroxidase isozymes. There is no obvious correlation in the number and activity of peroxidases with six or nutritional state. The variations defies rational genetic explanation, unless one postulates two different peroxidase loci, each with null alleles in significant frequency. The absence of these isozymes in pooled larval samples suggests that peroxidases becomes conspicuous only in the adult phase. Though some amphioxus adults have pink colour, which has been casually called myoglobin, no evidence for muscle haemoglobin could be found in these individuals. The peroxidase activity represents a true peroxidase and not haemoglobin.

Dehydrogenases of Triturus vulgaris, Rana esculenta, and R. temporaria

Newts, Triturus Vulgaris

In an initial survey of eight individual newts, four larvae and four adults, it was observed that both adults and larvae had only a single LDH isozymes in all tissues. One individual adult was observed with the typical five LDH isozymes expected of a heterozygote for a polymorphism. However, these studies were done utilizing only the pH 8 Ferguson-Wallace discontinuous buffer, which yields very sharp zones of LDH. When these samples were run in a pH 6.7 phosphate buffer (μ = 0.04), which also yields good resolution of LDH's, it was observed that all newts have distinct heart- and muscle-type lactate dehydrogenases. These are H_4 and M_4— and the intermediate isozymes, corresponding to H_3M, H_3M_2 and HM_3, typical of mammalian, avian, and some "lower' vertebrate LDH's, do not occur in certain amphibian, including the common newt. The individual newt-with five LDH zones in alkaline buffer turned out to have an M chain LDH polymorphism. Thus, in the newt hybrid isozymes do not form readily between H and M chains, although the heterozygote for the M chain polymorphism possessed the typical five zones as found in heterozygotes among birds and mammals.

No significant differences were observed between adults and larvae and, in the newt, it is clear that metamorphosis involves no changes in LDH gene expression. Similarly, larval and adult newts have similar glutamate dehydrogenases and the two zones of NAD-dependent malate dehydrogenase. Larvae have less G-6-P DH than adults but the position of the two zones is similar, even if larvae lack nearly all of the more anodal zone. No individual variation of these other dehydrogenases was observed among the small sample of eight efts.

Frog Lactate Dehydrogenases (LDH)

In contrast to the newt, the muscle and heart-type LDH's of *Rana esculenta* and *R. temporaria* are easily differentiated by electrophoresis in any of the conventional alkaline buffers. Tadpoles have primarily the H4, LDH in all tissues, whereas adults have M_4 as well, especially in muscle and liver. There are only traces of zones intermediate between H4 and M4. Thus, our results are identical to those of Sathe et al. studying *Rana pipens* and *R. palustris*.

Bullfrog *R. Catesbeiana* H chain has more of a tendency of combine with the M chain, although there is still only the H_2M_2 molecules formed and little, if any, of the H_3M and HM_3 tetramers. We have observed, though inconsistently, that refrozen and thawed samples containing H_4 and M_4 frog LDH's tend to increase in the amount of the trace intermediate isozymes although these never are as prominent as in extracts of appropriate avian or mammalian tissues. Electrophoresis of unfrozen frog tissues often fails to yield any intermediate LDH isozymes between H_4 and M_4, except for some H_2M_2 in bullfrogs. The bullfrog is also different from the other ranids in the more rapid electrophoretic mobility of the H_4 isozyme and that the M_4 isozyme occurs in appreciable amounts in tadpoles, especially in muscle and liver. Bullfrog adults also have a heterogeneity of the

H_2M_2 and M_4 LDH's that is not paralleled in *Rana pipiens*, *R. esculenta* and *R. temporaria*.

Genetic Polymorphism at the H Chain LDH Locus in *Rana esculenta*

The hypothesis that the H chain is the same in both tadpole and adult phases is proven by finding a particularly clear polymorphism at the H chain locus in *Rana esculenta*. The Pattern of variation is exactly the same as observed in LDH polymorphism of several vertebrates, including pigeons, mice, and men, for which the genetic interpretation has been confirmed by pedigree studies or appropriate matings. In both tadpole and adults of *R. esculenta*, there are three different H chain alleles. Heterozygotes have five zones, including three hybrid isozymes not found in homozygotes. Five of the six predicted phenotypes for a triallelic variation have been observed in the small sample of frogs so far investigated. It is of interest that the sample of eight adults yielded 1 AA, 1 AB and 6 AC, whereas the sample of eight tadpoles yielded 2 AB, 2 AC, 3 BB, and 1 BC. Thus, the estimated gene frequencies are in adults $p_A = 0.56$, $p_B = 0.06$, and $p_c = 0.38$; and in tadpoles $p_A = 0.2$, $p_B = -.59$, and $p_c = 0.19$.

Because of the small sample size, the difference in p_A in the two phases, as well as in p, are not statistically significant. However, the difference in gene frequency of the B allele, 0.56 in tadpoles versus 0.06 in frogs, in the two phases is statistically significant. The probability of such an unequal distribution occurring under the null

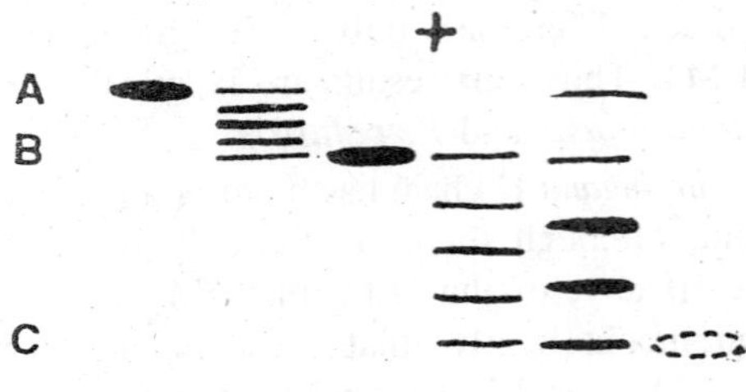

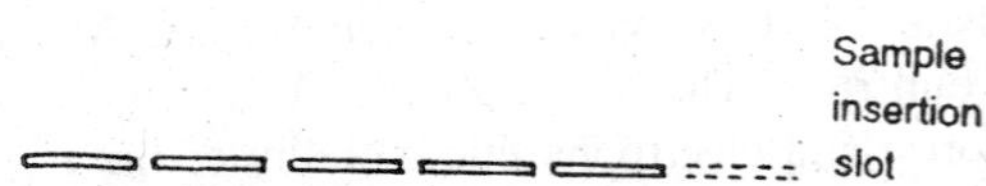

Fig. 16.11. Polymorphism of the H chain lactate dehydrogenase of tadpole or adult edible frogs, Rana esculenta.

hypothesis of no significant difference can be easily calculated – see Mather for discussion of the use of binomial and multinomial distributions for analysis of significance of differences in small samples. The total sample of 16 *R. esculenta* represents 32 alleles at the H chain locus. The probability of the 10 B alleles being distributed 9 in one phase and 1 in the other, or 10 in one phase and 0 in the other, is, from the appropriate binomial expansion:

$$p^{10} + 10\, p^{9}q + 10\, pq^{9} + q^{10}$$

where, because of the equal number of tadpoles and frogs in the sample, $p = q = {}^{1}/_{2}$. The exact value for the above sum of four terms is 22 $\times\ ({}^{1}/_{2})^{10}=0.0215$. The probability of observing such a difference in the frequency of the B allele in tadpoles and frogs due to chance is only about 2 in 100. Thus the data, in spite of the small sample size, are statistically significant at the $a = 0.05$ level.

There must be an extensive differential mortality of tadpoles and frogs to account for such a large gene frequency difference between individuals in different phases of the life cycle. However, frogs lay thousands of eggs. Many tadpoles fail to survive metamorphosis, and froglets are the food of many predators. There is plenty of numerical room for establishment of such different "genetic elites" in any species with a large number of zygotes. Survey of 6 adults and 44 tadpoles of *R. temporaria* failed to reveal any LDH variants at either the H chain or the M chain locus.

Frog Malate Dehydrogenases

There are two major NAD-dependent malate dehydrogenases in both tadpole and adult *R. esculenta* and *R. temporaria*. The tadpole enzymes migrate slightly more rapidly anodally than the frog enzymes in pH 8.0 discontinuous buffer. However, the resolution of these enzymes is poor and the electrophoretic mobility depends on the concentration of the extract and the organ extracted. Thus, the tadpole and frog differences are probably not significant. No clear individual variation was observed among 16 *R. esculenta* and 30 *R. temporaria*, although some type of individual variations exists in the position of the more slowly moving NAD-MDH isozyme in *R. esculenta*.

Frog Glutamate Dehydrogenases

In discontinuous buffer, pH 8.0 glutamate *dehydrogenase* of *R. temporaria* adults has two major zones. Tadpoles consistently have only the faster of these zones. However, in pH 6.7 phosphate buffer the tadpole and frog glutamate dehydrogenases present a different pattern of heterogeneity; thus it is possible that the tadpole and frog GDH's are completely different. No individual variation was observed among 30 *R. temporaria*. Thus, we cannot yet say whether the electrophoretic difference between the tadpole and frog GDH's is the result of different GDH genes being active in the tadpole and frog phases of the life cycle, or the result of some difference in chemical modification. It is well-known that vertebrate liver glutamate dehydrogenase is an allosteric enzyme.

Discussion

Enzyme Differentiation versus Adaptation

This comparative approaches indicates that in both amphioxus and amphibians, where metamorphosis is triggered (at least in amphibians) by thyroxine or similar iodinated compounds, some enzymes are unique only to one phase or the other of the life cycle. In general, as differentiation proceeds there is an increase in the number of different enzymes detectable after electrophoresis of extracts, e.g., larval amphioxus have only one possibly unique larval esterase, whereas adult amphioxus have at least seven different unique adult esterases. Similarly, in *Rana temporaria* and *R. esculenta*, though not in *R. catesbeiana*, only traces of the M LDH chain are synthesized prior to metamorphosis.

In the newt, however, we have observed no differences between larvae and adults, whether the comparison involves H chain LDH, M chain LDH malate dehydrogenases, or glutamate dehydrogenase. Additional studies on the same eight individuals have revealed neither any significant haemoglobin differences, nor differences in phosphatases; adults have two sets of esterases present in significant levels of activity which are not present, or barley detectable, in larvae; however, at least six other esterases are identical in adults and larvae. Thus, most of the newt enzymes are amphienzymes and are unaffected by metamorphosis. The patterns of enzyme differentiation show some degree of variation between species of the same genus.

The best example in the present studies is the much greater activity of the M chain locus in bullfrog tadpoles as compared with-tadpoles of other related ranids. Similar differences suggesting species differences in "control genes" have been observed in haemoglobin differentiation—e.g., the persistence of larval haemoglobin in adult lampreys of the semineotenous species *Entosphenus lamottei*, the difference in the time of appearance of the major and minor adult haemoglobins in the embryos of galliform birds, the relative differentiation an absolute differentiation of coelomic and water vascular haemoglobins in sea cucumbers, and the differences in the presence or absence of haemoglobin components, as well as their number, in embryonic, fetal, and adult mammals. Although enzyme changes in different phases of a life cycle are often studied as examples of hormonal influence on the activation of genes—or at least the activation of messenger RNA-ribosome complexes—it must not be forgotten that some of these differences in protein phenotype reflect differences in the environment. To what extent are such examples of enzyme differentiation altered by differences in the environment?

It is known that the relative amounts of the H and M chains of lactate dehydrogenase can be altered by oxygen level independently of other embryological events. Accordingly, it is just possible that the tadpole-frog difference in the amount of the M_4 isozyme reflects differences in the availability of oxygen and would be reduced or eliminated if tadpoles and frogs were kept at identical oxygen levels. A further paradox is posed by the fact that tadpoles of some species have primarily the H LDH chain type, which is considered to be more suited in its kinetic properties to metabolism in obligatory aerobic tissues. However, tadpoles live in environments of variable, and often low, levels of oxygen—for which the absence of a Bohr effect in tadpole haemoglobin is ideally suited. Thus, at our present state of understanding—or misunderstanding—tadpoles are better adapted anaerobically than frogs in respect to haemoglobin, but better adapted aerobically than frogs in respect to lactate dehydrogenase. In general, the greater the morphological difference between phases of a life cycle, the greater the difference in the habitats of larvae and adults—although the correlation is not perfect, especially among the some insects and amphibian.

It is of interest that the highest frequency of amphienzymes is found in the newt, where adults spend considerable time in the same aquatic environment as the larvae. It may well be that some of the specific enzymes being studied as examples of differentiation are, in fact, really or more significance in terms of adaptation to the difference in larval and adult environment. As suggested earlier, from the stand point of changes in morphology, the major soluble and insoluble proteins are particular importance, even though *de novo* synthesis, or activation, of specific modifying enzymes is also necessary for those morphological changes—e.g., the collagenase of tadpole tails. It is worth bearing in mind that metamorphosis in some marine invertebrates, e.g., tunicates, barnacles, and polychaetes, can be initiated and finished within a few hours, in some cases as little as 15 minutes.

It is unlikely that all the necessary protein changes are accomplished by *de novo* protein synthesis. Rather, reliance must be placed more on synthesis or activation of a few key modifying enzymes, which in turn can alter certain major proteins, which might be additional enzymes or structural proteins. Very likely there is a series of such activation steps involving successive amplification, as in the activation of the "factors" in blood clotting. It is proposed to call such changes in development that involve the chemical modification of already synthesized proteins transubstantiation. Transubstantiative changes may involve proteolytic modification, as in the case of collagenase, addition or removal of non-protein moieties, such as the carbohydrate changes in transferrin, or modification of configuration by combination with a small molecule, e.g., an allosteric effect. The very unusual difference between adult and fetal shark *Squalus* haemoglobin may reflect a transubstantiative change which results in the loss of two tryptic or seven chymotrypitic peptides of what are otherwise identical proteins.

Consequences of Individual Variation

In biochemical and endocrinological studies it has often been the practice to combine a given organ or gland from many individuals in order to obtain sufficient protein. In relation to comparison of tadpoles and adult frogs, the electrophoretic studies warm that such a procedure may give rise to differences in enzyme activity or other properties even when the enzymes are fundamentally identical in both phases of the life cycle. This can result in three ways:

1. Pooling of a small number of individuals may result, purely by chance, in combining different numbers of various variants for a protein polymorphism common to both phases. This can be minimized by increasing the size of the pooled sample.
2. When there is a change in the gene frequencies of alleles for a polymorphic locus active in both phases of the life cycle, this difference in variants will manifest itself in a pooled sample no matter how large that sample is. The data on *R. esculenta* B allele of the H chain LDH locus indicates that considerable differences in frequency exist in a sample of tadpoles and frogs from the same locality. This could result from seasonal differences in selection pressure, for adults will have been exposed longer to any progressive changes than will tadpoles. O, this could result from differential selection arising from the difference in the tadpole

and frog environments. The large number of eggs laid by frogs makes it perfectly possible to have very different "genetic elites" in the two phases, the only restrictions being those imposed by the fact that any unusual distribution in frogs must arise from what is already present in tadpoles and vice versa. Such a system could stabilize polymorphism which would otherwise be unstable. If the different H chain variants of *R. esculenta* differ in their enzyme kinetics, a point which has not been investigated, then pooled samples of tadpoles and frogs would show differences in kinetics because of the differences in the frequency of the variants, although fundamentally the same LDH genes were active in both phases of the life cycle. Similarly, a pooled sample of tadpoles or frogs would yield somewhat different electrophoretic patterns or chromatographic elution profiles.

3. Another observation is that protein polymorphism are by no means rare. In some species from 20 to 73 per cent of various proteins show genetically typeable variation. Furthermore, not all such polymorphisms involve only two or three alleles. Several protein

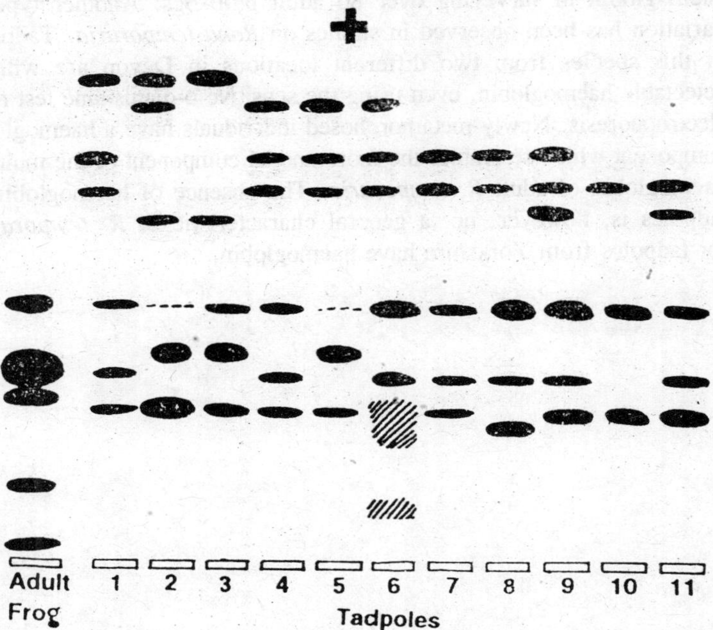

Fig. 16.12. Individual variation of the hemoglobins of 11 bullfrog Rana catesbeiana tadpoles out of a sample of 13.

polymorphism are known where from 6 to 13 different alleles all occur in relatively similar gene frequencies, probably representing natural selection for diversity. A small sample from such a population will invitably miss many of the possible phenotypes. What is worse is that two different samples will tend to contain quite different phenotypes, simply by probability theory and without any necessity for differential selection. Consider the amphioxus Locus I and Locus II-G-6-P DH's. With a total of five different alleles at one locus and four alleles at another, there are 15 × 10 = 150 different possible phenotypes of this one enzyme.

In view of the number of studies which have compared tadpole and adult frog haemoglobins, especially for determination of the number of polypeptide chains present, it is instructive to see the diversity of variation that occurs in one population of the bullfrog *Rana catesbeiana*. A diagram of the 11 different haemoglobin patterns observed among 13 tadpoles by *Elzinga*. In addition, many tadpoles have appreciable amounts of non-haemoglobin protein in hemolysates. By way of contrast, neither Elzinga nor the authors found any individual variation of haemoglobin in surveying over 80 adult bullfrogs. Another type of variation has been observed in studies on *Rana temporaria*. Tadpoles of this species from two different locations in Devon are without detectable haemoglobin, even using the sensitive o-dianisidine test after electrophoresis. Newly metamorphosed individuals have a haemoglobin component which resembles the more anodal component of the multiple haemoglobin of adult *R. temporaria*. The absence of haemoglobin in tadpoles is, however, not a general characteristic of *R. temporaria*, for tadpoles from Yorkshire have haemoglobin.

17

RECOVERY OF LOST PARTS

Many kinds of living organisms regrow appendages that are crushed or torn off in the mishaps of an active life. Human beings have scarcely any abilities of this sort, a fact which contributes to their curiosity about the mechanisms of regeneration in more resilient organisms. This curiosity runs deeper because regeneration in many ways resembles the initial normal development of an animal's structures. Normal development plus regeneration, collectively called *morphogenesis*, presumably operates by some general rules that we might at least elucidate empirically as a prelude to ferreting out deeper mechanisms. Yet for all the imaginative and neticulous efforts of at least four generations of developmental biologists, few general rules have stood the test of time.

If principles of widespread applicability exist they remain tantalizing obscure. In this situation, the whimsically dubbed *clockface* ("*clock-phase*") model of regeneration created a stir among developmentalists of both experimental and theoretical bents. Its most fundamental and radical tenet is that positional information is stored in cells as a "phase" or an "angle" analogous to the hour-hand position on a face of a clock, a quantity topologically unlike the conventional scalar variables of physiology. This has several startling implications which have been verified experimentally in several contexts, so it seems to call four some interpretation, perhaps involving a rhythmic or somehow periodic process. It also turns out that this model, unless adorned with special hypotheses to cover for exceptions, makes other predictions that are not observed. Moreover, a less original interpretation suffices to account for the remarkable regularities that gave the clockface model its first appeal. This relatively prosaic rendering assumes that positional information is encoded, not in a single angle or phase, but in a pair of transverse concentration gradients. This is a completely ordinary idea, essential qualitative facts about regeneration in epimorphic fields.

The Clockface Model

The clockface model was contrived as the vehicle for a brilliant resynthesis of observations by Vernon French, Susan Bryant, and Peter Bryant. They draw on experiments with the fruit-fly larva's "imaginal disks" (the precursors of the adult fly's various appendages), with the legs of cockroaches, and with amphibian limbs. Their original paper is a *tour de* force of focused experimentation and organization of data. They come up with two principles which together account for an extraordinary diversity of peculiar experimental results. Let me first remark that what one takes to be "the principles" is presently subject to so much subjective interpretation that there are as many versions as there are people retelling this story I will retell it based on four principles. The reader will want to examine the bibliography for other versions and for more detailed references to the original experimental papers.

It should also be noted that the experiments under consideration here do not span the whole diversity of regenerative processes. We focus here on situations in which cells seem to have no intrinsic directional polarity, seem to retain their differentiated identity, and seem to impose on adjacent cells (possibly newly created from dividing populations) to adopt a similar identity. The main point introduced by French et al. is that each tiny patch of tissue is somehow labeled permanently with an unchanging intensive quantity—a tissue specificity—that behaves like time in a cycle or like a hue or like an angle in that any one of them denotes a point on some abstract circle of states. French et al. (4) placed the digits 1-12 as indicators of local tissue specificity around the circumference of a limb or other developmental field. Many of their diagrams thus resemble the face of a clock. Although this notation gives the model its name, it unfortunately also requires a numerical discontinuity where none is intended biologically.

A more apt analogy is provided by the seasonal states of an ecosystem: spring grades into summer grades into fall grades into winter grades into spring with no discontinuities. The numerical analogy is unfortunate, but the principle is clear and tantalizingly suggests something oscillatory, somehow periodic, underlying. Without committing ourselves to any such interpretation, it will still be convenient to accept the "clockface" metaphor and refer to the quantity denoted around its rim as a "phase. Independent of its circlelike topology, the import of a cell's neighbourhood bearing this putative state label is that an appropriate stage in regeneration, tissue will develop structures (sensory hairs, muscle attachments, colour patches) corresponding to the local label. This is an application of Wolpert's principle that tissue specificity, encoded in a local quantity of unknown origin, is conceptually distinct from the cell's interpretation of that "positional" specificity and from other convert internal states such as "determination". The following simple rules make use of the putatively *circular* label of cell states, and suffice to systematize a lot of otherwise very perplexing experimental results.

Rule 1: Each little patch of cells is labeled with a phase which is part of a smooth phase gradient across the tissue. Thus there exists a smooth map from the tissue to an abstract ring of biochemical specificities. Once established, this map does not change. (Some independent second label, e.g., the proximo-distal level of cell specificity, is implicitly assumed, whereby to distinguish cell types on the two dimensions of an organism's surface.

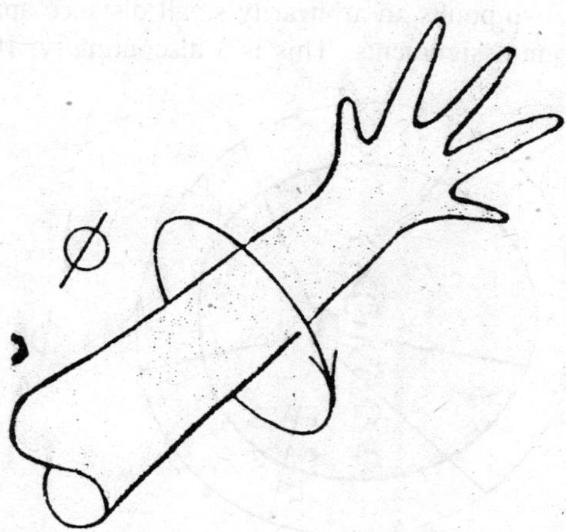

Fig. 17.1. The circular dimension around an appendage is putatively enclosed in the cells by a time-independent quantity that increases smoothly from an initial value and returns to it without ever decreasing.

Rule 2 introduces an exceptional point near which this map is not smooth.

Rule 2: In the normal limb, the phase values are supposed to run one full cycle around the limb axis (the azimuthal direction). If a limb were an open-ended cylindrical surface, this would pose no problem. But besides having an inside, here discretely ignored, a limb has a foot or a hand at the end: its surface resembles the glove with phase increasing full cycle around the open end. Herein lies the tantalizing nucleus of this whole subject. It phase increases smoothly through a cycle around the perimeter of a two-dimensional patch of tissue, let us say that the patch has "winding number" $W = 1$ (or 2, should phase increase through *two* full cycles, etc.). It is an unfortunate fact of geometry that only if the winding number is zero can a distinct phase value be attributed to every point inside. To convince yourself of this, try to sketch on the tissue a curve along which phase = 1, and an adjacent curve along which phase = 2, and so forth. All these curves start from the corresponding points along the tissue's boundary, and extend inward. If the winding number of phase along that border is zero, then each curve that enters the patch from one point on the border can again exit the patch at another point where the phase value is the same.

But if the winding number is not zero, then that integer number of full sets of phase contours enter and can't get back out. Wherever they converge, two points an arbitrarily small distance apart may have radically different assignments. This is a discontinuity. If W=1 there

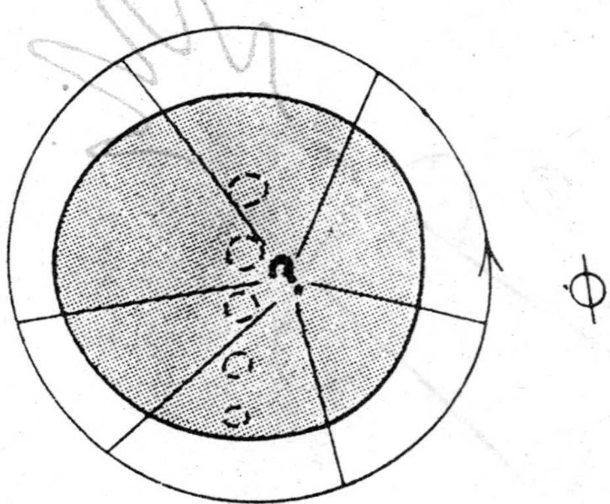

Fig. 17.2. The epidermis is stretched flat. The dashed rings indicate fingernails. From the five points on the abstract circle of phase values contour lines cross the epidermis, joining cells of like value. If the perimeter spans a full cycle, then these contour lines inevitably converge on the borders of an internal hole, or, if there is none, on a region of ambiguous phase (a "phase singularity").

must be a discontinuity in the assignment of phase values somewhere; the tissue could evade this discontinuity only by placing it in a hole (a region devoid of living, labeled cells). One might suppose that the discontinuity would appear along some "seam" running the length of the leg (the proximo-distal direction) like an International Date Line, but none was found. The discontinuity is apparently more compactly localized than that. It might be an isolated phase "singularity". Alternatively, there could be several isolated phase singularities, e.g., three, of which two have opposite handedness and cancel out.

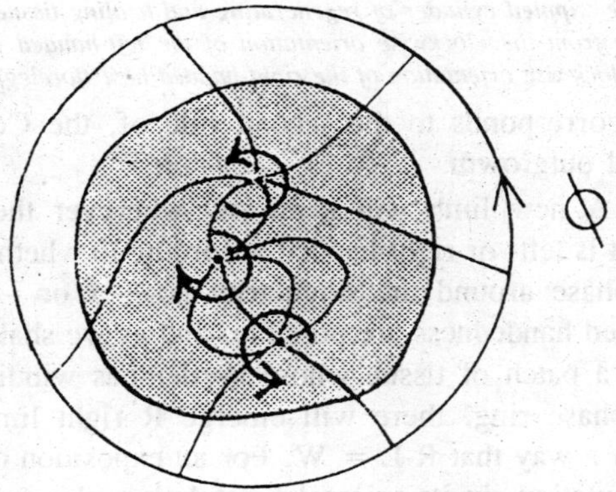

Fig. 17.3. As in but showing that any picture may be supplemented by any number of additional paired singularities of opposite handedness without violating the boundary conditions.

Rule 3 corresponds to the Rule of Intercalation.

Rule 3: So long as the phase gradient is shallow enough (if adjacent cells are sufficiently similar in phase), cells divide only to replace those that happen to perish. But if cells with normally non-adjacent phases are juxtaposed (e.g., by cutting off a leg, rotating it, and sticking it back on), then those cells begin to divide in earnest. As proliferation continues, the new cells take on phase values intermediate "between" their immediate neighbours. Thus the phase discontinuity is soon bridged through newly regenerated tissue. Proliferation continues until it has restored the initial shallowness of the phase gradient in space. In some versions of the clockface model, intercalation of intermediate phase values might, without an additional rule, go either of two ways: there are two arcs of phase values "between" any pair of points of a circle. We defer this puzzle until it can be seen in its more general aspect.

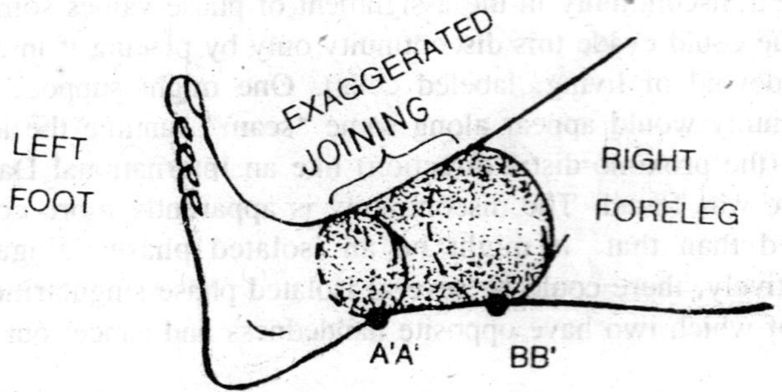

Fig. 17.4. In the stippled cylinder of regenerating and healing tissue, the transition is made from the clockwise orientation of the left-handed graft (foot) to the anticlockwise orientation of the right-handed host (foreleg).

Rule 4 corresponds to the second rule of, the Complete Circle Rule of distal outgrowth.

Rule 4: A new limb will grow out wherever there is a phase singularity. It is left- or right-handed according to whether the winding number of phase around the singularity is + 1 or − 1. Additional limbs of paired handedness are possible, and as we shall see, occur in fact. Within a patch of tissue whose border has winding number W around the phase ring, there will emerge R right limbs and L left limbs in such a way that R-L = W. For an exposition of this winding number formulation. In its original formulation, this rule was worked as though cells make a local decision based on fulfillment of a global criterion (non-zero integer winding number on a distant ring of tissue).

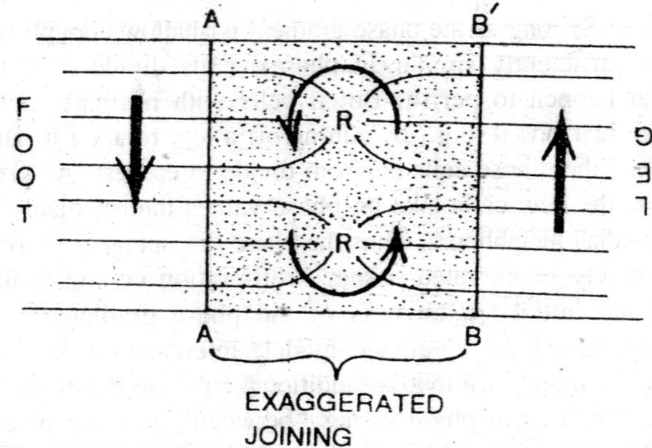

Fig. 17.5. The stippled cylinder is slit open along the line from AA′ to BB′. Lines A′B′ and AB are the same.

This is implausible physiologically, and lends itself to the construction of paradoxes, besides.

In the approach taken here, emphasizing strictly local principles of cell interaction, something like the Complete Circle Rule emerges as a mathematical artifact, valid in especially simple situations. These difficulties prompted several independent analyses that converged on substantially the same ideas as presented next. It was not long before the contradictions implicit in the Complete Circle Rule made themselves felt in published experiments, too: an amended version of the clockface model subsequently appeared in which a local rule replaces the former global rule.

Examples of the Foregoing

Applying the polar co-ordinate rules by way of illustration, consider the regeneration of a served foot. A disorganized layer of vaguely differentiated tissue, a blastema, grows from the cylindrical edge to cover the stamp. By Rules 1 and 3, it must contain a phase singularity of the same handedness as the original foot. So by Rule 4 a new foot of the same sort replaces the old. In terms of winding numbers, W = 1 (or −1), so in the simplest case $R = 1$ and $L = 0$ (or $L = 1$ and $R = 0$). Consider a second example in which a right foot is cut off and replaced by a left foot served from the other leg.

The cylinder of new tissue proliferating in the junction according to Rule 2 is bounded by two complete circles of phase, as indicated. This two - part border has winding number $W = 2$ around the cylinder of skin it encloses. This is shown by slitting the stippled skin along the line AB and laying it flat: along path ABB 'A'A the phase increases by 0 + 1 + 0 + 1 cycle. Thus we expect two additional new right feet to emerge. In animals capable of regeneration, the actual result is indeed two additional right feet. A Chinese woman suffered this very operation in 1973, following piecemeal destruction of both limbs in a railroad accident (15). The present inability of humans to regenerate whole limbs presumably spared embarrassment to all concerned.

As a last example consider a left foot cut off and replaced with a 180° rotation. By the same argument, we expect no supernumerary limbs, or else any number of left-right pairs. A common result is a right foot and a left foot. The other common result is no new limbs. An amusing implication of all this, incidentally, is that no organism capable of regeneration in this mode can have an odd number of asymmetric appendages, and its even number must occur in left-right pairs. A five-armed starfish, for example must be a covert tetrapod

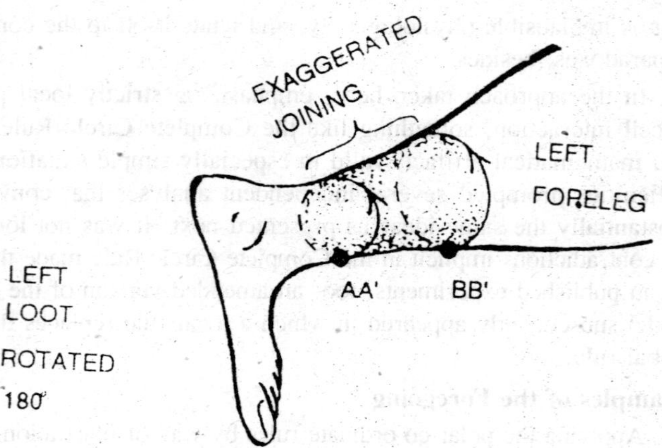

Fig. 17.6. As the left foot is rotated 180° rather than suffering removal to a right stump.

with a prehensile tail (a bilaterally fused appendage)-as it is in fact, the echinoderms being descended from bilaterally symmetric ancestors.

Alternative Formulations

Superficially, the data compiled by French et al., would seem to argue forcibly that phase singularities do play a central role in the morphogenesis of higher animals, much as they do in wave-conducting sheets of oscillators such as the social amoebae, the ascomycete fungi, and certain chemical reactions. This is a distinct possibility that I personally find very exciting, but it does have difficulties. These arise when we ask, as in the other organisms exhibiting phase singularities, what happens at the singularity? There is no abrupt discontinuity in cell type, no unique structure, and in apparent violation of Rule 2; cells do not keep proliferating indefinitely. All this suggests that the proximo-distal aspect of tissue specificity interacts in an essential way with the azimuthal tissue specifieity represented on the phase circle.

Although the original paper does not dwell on this interaction, it is implicit in the polar co-ordinate diagram in which the proximo-distal aspect of tissue specificity is represented radially with the most distal types (making toes, fingers) at the center. Sibatani gives a full explicit analysis of the peculiar implications of this diagram. The landmark contribution that came with the clockface model consisted primarily of a clear statement of correlated phenomena.

There are some astonishing regularities spanning diverse phyla, never before perceived so lucidly. The model itself had less appeal, its own internal contradictions lying so near the surface. The inevitable flurry of counter-experiments and *ad hoc* amendments was well summarized, but the ostensible elegance of 1976 was irretrievable once the subject had been glimpsed from another perspective. At present it appears that the reported phenomenology is best understood as a consequence of little more than the continuity properties of any topologically suitable representation of "positional information" or (to avoid confusion with physical position) "tissue specificity" underlying visible differentiation. The most tantalizing feature of the clockface model was its use of a *circular* co-ordinate for measuring cell types. It cells have two interdependent aspects of tissue specificity then one must ask exactly what motivates the choice of one co-ordinate system over another on this two-dimensional state space. Of course, one would prefer a co-ordinate system natural to the topology of the state space.

An Inverse Fate Map

A simple co-ordinate-free representation might be constructed as follows. Let us imagine a tissue-specificity space (TSS). We endow it with enough dimensions (two) to distinguish cell types on a two-dimensional surface such as the surface of a leg. We depict as a *place* in TSS the latent tissue specificity of a cell or a patch of neighbouring cells. To each region in TSS there corresponds a type of structure that cells in that region will make when they mature. Map the creature, organ, or tissue into TSS. There is nothing new in this procedure. This is only drawing a fate map, inside out as it were: instead of drawing the organism in the real world and writing tissue names on an overlay, we write the tissue names at fixed places in tissue-specificity space and then draw the organism as an overlay, distorted as required to put places on their corresponding names. For example, this "inverse fate map" of any bilaterally symmetric organ is folded so that two patches of cells symmetrically disposed to the left and right of a mirror plane both map to the same place in TSS. The mirror axis maps to a fold line. Now consider two pieces of an organism, normally not adjacent in the intact, mature stable organism.

There appear in TSS as two islands of tissue. If they are now physically juxtaposed, tissue specificities are not initially affected, so fine lines must be used to connect the cells that are quite separate in TSS though physically adjacent along the surface of contact. In experiments on the crayfish, Mittenthal shows that these "in-betweenness" connections *cannot* always be made straight by suitable rubber-sheet rearrangement of TSS, contrary to my conjecture of 1977. This may be a consequence of nothing more exotic than *curvature* of TSS:

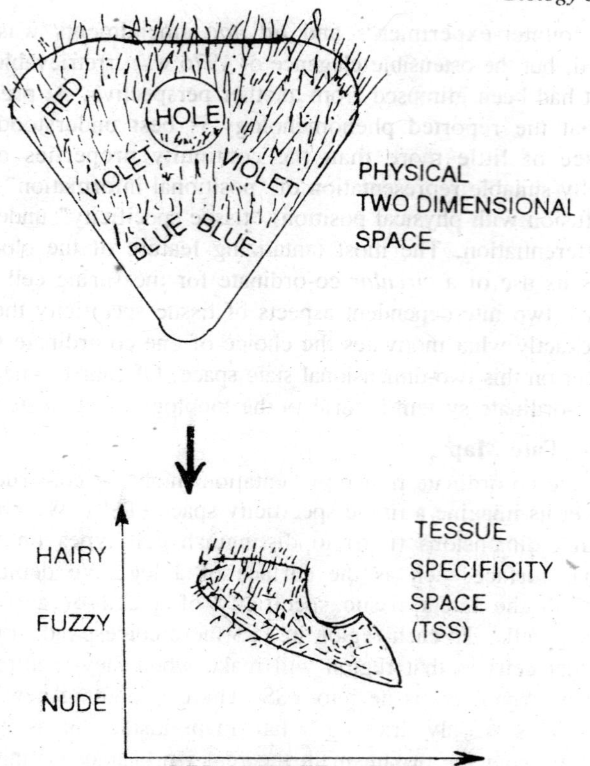

Fig. 17.7. An imaginary two-dimensional organism and its image in tissue specificity space.

if TSS were like a steep hill on an otherwise flat plane, then between any two points astride the hill there would be two shortest paths (left and right along the hill's flanks) from one to the other through "in-between" cell states. Any representation that purports to straighten all such curves would thereby inadvertently superimpose these two, though they consist of distinct cell types.

So a two-dimensional TSS cannot always be so defined that the linking curves appear straight. Nonetheless, straight-line connections in two-dimensional-space appear to suffice for the purposes of this Chapter, obviating the need to adopt a special "shortest arc" rule and the implicit metric to decide which path to take in a one-dimensional space. We rewrite the rules now in three parts using this geometric language:

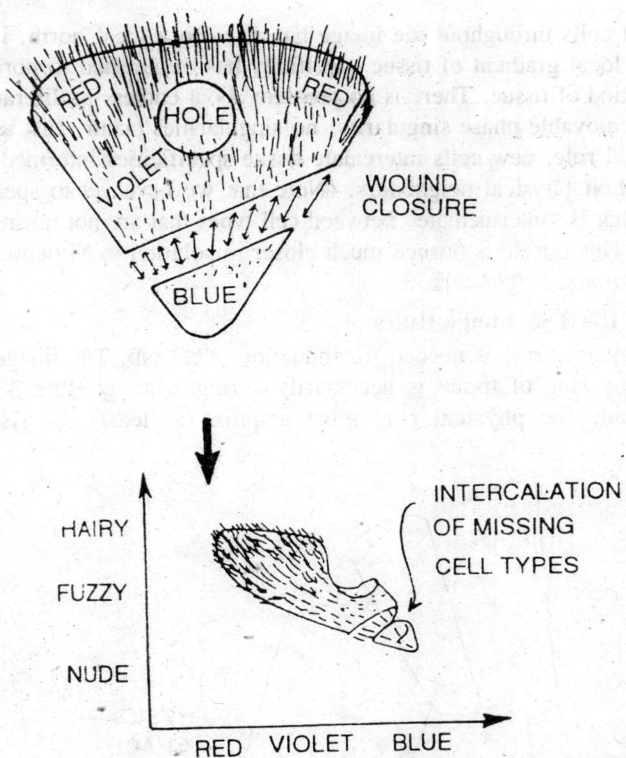

Fig. 17.8. A strip of tissue is removed from the organism and its image without affecting adjacent tissues.

Rule 1: Each little patch of cells is labeled with a state which is part of a smooth gradient of states across the tissue. This state is a point in a two-dimensional space of biochemical specificities topologically equivalent to a plane. Once established, this smooth map does not change. The two components of this "state" specification might be, e.g., the concentrations of two kinds adhesive site on the outer cell membrane.

Rule 2: No exceptional point or state exists.

Rule 3: Cells proliferate at a rate determined by the separation of adjacent cells in TSS (the unlikeness of surface properties, for example). If they are initially quite different in tissue specificity (e.g., cells on the interface, connected by fine lines, then they begin mitosis. The tissue expands in physical space while its image in TSS remains unmoved but becomes denser with cells. Proliferation stops when the density of cells throughout the image has risen to a local norm, i.e., when the local gradient of tissue specificity has diminished to normal for each kind of tissue. There is no problem about endless proliferation at an unremovable phase singularity: no singularities occur. Just as in the original rule, new cells intercalate tissue specificities intermediate between their physical neighbours. (Note that we have yet to specify exactly what is "intermediate" between cell types that are not normally adjacent. This puzzle is pushed much closer to solution in Mittenthal's *Rule of Normal Neighbours*.

Applying the TSS Image Rules

No separate rule is needed for induction of a limb. The image in TSS of any ring of tissue is necessarily a ring, and by Rule 3 the tissue inside the physical ring must acquire (at least) the tissue

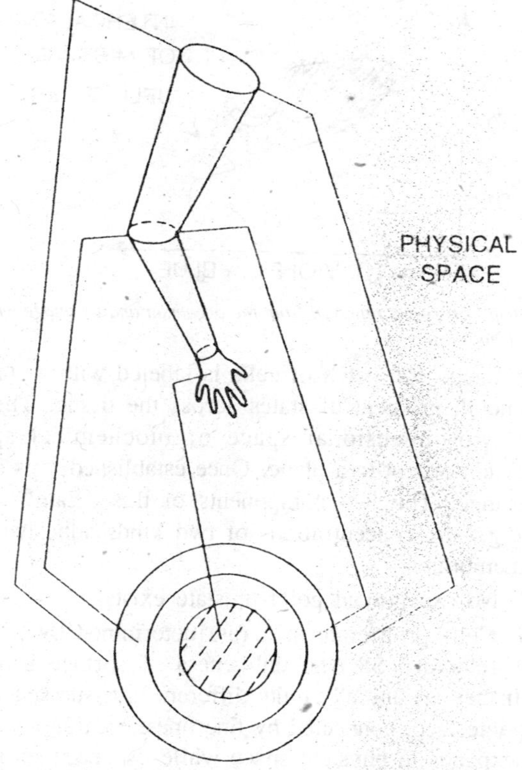

Fig. 17.9. As in, a limb maps to its image in tissue specificity space. More distal structures are believed to map concentrically interior to move proximal structures.

specificities inside the ring's image. At an appropriate stage in development, cells will develop the structures by which their positions in TSS were named. If those should happen to include the distal parts of a limb (fingers, for example), then such structures (among others) arise when tissue specificities are interpreted biochemically, and we have a "limb". If the most distal structures are normally internal to more proximal structures in TSS, then amputation corresponds to ablation of a disk in TSS, leaving a ring.

Blastema formation followed by wound healing corresponds to stitching across the empty disk fine lines along which new cells take up the missing tissue specificities, restoring the more distal organs. Replacing a left hand (L) by a right hand (R) on a left stump corresponds to cutting out the disk in TSS and replacing it with an identical disk. But remember that because of the inevitable L-R mismatch in physical space, tissue connections from the (R) central disk reach across to the opposite side of the (L) hole. We thus have a three-layered map across the distal core of TSS. The three layers are the existing hand and the cylinder of wound along the wrist that joins hand to forearm. The two new layers (the cylinder of wound proliferation) have the same orientation, so that we get two replicate left hands. Or imagine that the arm is amputated at both ends. Both the proximal and distal blastemas must then span the middle ("hand") region of TSS.

So mirror image hands must emerge (and do) at each end. This is the simple geometric essence of the rule of distal outgrowth. According to this rule, more distal structures regenerate from any stump even at the proximal end where, if regeneration were naively expected, a shoulder should grow. This picture in TSS is the same (not inverted) for right or left limbs. Thus replacing a left hand (L) by a right hand (R) on a left stump corresponds to cutting out the (left-hand, L) disk in TSS and replacing it with another (right-hand, R) disk. The new disk looks exactly the same as the old in TSS: it covers all the same regions of the "inverse fate map," since it has all the same cell types in the same neighbourly arrangement. But the reversal of handedness corresponds to a reversal in the correspondence between cell types and points in real physical space.

This only becomes important at the boundary where the graft and host are trying to fit together: in real physical space, that ring of cell types in TSS corresponds to a clockwise ring of host tissue, but to an anticlockwise ring of graft tissue. Because of this inevitable L-R mismatch in physical space, tissue connections from the (R) central disk must reach across to the opposite side of the (L) hole. Note that this is not quite crazy: to switch handedness, one must invert an organ through some axis, commonly the medio-lateral axis as in, but often (and with comparable effects) the antero-posterior axis or any intermediate. Each cell type at the junction finds itself adjacent to cells of the same antero-posteriorness, but opposite medio-lateralness.

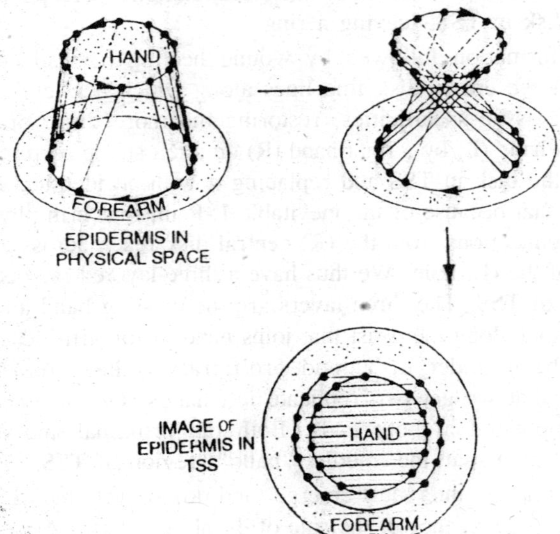

Fig. 17.10. The experiment is represented in the terms, rather than in terms of phase contours.

The new cells arising in the junction between graft and host must accordingly cover the range of medio-lateral positional values (tissue specificities) between opposite extremes. This range is spanned by the new cells on each side (and is of course already spanned within the graft itself). We thus have a three-layered map across the distal core of TSS. The three layers are the existing hand and the cylinder of wound along the wrist that joins hand to forearm. The two new layers (the cylinder of wound proliferation) have the same orientation, so that we get two replicate hands. Since that orientation is upside-down relative to the (R) graft, they are of opposite (L) type, the same as the host. The parity-conserving final result is a left limb with a left hand, but with a complementary L-R pair in addition. This strange, and (originally) unexpected experimental result derives from nothing more than an assumption of continuity(!).

Equivalent Metaphors and a Coat of Many Colours

Another way to articulate the TSS rules uses the language of *colour*. The diagrams in every paper about the clockface model would be much simplified, and both design and interpretation of experiments would be made transparent if only we could publish airbrushed colour diagrams as follows. Let the phaselike, circular aspect of tissue specificity be represented not in digits 1- 12, nor in contour lines, but in *hues* taken from a colour wheel. On a colour wheel, red grades into orange grades into yellow grades into green grades into blue grades into indigo grades into violet grades into red grades into orange, and so on: without backtracking, one can progress smoothly and unidirectionally through a cycle of tissue states, as required. Proximo-distal-ness may then be represented by *saturation* of the hue, the most distal tissues being least saturated (gray) so that all hues nearby are compatible without discontinuity. Our labeling of tissue types then consists only of assigning to each tissue type (and so to each place on the organism) a colour: have and its saturation. Every organism, in this view, wears a coat of many colours.

The colour is permanently imbued in each cell at mitosis, in cases of strict epimorphosis. Surgical interventions consist of abutting together colours that were previously separated by gradations of colour. If the colours now "run" and blend (as in morphallaxis), or if new fabric intercalates itself, adopting hybrid colouration compatible with its boundaries (as in epimorphosis), then we end up with arrangements of colour that satisfy all the phenomenology of regenerating and duplicating tissues, and, incidentally, the mathematical phenomenology of phases 1-12 (hues) and winding numbers thereof (complete colour wheels) and singularities (gray spots). The phase circles of Glass may be traced through full cycles of hue, ignoring saturation. They can be constructed by drawing circles on the physical tissue and then following the circles' images (rings) in TSS. The number of singularities corresponds to the winding number of the image about an arbitrarily chosen origin for the angular co-ordinate system.

In the coat-of-many-colours metaphor, the usual choice of origin would be gray, but this is an arbitrarily convenience dependent on ambient lighting. Biologically, the usual choice of origin is the distal-most tissue. From my point of view, the arguments of French *et al.* and of Glass about circles, phase maps, and winding numbers amount to using a circle embossed on the organism as a means of book-keeping the folds and rotations of the regenerating tissue's image in

TSS. Glass's calculational procedure is very useful here, because until one has a little practice, it seems awkward to visualize handedness in TSS maps.

The essential feature common to all these metaphors is the mapping from a physically distributed tissue to an abstract state space of two dimensions with continuity properties like ordinary two-dimensional Euclidean space. The most general underpinning for such a mapping is a set of chemical substances arranged in the tissue along transverse concentration gradients. To take but one example, the local concentrations of any three distinct pigments define a three-dimensional state space. Any surface coloured by mixtures of those pigments is thereby mapped into that space. If the total amount of pigmentation (the sum of the three concentrations) is limited to a fixed quantity, then the surface is confined to the two-dimensional triangular plane of standard colorimetric charts.

Colour mixtures happen much as in. If the "pigments" (or cell-surface antigens, etc.) should happen to mutually influence their own rates of synthesis and degradation then, as is well-known from the classical theory of reaction/diffusion structures, the chemical gradients may become self-stabilizing. The only experimentally demonstrable

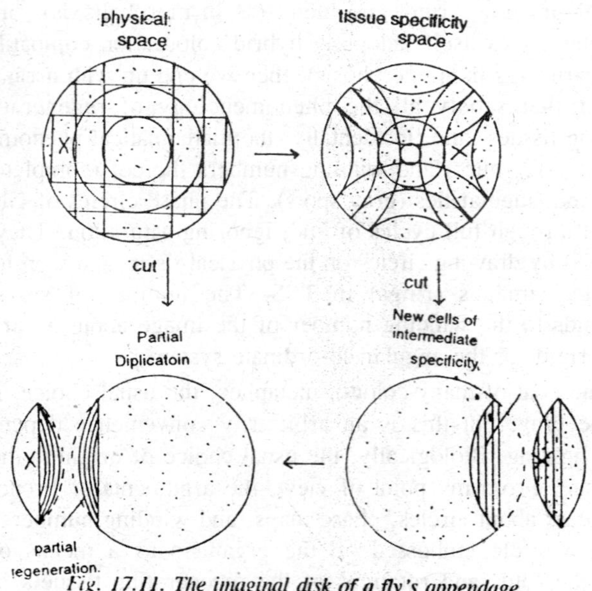

Fig. 17.11. The imaginal disk of a fly's appendage.

reaction/diffusion structure at present is the "rotor" in Belousov-Zhabotinsky reagent. Its structure, dynamics, and reaction to trauma may be expounded exactly in the manner of this chapter, its development into three dimensions resembles nothing so much as the embryology of simple organisms. From the literature surrounding the clockface model (1), one gains the impression that proper understanding of the phenomena of regeneration somehow repudiates (in favour of the more modern preoccupation with cell surface chemistry) the classical and apparently sterile notion that chemical concentration gradients and reaction/diffusion structures underlie it all. Perhaps it is indeed time for a fresh perspective on these unanswered riddles, as old as the science of biology. But to my mind, nothing more resembles the phenomenology of the (known and nonlinear, as opposed to theorized and linear) reaction/diffusion structure than contemporary reports on regeneration in limb fields of comparable size.

Experiments Needed

Certain implications of the geometric viewpoint adopted above may allow critical testing:

1. This style of mapping physical media or organisms to a state space or a TSS differ its better-established prototypes in that no "dynamic" is postulated in state space. Applied to non-dividing or relatively slowly dividing tissue, this is the assumption of epimorphosis. But dynamics may play an essential role in more rapidly dividing young or regenerating tissue. As Goodwin was first to emphasize, this is a deficiency that needs attention. For example, consider the following:

(a) Every tissue starts as a tiny patch of indistinguishable cells which, as they grow, acquire divergent specificities. Developmental fields originate, like the "big-bang" model of the expanding universe, from a singularity in TSS. At least at that early stage of development a "dynamic" *dominates* TSS. Kauffman *et al.* provide an alluring linear theory of this "dynamic."

(b) Although French *et al.* address themselves exclusively to epimorphosis (old cells retaining their specificity), some amount of morphallaxis (respecification) may commonly occur in regenerating limbs. According to Maden's interpretation, this fact is assimilated by simply allowing our Rule 3 to govern wounded tissues even prior to cell division. This scarcely alters our diagrams. However experiments to elucidate the following two points are potentially more subversive.

(c) Observations on transected imaginal disks of the fruit-fly argue for a strong tendency of tissue specificity to evolve more proximally. The qualitative observation is that both edges of the cuts produce the same new structures: one piece exactly duplicates and the complementary piece regenerates completely. The knife line though straight in the real world, is generally curved on the image of a disk in TSS, TSS being defined in such a way that inbetweeness follows straight lines so far as possible. No matter how the cut edge heals onto itself, the proliferating cells will adopt tissue specificities intermediate (parallel lines) between points on the knife cut's image in TSS. These intermediates cannot include the whole domain occupied by the excised piece if the disk's image is convex in this TSS- and it must be, or else intercalation between protrusive extremes of the image would produce tissue types outside the image, i.e., not normally present. So developmental fields are *convex* and cutting the edge off one leaves each part completely without access to some region of TSS occupied by the other part. Without a "*dynamic*," one piece must then fail to *fully* regenerate while the other fails to *fully* reduplicate. Transection experiments may thus provide the means to verify the existence of a "dynamic" and to quantitate the rates at which tissue specificity can change during normal growth and during regeneration.

Without a proximalizing "dynamic," there must be parts of any organism that *cannot* be regenerated: the adjacent most proximal parts of neighbouring developmental fields, located on the common rim of their convex images in TSS. By the same token, the kind of regeneration here considered cannot proceed beyond any symmetry plane: a bilaterally symmetric organism, for example, could regenerate at most half of its complete body. How do these limitations contrast with the facts of regeneration in radially symmetric organisms and in morphallactically flexible organisms such as *Planaria*? Recent experiments using artifically constructed bilaterally symmetric limbs are particularly intriguing in this context. An unexpected progressive loss or gain circumferential positional values (hues, clock-phase values, TSS regions occupied). This may betray a slow underlying dynamic such as we inferred previously.

2. According to this coordinate-free representation, ablation of a feature on the symmetry plane of a bilaterally symmetrical organism (a tongue, a nose, a penis, a tail, the genital disk of a fly) corresponds to cutting a hole out of the fold of the two-layered image in TSS. Such a hole can heal in two quite distinct ways.

(a) Closing horizontally (top-to-bottom contacts), the hole in TSS would be filled again. Thus we should anticipate complete regeneration.

(b) In contrast, a vertical closure (side-to-side contacts) would join mirror image tissues. Thus, no proliferation should ensue. Even if it did, the missing tissue specificities would not be recovered.

The results of diagonal closure must vary between the external results (a) and (b) according to the angle of the diagonal. This series of experiments may help to distinguish between geometric interpretations of tissues specificity. For discussion of experiments recently undertaken along these lines, consult. One source of ambiguity at present is the lingering uncertainty about the actual (uncontrolled and unobserved) geometry of wound closure.

3. If one had to guess before observing, it might reasonably be supposed that an organism would respond to surgical challenges by intercalating positional values between given boundaries in the smoothest way. This surmise comes from the observation that paired phase singularities (left-right pairs of supernumerary appendages) are uncommon where uniformity of phase might alternatively prevail.

4. It seems not entirely proven by experiment that apposition of dissimilar tissues is required to elicit regeneration. It is possible that the external medium may under some circumstances serve as a "tissue type"? Where is its point in TSS, if so?

5. It would be of interest to treat a strictly non-living chemical "organizing center" as though it were, for example, a limb field, repeating many of the physical manipulations of classical embryology. Some of the practical difficulties are similar: Accurate separation and recombination of microscopic chunks of wet gel that tend to adhere through surface tension or to float apart when submerged; accurate observation of outcomes that consist of garbled and sometimes subtle patterns, constantly changing; limited reproducibility due to uncertainty of initial conditions. The outcomes might also bear sufficient formal resemblance to provoke stricter attention to strategic issues: What questions are we asking? How will the answers aid understanding and control? Since the morphogenesis of organizing centers in Belousov-Zhabotinsky reagent can be understood in exact Physical-chemical detail, it might be illuminating to know how nearly it really does resemble cytochemical morphogenesis.

Regeneration in Amphibian Limbs

Many organism or parts of organisms are capable of pattern regulation not only during embryonic stage, but throughout life. The domain within which regulation can occur has been designated a 'field' and it has been recognised for many years that field can be roughly categorised according to their mode or regulation following surgical intervention. One category contain those systems which display morphallactic regulation, where a part of the pattern can reorganise to form a miniature but complete pattern. The other category consists of those systems showing epimorphic regulation, where a part of the pattern can add on new pattern elements trough growth. The new elements formed by epimorphic regulation may be the same as those

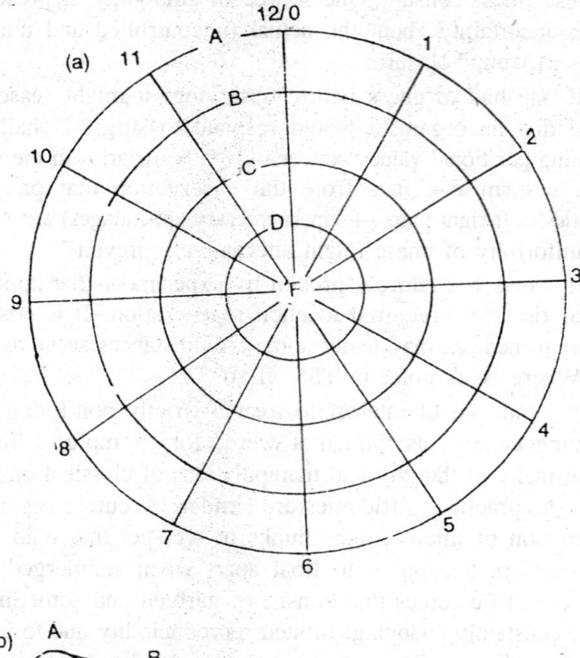

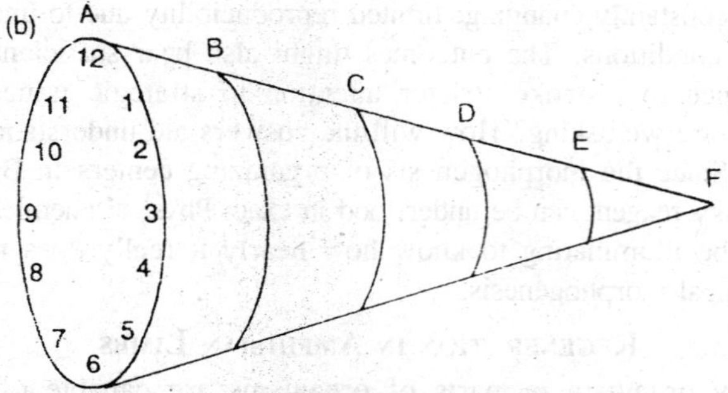

Fig. 17.12. (a) Polar co-ordinates of positional information in an epimorphic field. (b) Polar co-ordinates of positional information in a leg field.

which were removed, in which case together with the regulating fragment they make up the complete pattern, or they may be the same as structures already present in the regulating fragment, leading to pattern duplication. French, Bryant & Bryant (1976) have recently developed a formal interpretation of epimorphic regulation based on the results of experiments on the legs of hemimetabolous insects, the imaginal discs of *Drosophila*, and amphibian limbs. This interpretation assumes that positional information in field which exhibit epimorphic regulation is specified in terms of polar co-ordinates where one component of positional information corresponds to position on a radius and the other to position on circle (0-12 sequence). In amphibian limbs, the radial sequence is in the proximo-distal axis, and the outer circular sequence around the base of the limb. It is proposed that cells in such fields respond to surgical intervention according to the following principles.

Shortest Intercalation

When cell with normally non-adjacent positional values in either the radial or circular sequences are brought together (as in grafting experiments or as a results of wound healing) growth occurs to intercalate the missing positional values. The circular sequence is continuous, hence there are two possible sets of intermediate values between any two non-adjacent values. It is the shorter of the two intervening set of positional values which is intercalated.

Complete Circle

From any given radial positional value, transformation to from all more central (distal) radial values can occur provided that a complete set of positional values in the circular sequence is either exposed by amputation, or generated by intercalation.

Supernumerary Limbs

The shortest intercalation and complete circle rules can together account for the occurrence, number, location, handedness and orientation of the supernumerary limbs which form following certain grafting operations in regeneration newt limbs. Iten & Bryant (1975) showed that when forelimb regeneration blastemas at stages from medium bud to early digits for normal stages were transplanted to the same or a more distal level of the contralateral forelimb stump from which a blastema had been removed with anterior and posterior positions of graft and stump apposed, supernumerary limbs formed in the anterior or posterior of the graft junction. At the earliest stage studied, early bud, separately identifiable supernumerary limbs were rarely formed,

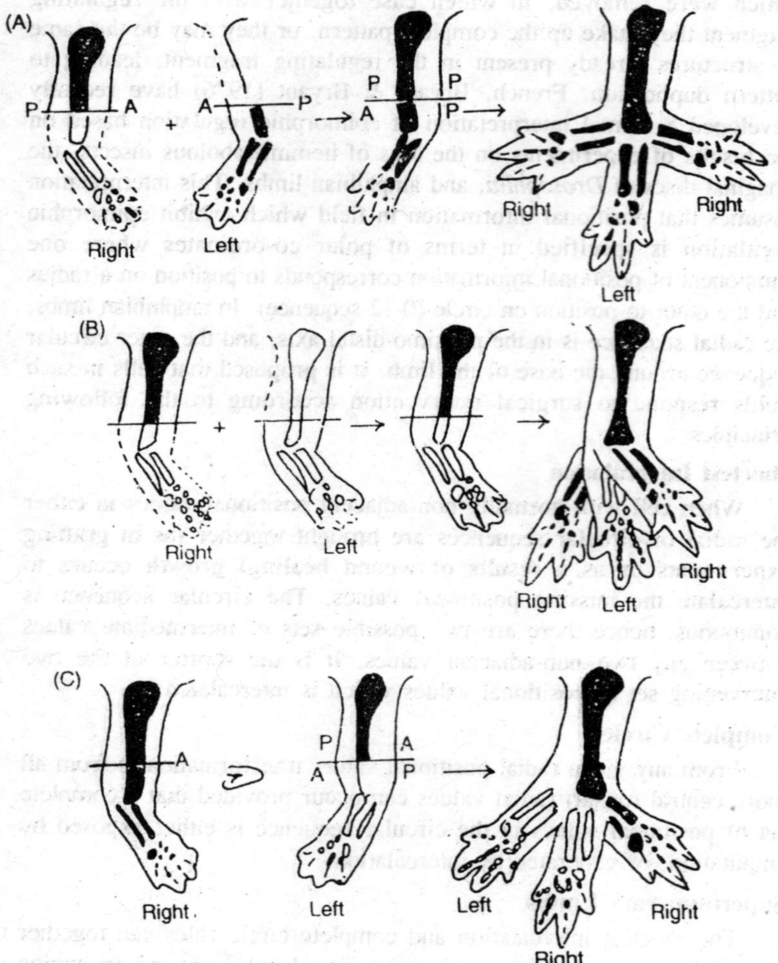

Fig. 17.13. Diagram of the experimentation which a left blastema was transplanted to the contralateral limb stump with anterior and posterior position of graft and stump apposed.

but often the limb had multiple distal structures, suggesting the formation of anterior or posterior supernumerary elements which remained closely associated with the structures originating from the transplant.

In a subsequent study, Bryant & Iten (1976) showed that dorsal or ventral supernumerary limbs would form following transplatation of a blastema at medium bud or early digits stages to the contralateral limb stump with dorsal and ventral positions of graft and stump apposed. Thus it appeared that supernumerary limbs could form wherever anterior and posterior or dorsal and ventral tissues were in contact. However, when both dorsal and ventral as well as anterior and posterior positions of graft and stump were apposed by ipsilateral grafting of a blastema rotated by 180°, a maximum of two supernumerary limbs were formed, not four as might have been expected on the basis of the results from contralateral grafts.

In both the contralateral and ipsilateral grafting operations the location, handedness and orientation of the supernumerary limbs are reproducible. When anterior and posterior positions were apposed in grafting, supernumerary limb arose in the anterior and posterior and were arranged in mirror-immage symmetry with the graft. Similarly, when dorsal ventral positions were apposed, the dorsal and ventral supernumerary limbs were mirror images of the graft. However, when both dorsal and ventral as well as anterior and posterior positions were apposed by rotation of the graft through 180°, supernumerary limbs formed either in a posterior-dosral location or in an anterior-ventral location, with respect to the stump. The posterior-dorsal supernumerary limbs were mirror images of the grafted limb, whereas the anterior-ventral supernumerary limbs were of the same handedness as the grated limb.

As has been pointed out by French *et al.* (1976) and Bryant & Iten (1976), all of these results can be understood as a consequence of the shortest intercalation and complete circle rules, if it is assumed that in the regeneration appendage positional values in the circular sequence are slightly non-uniformly distributed. Intercalation between graft and stump can occur where normally non-adjacent positional values in the circular sequence are juxtaposed. For each type of graft illustrated, there are two positions around the graft junction where the two possible routes of intercalation involves the same number of positional values. Intercalation on either side of these locations is in opposite directions, and a complete circle is generated where the two half-circumferences meet. Distal transformation can then occur from these complete circles, but not form any other position in the graft junction. The handedness and orientation of the supernumerary limbs so produced is controlled by the direction of intercalation in the regions of the graft junction adjacent to them.

Recent studies have shown that the ability to form supernumerary limbs is not limited to regeneration limbs. When mature, non-regeneration hands were transplated to the contralateral limb stump, supernumerary limbs in the expected locations, and with the predicted handedness, sometimes formed. The incidence of supernumerary limb formation in mature limbs was greater if the graft was made to a more proximal level on the contralateral limb, rather than to the same level. The reason for this difference in incidence is not understood, but it may indicate that mature, fully differentiated tissues do not readily respond to discontinuities of positional values in a single sequence. When mature limb tissues were grafted to create double half-limbs (see below), a single supernumerary limb sometimes formed at the proximal junction between graft and limb stump. This is the position at which a complete circle of positional values in created by the grafting operation on. The fact that the formation of a supernumerary limb in this location was often not apparent until after the limb had been amputated through the distal region of the graft and dedifferentiation of the graft had been initiated, suggests that dedifferentiated cells may be better able to respond to discontinuities in positional values than are differentiated cells.

In addition to the regeneration and non-regnerating limbs of adult urodeles, the developing limbs of larval urodeles are also known to be capable of producing supernumerary limbs following certain types of orthotopic and heterotopic transplantations. The results of many of the pioneering experiments of Harrison and Swett as well as those of other workers, can be understood in terms of the polar co-ordinate model.

Double Half-Limbs

Further support for the complete circle rule comes from experiments on double half-limbs. To make double anterior or double posterior limb stumps, half of one upper arm was removed and replaced by the reciprocal half from the contralateral upper arm. Similar operations were performed to create double dorsal and double ventral stumps. The grafted pieces were allowed to heal into place on the unamputated limbs for 3 to 4 weeks, after which time amputation was performed through the distal region of the graft.

All 44 control limbs, in which either an anterior, posterior, dorsal or ventral half of an upper arm was removed and replaced *in situ*, regenerated normally (six of these had only three digits rather than the normal four, a common feature of normally regeneration limbs). However, the double half-limb stumps usually either failed to regenerate distally or produced abortive, incomplete outgrowth. Of 12 double

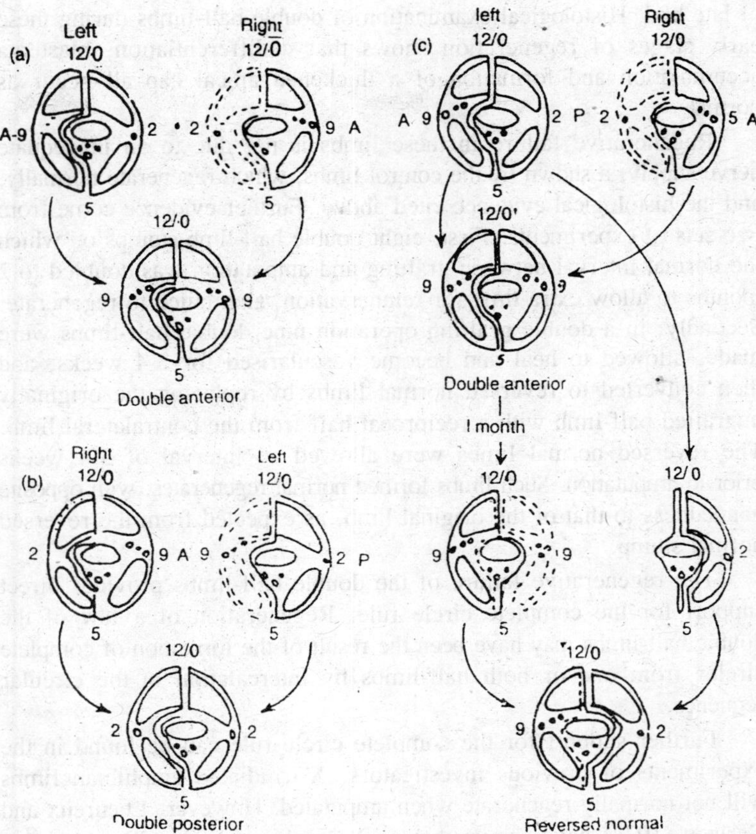

Fig. 17.14. Schematic diagram of operations to produce double half and reversed normal limbs.

anterior limbs amputated, none regenerated normally and of 14 double posterior limb, only one formed a complete regenerate. The results with double dorsal and double ventral limbs were similar; two of the 15 double dorsal limbs regenerated normally, and of 12 double ventral limbs, only one formed complete distal regenerates. Histological examination of double half-limb stumps at the time of amputation showed that they were innervated, and comparison with control limb stumps in which half of the upper arm had been removed and replaced, revealed no major differences in appearance which might account for the difference in regenerative ability. In addition, double half-limb of all types initiated regeneration, and several proceeded as far as the stage of late bud. Histological examination of double half-limbs during these early stages of regeneration shows that dedifferentiation, blastema accumulation and formation of a thickened apical cap all occur as normal.

Regenerative failure in these limbs is not due to an inadequate nerve supply, a shown by the control limbs, which regenerate normally, and the histological evidence cited above. Further evidence come from two sets of experiments. First, eight double half-limb stumps on which the normal interval between grafting and amputation was doubled to 2 months to allow extra time for reinnervation, also failed to regenerate. Secondly, in a double grafting operation nine double half-limbs were made, allowed to heal and become vascularised for 3-4 weeks, and then converted to reversed normal limbs by replacing the originally ungrafted half-limb with a reciprocal half from the contralateral limb. The reversed normal limbs were allowed an interval of 3-4 weeks prior to amputation. Such limbs formed normal regenerates, with opposite handedness to that of the original limb, as expected from the reversed normal stump.

The regenerative failure of the double half-limbs provides direct support for the complete circle rule. Regeneration of a few of the double half-limbs may have been the result of the formation of complete circles from one or both half-limbs by intercalation in the circular sequence.

Further support for the complete circle rule can be found in the experiments of previous investigators. X-irradiated amphibian limbs will not normally regenerate when amputated. However, Lheureux and Carlson (1974) have shown that an X-irradiated limb will regenerate if it is provided with a normally oriented cuff of non-irradiated skin. Regeneration does not occur if a cuff of skin is taken from one surface of the limb and is oriented so that only one 'quality' (dorsal, ventral, anterior or posterior) is present at the amputation site. These results suggest that regenerative ability can be restored in X-irradiated limbs by a skin cuff containing all positional values in the circular sequence, but cannot be restored of only part of the circular sequence is present.

Half-Limbs

Previous studies have shown that longitudinal half-limbs can regenerate normally or relatively normally. In parallel with the double half-limb experiments described in the previous section, we have also looked at the regenerative ability of half-limbs stumps under two different conditions: first, removal of half of the upper arm, followed by immediate amputation through the distal region of the half-limb; secondly, as above, but with amputation 1 to 2 months after the removal of half the limb.

Of the 12 half posterior limbs amputated immediately, 10 formed normal regenerates (two had only three digits), and two failed to regenerator. Of the 12 half anterior limbs amputated immediately, 10 regenerated normally (one had only three digits) and two failed to regenerate. The regenerates originated over the junction between the distal and lateral would surfaces, but during the later stages of regeneration most became more normally oriented on the distal half-limb stump. A tentative explanation for these results is that the missing positional values can be restored along the lateral wound surface by migration of cells during wound healing from the proximal half-limb stump. A complete çircle of positional values can therefore be created at the edge between the distal and lateral wound surfaces, and distal transformation can occur from this complete circle. A similar explanation can account for the results of experiments in which half-blastemas were removed at different stages of regeneration. At all stages prior to early digits, such half-blastemas developed into normal limbs. An explanation for this result is that the remaining half-blastemas, together with cells migrating during healing from the proximal stump, will constitute a complete circle of positional values, leading to the formation of a normal regenerate.

When amputation of half-limbs is performed 1 to 2 months after removal of half of the limb, the result differ in some respects from those following immediate amputation. Of 20 half posterior limbs amputated after 1 or 2 months, 17 regenerated normally (two had only two digits and one had three digits). A further three half posterior limbs formed a regenerate from the distal wound surface, and supernumerary elements from the lateral would surface, leading to a total of five or six digit per half-limb. Of 19 half interior limbs amputated after 1 to 2 months, 17 regenerated normally and two failed to regenerate. All of the 14 half ventral limbs amputated after 1 month regenerated normally. Of 14 half dorsal limbs amputated after 1 month, 10 regenerated normally (two had less than four digits) and three formed a regenerate from the distal would surface, and additional structure from the lateral wound surface. The remaining limb formed structures from the lateral wound surface but not from the distal amputation surface. The formation of a normal regenerate from the distal surface. The formation of a normal regenerate from the distal surface of the half-limb can be understood if it is assumed that the missing positional values are replaced along the distal region of the lateral wound surface. The occurrence of multiple regenerates on some half posterior and half dorsal limbs but not on half anterior and half ventral limbs, suggests the possibility that shortest intercalation may occur in the central regions of the long lateral wound prior to wound healing. If positional values around the circle are slightly non-uniformly spaced, as we have suggested from the supernumerary limb results, then shortest intercalation on the lateral wound surface of a half ventral or a half anterior limb stump will not introduce any positional values which are not already provided by migration from the distal and proximal wound surfaces. Hence, complete circles of positional values will not form on the lateral wound surfaces of such half-limbs. However, shortest intercalation on the lateral surface of a half dorsal or half posterior limb stump could introduce new positional values which are not provided by migration and could lead to the formation of two complete circles on the lateral surface. Distal transformation could then occur from one or the other of these complete circles. The fact that the structures forming on this surface can sometimes be recognised as being of the same handedness as the normal distal regenerate, and sometimes of opposite handedness, supports this suggestion. The opportunity for intercalation in these half-limbs may depend on the rate of wound healing from the proximal and distal stumps. Half-limbs with delayed amputation will have a greater opportunity for intercalation compared to half-limbs amputated immediately, because the length of the lateral wound is greater in the former than in the later. In the latter limbs, the size of the lateral wound is reduced due to amputation, wound trimming and subsequent dedifferentiation of the stump.

The relationship of wound healing to the results obtained with half-limb stumps is being explored further, using scanning electron microscopy.

Regeneration of the Axolotl Limb

The development in *Ambystoma* is indirect. Its larva is the *axolotl* and wherever it develops in the absence of iodine it becomes sexually value and starts reproducing sexually. This breeding behaviour is called neotenry. These neotenic axolotls spend all of their normal life as aquatic larvae. As is the case with all larval urodele amphibians, axolotls are capable of regenerating their appendages remarkably well. In recent years, they have become increasingly important as laboratory animals because they are easy to maintain and breed readily over the

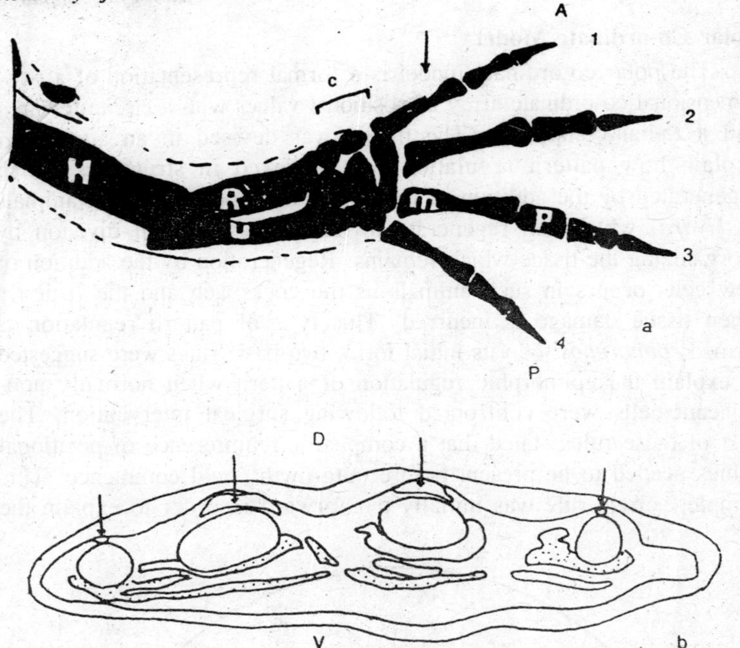

Fig. 17.15. The axolotl forelimb bears numerous clear structural markers that can be used to analyse the results of tissue level grafting experiments.

winter months. The limbs of the larvae are particularly well suited to studies of pattern regulation because normal limbs faithfully replace the limb parts distal to the level of amputation in an ordered histological sequence, and the skeleton, muscles, and skin provide clear anatomical markers for the results of tissue level manipulations.

In this discussion, a theoretical framework for the control of pattern regulation will be analyzed in terms of a series of recent experiments. The intention of the discussion is twofold; the experiments to be outlined will reveal some of what we know of the cellular interactions occurring during distal outgrowth and will indicate how a theoretical framework for these interactions molds the design of experiments and evolves as new data concerning specific tenets of the model become available. Some of the basic features of the polar co-ordinate model have remained unscathed during the last six years, whereas others have been discarded or have been refocused. Prior to describing the series of experiments that will form the bulk of the discussion, the model itself will be outlined.

Polar Co-ordinate Model

The polar co-ordinate model is a formal representation of a two-dimensional co-ordinate array of positional values with a circumferential and a radial component. The model was devised in an attempt to explain how pattern regulation was achieved in structures which regenerated by the addition of new cells, as opposed to such animals as *Hydra*, which can regenerate in the absence of cell division by reorganizing the tissue which remains. Regeneration by the addition of new cells occurs in such animals as the cockroach and the fruit fly when tissue damage is incurred. This type of pattern regulation is termed *epimorphosis*. In its initial form, two basic rules were suggested to explain the epimorphic regulation of pattern when normally non-adjacent cells were confronted following surgical intervention. The first of these rules stated that a complete circumference of positional values needed to be present before outgrowth could commence. This complete circle rule was initially put forward in order to explain the

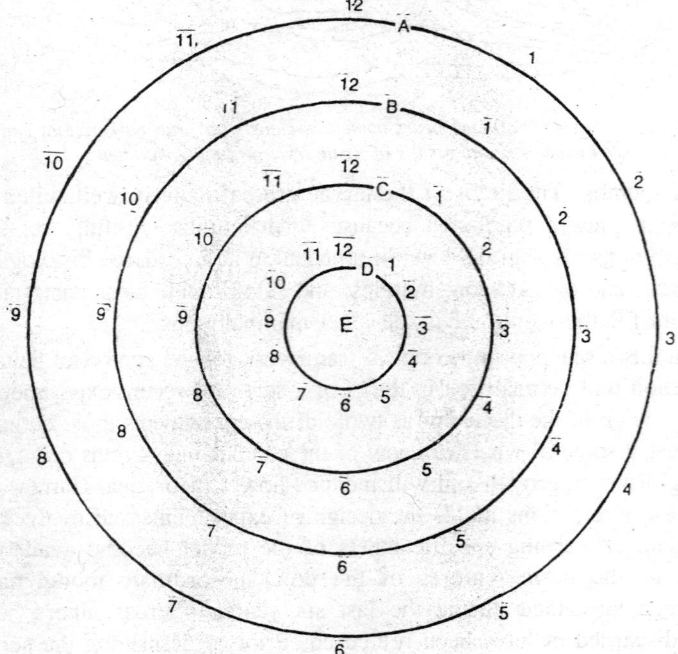

Fig. 17.16. The polar co-ordinate model can be formally represented as a two-dimensional array of positional values.

appearance of supernumerary limbs that appear after axial misalignment of blastemas and stumps. In a sense, the rule was in conflict with the remainder of the model because it relied on a global realization within the limb that a complete asymmetrical circle was present before outgrowth could begin. The remainder of the model relies solely on local cell-cell contacts and the intercalation of new cells by cell division that restore circumferential or radial continuity after the confrontation of normally non-adjacent cells.

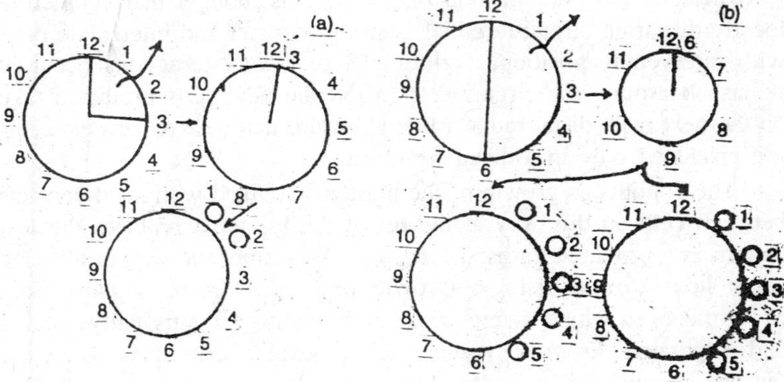

Fig. 17.17. Diagrammatic representation of the shortest route intercalation rule that governs the positional values held by new cells produced following contact of cells from normally non-adjacent positions.

The polarity of restoration of continuity by intercalation is controlled by the second rule which states that this always occurs by the shortest possible route. In this way, the actual positional values intercalated should always be predicted except when maximum points of discontinuity are confronted when a choice of two paths of intercalation are evident. Soon after the model first appeared, numerous experimental results demonstrated that the complete circle rule was not required to explain pattern regulation. It is now clear that tissues bearing incomplete asymmetrical circles or symmetrical circles can commence outgrowth. This is evident from experiments concerned with both the outgrowth of stumps bearing symmetrical tissues and from the structure of supernumerary limbs formed following ipsilateral blastemal rotations. In the first section of experimental results presented, the likely controls of normal outgrowth will be discussed. The possible interactions leading to the formation of supernumerary limbs will be outlined later on.

Control of Distal Outgrowth

Following amputation of the limb, differentiated cells in the stump close to the amputation plane begin to divide and undergo a morphological transformation resulting in the accumulation of many cells that appear to the undifferentiated. This process occurs beneath a thin would epidermis which covers the forming blastema throughout the process of pattern regulation. The initial accumulation of blastemal cells forms a mound known as the stage of medium bud. During the formation of the medium bud blastema, it is thought that cells from locally disparate circumferential positions contact and intercalate cells with intervening positional values. These local interactions occur at points all around the circumference and the new cells produced take on the next most distal radial value to ensure that new pattern elements are produced only in a distal direction.

The continuous growth of the blastema coupled with local contacts between cells in this way at the tip of the blastema replaces the lost pattern in a proximal to distal sequence. This simplistic way of thinking about how normal distal outgrowth may occur gains support from experiments in which symmetrical limb stumps are surgically created and amputated. From the results of these experiments, it is now clear that the amount of distal outgrowth involving the sequential production of new radial levels is intimately linked to both the nature of the circumferential sequence at the amputation plane, and the mode of healing that controls which circumferential cells actually make contact during outgrowth. Symmetrical limb stumps have been created surgically

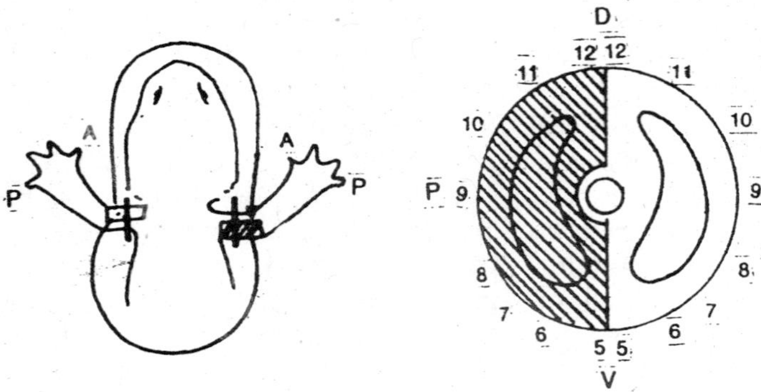

Fig. 17.18. (a) Diagrammatic representation of the operation performed to create double anterior and double posterior symmetrical stumps. (b) An end-on view of the amputation plane of a double posterior stump.

in both the upper and lower limb regions. The basic experiment involves cutting one-half of the limb out from upper or lower limb positions and exchanging tissues from contralateral limbs. Double anterior and double posterior constructions have been made in most cases, although double dorsal and double ventral constructions have also been examined.

The amputation of such limbs produces a variation of structures ranging from extensive symmetrical regenerates to nothing at all. The very fact that any regeneration is achieved is evidence that the complete circle rule is not upheld. Several other directives emerge from an analysis of the data from these experiments. It is clear from operations performed on the upper forelimb that double posterior constructions consistently regenerate more structure than double anterior constructions. Furthermore, the degree of distal outgrowth is dependent upon the length of time that the initial graft is allowed to heal prior to amputation. The longer the graft healing time, the less the regeneration that is seen following amputation. This effect has also been detected in similar experiments in the thigh region of the axolotl leg. Two further conclusions have been drawn from these results. The reason for posterior tissues being more productive than anterior tissues in symmetrical combinations is assumed to be due to there being greater than half the number of circumferential positional values on the posterior side of the upper limb.

The role of tissue healing is thought to involve the structural constraint of contacts between cells on either side of the wound created during surgery. This last point can be most clearly thought of in terms of the simple model discussed above for normal limb regrowth. It is thought that after long periods of healing, cells on the two sides of the wound can interact freely. In this case, cells will interact locally as in normal outgrowth and, because of the symmetry in the experimental cases, this will result in a number of cell contracts occurring between cells with the same or normally adjacent circumferential value. This type of contact will not stimulate intercalation. In addition, no local contact around the circle will result in the intercalation of the mid-line circumferential values at the next most distal level to that of amputation. The outcome of these two properties of symmetrical circles will be a tapered symmetrical outgrowth that forms from a blastema containing few cells.

In contrast to this effect, which occurs after long periods of healing, short healing grafts regenerate far more structure. This is assumed to

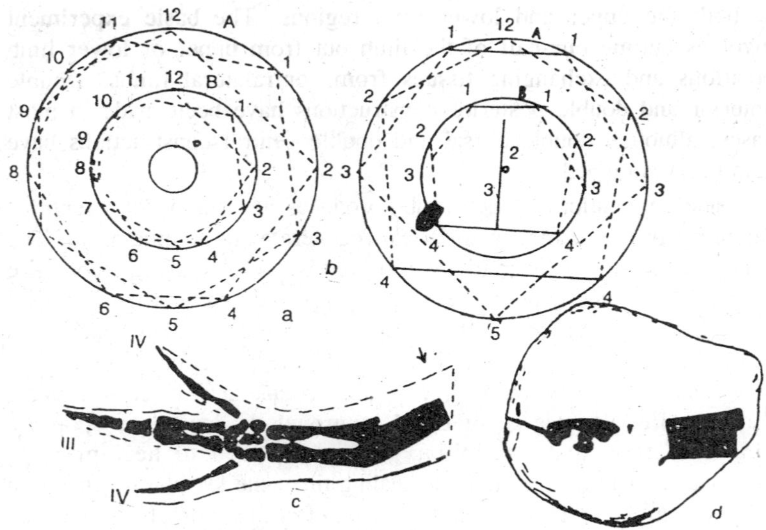

Fig. 17.19. Representation of the theory of short arc intercalation.

be due to the prevention of contacts between cells on either side of the central wound, and consequently the prevention of non-productive cell interactions. The result of this will be more cells available for intercalation at the next most distal level. The result of amputating such short healing time limbs should be a distally tapering structure that regenerates to a more distal pattern level than the longer healing type of graft. In many cases, regenerates of this class distally transform to the digits. Thus far we have examined only symmetrical outgrowths from upper arm stumps. Two further experiments have been performed on axolotls which examine the regenerative ability of symmetrical double posterior and double anterior lower arms.

The first of these experiments clearly suggested that lower arm symmetrical constructs behave differently from comparable upper arm constructions, in two ways. First, no clear healing time effect is seen in the axolotl lower arm. Amputations of symmetrical stumps immediately after surgery produces similar degrees of outgrowth of those amputated after a month of healing. Secondly, the degree of regeneration of double posteriors and double anteriors is comparable. In the second experiment, double posterior limbs that had regenerated from initial short healing time upper arm amputations of symmetrical double posterior constructions produce expanded patterns when amputated in the lower arm. These types of results strongly suggested that the lower arm was behaving differently from the upper arm in some fundamental way. Not only is the distribution of values around the circumference thought to be more even, but the constraints on contacts between the cells around the circumference seem to be different.

This point can be clearly demonstrated by examining how distal expansion of a double posterior limb amputated in the lower arm can occur. In this majority of cases, double posterior upper arms amputated immediately make symmetrical limbs bearing between three and six digits. When reamputated in the lower arm, limbs bearing up to seven digits regenerate with a normally anterior digit appearing in the middle of the pattern. In terms of shortest route intercalation, this can only occur if cells with the dorsal and ventral extremes of the circumferential set of values interact. The question at hand is why should this occur in the lower arm but not in the supper arm? The simplest conclusion, if local cell-cell contacts are requried, is that the dorsal and ventral sides of the blastema are closer together in the lower arm than in the upper arm; that is, the blastemas formed in these two different limb regions are of different shape. The same conclusion can be drawn from the observation that symmetrical double posterior and double anterior lower arm stumps do not exhibit a healing time effect. This will occur if cells preferentially contact from dorsal to ventral sides of the blastema and not from anterior to posterior sides.

In conclusion, several points can be drawn from the results of amputating double anterior and double posterior limb stumps.

1. It appears that the spacing of positional values around the circumference is not even in the upper arm of both newts and axolotls.
2. Interactions between cells can be modulated by physical constraints, such as lines of healing or by the shape of blastemal populations (this point will be demonstrated further below).
3. The amount of distal outgrowth (radial value production) is directly related to the number of circumferential positional value present at the amputation plane. In short, the amount of growth that is achieved during regeneration is intimately linked to the pattern formation process and the constraints on cell contacts.

Blastemal Shape: Its Control and Significance

Blastemal shape has recently been quantitatively described for different morphological stages of blastemas derived from different limb levels of both axolotl hindlimb and forelimb. It has clearly been demonstrated that upper arm and leg blastemas are round in cross-section whereas lower arm and leg blastemas are elliptical. The observation is entirely consistent with the predictions made from the results of the experiments just discussed.

This analysis of blastemal shape was based upon a simple morphometric analysis. Limbs were amputated at the required level and allowed to regenerate to the stage of medium bud, medium to late bud, and palette. At the requried stage, limbs were fixed, wax embedded, and serially sectioned. The distance of the main anterior to posterior and dorsal to ventral axes were measured at 100-m levels from the distal tip and a simple ratio of these values calculated. This ratio, r, was then plotted for each blastemal stage at each level of amputation. It is clear from these graphs that the lower limb blastemas are significantly more elliptical than the round upper limb blastemas. It should be emphasized that this shape analysis was carried out subsequent to the surgical symmetrical limb stump experiments and indicate therefore that structural constraints can be predicted from such an approach.

At the same time, however, blastemal shape may be used to examine the problem of pattern formation. If the control of shape can be established, then it should be possible to alter blastemal shape and the critical relationship between shape of tissue fields and pattern formation can be further examined. For this reason, a further set of experiments were performed in an attempt to elucidate the tissue controls of blastemal shape. At the outset of these experiments, the simple questions to be examined involved the role of the stump in the control of blastemal shape. Clearly the stump is of maximum importance during the early phases of regeneration because all blastemal cells derive from it. However, it is already known that by the stage of MB the pattern regulation mechanisms of the regenerate are autonomously controlled by the blastema and are not influenced by the stump from which it derives. We wished to know if the same was true of the control of blastemal shape. An indication that this was the case came from observing a change in shape from round to elliptical during outgrowth of a palette staged upper arm blastema.

In this case, the flattening was occurring spontaneously within the blastema because it was forming on an upper arm stump. Further evidence for an autonomous blastemal control came from surgical experiments where specific tissues of the stump were removed prior to amputation to see if an abnormal stump would produce misshapen blastemas. In these experiments, individual fore and hindlimbs had

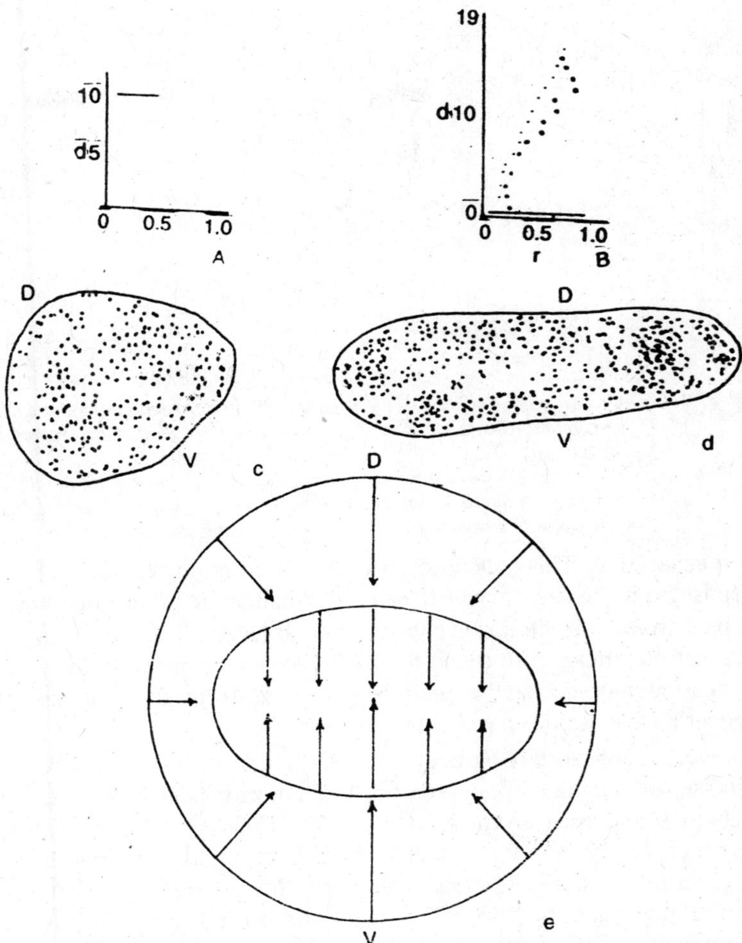

Fig. 17.20. The shape of blastemas formed at different proximal to distal limb levels and at different stages. r=shape ratio; d=ditance from the blastemal tip in 100 micron steps, D=dorsal; V=ventral. The arrows on the graphs represent 95% confidence limits for the range of r values for each blastema population.

either the cartilage, muscle, or dermis removed. Removal of one bone, either the radius or ulna, from the lower forelimb and removal of stump musculature from the thigh region of the leg had little effect upon blastemal shape as compared with sham operated controls. These results suggested that the role of the stump in governing blastemal shape involved cryptic, level specific information rather than a purely

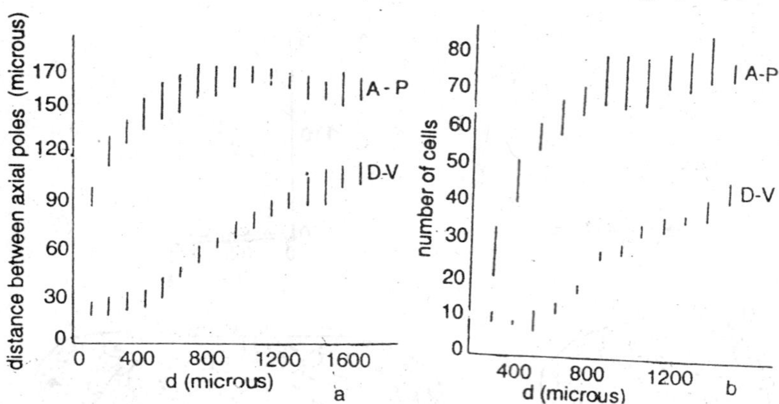

Fig. 17.21. (a) Graphic representation of the distance (in microns) separating the poles of the cardinal axes in a palette staged blastema formed after amputation in the upper arm. The anterior and posterior (A-P) poles are clearly wider apart than the dorsal and ventral poles (D-V). d is distance from the distal blastema tip in microns. (b) The number of cells separating the cardinal axes in a palette staged blastema formed after amputation in the upper arm.

mechanical role. This conclusion was supported by the removal of the dermis from the stump which caused a flattening of the blastemas derived from both the upper and lower limb levels. The alterations in shape of blastemas from round to elliptical is consistent with the mode of healing and interaction that is predicted from the analysis of regeneration of symmetrical limb stumps.

With regard to this particular point, an analysis of the number of cells separating the dorsal-ventral and anterior-posterior poles of blastemas and measurement of the dimentions of blastemas with reference to these axes demonstrates that these facets of blastema outgrowth are entirely consistent with the process of short arc intercalation and the proposed constraints on healing following amputation at different limb levels. Geraudie and Singer have recently demonstrated, using the scanning electron microscope, that newt limb blastema cells have cellular processes that may be up to 50m long. In the distal region of a palette staged blastema formed after upper arm amputation, the dimensions of and cell counts along the cardinal axes are consistent with the possibility of cells from dorsal and ventral sides of the blastema contacting directly. For example, in the distal 700m of such blastemas, which is the region forming the limb region distal to the elbow, the distance from dorsal to ventral extremes varies from 20 to 50m and are separated by between means of 7 and 26 cells. In contrast, the anterior and posterior poles are between 84 and 151m apart and are separated by between means of 25 and 67 cells. Thus the likelihood of cells contacting preferentially from dorsal to ventral sides of the blastema is higher than the likelihood of anterior and posterior cells contacting and the dimensions are consistent with a local cell contact model. Furthermore, dimensions and cell counts analyzed in upper arm MB blastemas are consistent with a radial healing mode.

Conclusions: The Importance of Moderators of Cell-Cell Contact

The data presented in this discussion are consistent with a short range cell-cell contact model controlling pattern regulation in the amphibian limb. It is clear, however, that trying to explain the results solely in terms of the formal representation of the polar co-ordinate model is inadequate and that secondary considerations, such as healing constraints and blastemal shape are needed to make the explanations more complete. This point is further emphasized by recent experimental findings where regenerating limbs bearing clear structural discontinuities have been demonstrated. These experiments will be briefly outlined as the conclusion to this discussion to demonstrate the problems that face attempts to elucidate the basis of pattern regulation in the urodeles. The first demonstrations that limbs could form bearing structural discontinuities were made by Maden and Maden and Mustafa who analyzed the muscle patterns in supernumerary limbs produced following ipsilateral blastema rotations in the axolotl.

It is now clear that these extra limbs bear a variety of dorsal-ventral pattern characteristics despite the fact that the anterior-posterior pattern appears normally asymmetric. Four basic limb anatomies have been found. These include double dorsal and double ventral limbs, limbs with part of their pattern symmetrical (double dorsal and double ventral) and part asymmetrical, limbs with mixed handed patterns where half of the pattern is dorso-ventrally reversed compared with the other half, and normal asymmetrical limbs. The supermumerary limbs comprising part symmetrical and part asymmetrical patterns or mixed-handed patterns have clear structural discontinuities within them.

The formal polar co-ordiante model and any other model in which clear discontinuities should be smoothed out fail to explain how such limb patterns can occur. The same is true of the recent descriptions of co-ordiante-free models based simply on rules of pattern continuity that do not discuss likely cellular mechanisms. An additional problem is that these supernumerary limbs are derived from interactions between positionally mismatched blastemas and stumps and no way is currently available for elucidating the details of these interactions. For this reason, we have recently surgically created forearms that are anatomically equivalent to the mixed-handed supernumerary limbs. Such forearms were made by splitting the axolotl limb between digits 2 and 3 and separating one complete half of the forearm to the elbow between the radius and ulna. Anterior or posterior halves were then exchanged between left and right limbs and sutured into the vacant lower arm position such that the dorsal and ventral axes of graft and host limb tissues were opposed but the anterior to posterior axes were normal. Such constructions were allowed to heal for one month before amputation in the mid forearm position or amputated immediately. In both instances, such mixed-handed limbs regenerated essentially the same mixed-handed pattern that was removed. This fact was established by the analysis of muscle patterns in the regenerates. Thus in these surgically created cases, as in the supernumerary limbs, clear structural discontinuities in the limb are maintained during outgrowth and apparently no attempt is made to smooth out the discontinuity. No extra pattern elements are found in these regenerates. The faithful regeneration of mixed-handed limbs also occurs following amputation of such limbs formed following ipsilateal blastema rotation.

Maden has recently suggested that the complete range of anatomies in supernumerary limbs initially formed following this operation can be explained by the fusion of initially discrete blastema populations around the graft-host junction. This idea may well explain the first formation of these limbs but no such fusion can occur following their amputation. The regeneration of mixed-handed and partly symmetrical, partly asymmetrical limbs has as yet only been observed following amputation of these structures in the lower arm. One clue as to the reason why they are able to regenerate faithful copies may be the elliptical shape of the lower arm blastema that was illustrated earlier in this discussion.

It is possible that the preferential dorsal to ventral healing constraint within the forearm blastema prevents, or at least restricts, contact between cells across the anterior-posterior axis and thus allows the outgrowth of limbs bearing discontinuities. This is essentially the same explanation as that put forward to explain the lack of a healing time effect following amputation of double anterior and double posterior axolotl forearms. It is important, therefore, to examine the regenerative potential of mixed-handed tissues constructed in the upper arm where the radial healing mode should ensure that the discontinuities in the pattern are smoothed out by intercalation. In these experimental limbs, extra structures would be predicted. This particular category of results reveals that we are yet far from an adequate explanation for the cellular basis of pattern regulation in urodele limbs. It is clear, however, that the use of formal models that are clearly predictive remain essential as tools with which to further our knowledge. We must proceed with the hope that in the near future enough data will be available to allow the move to the analysis of pattern regulation at the cellular level. A clear understanding of this complex problem will only emerge when information from these different levels of organization can be held together. The urodele limb may be one of the systems in which this daunting task will be achieved.

INDEX

C

D